Efficacy, Safety, and Immunogenicity of Vaccines against Viruses: From Network Medicine to Clinical Experimentation

Efficacy, Safety, and Immunogenicity of Vaccines against Viruses: From Network Medicine to Clinical Experimentation

Editors

Pietro Hiram Guzzi
Marianna Milano
Jayanta Kumar Das

MDPI

Basel • Beijing • Wuhan • Barcelona • Belgrade • Novi Sad • Cluj • Manchester

Editors
Pietro Hiram Guzzi
Surgical and Medical Sciences
Magna Graecia University of Catanzaro
Catanzaro
Italy

Marianna Milano
Experimental and Clinical Medicine
University of Catanzaro
Catanzaro
Italy

Jayanta Kumar Das
National Institutes of Health
National Institute on Aging
Biomedical Research Centre
Baltimore MD
United States

Editorial Office
MDPI
St. Alban-Anlage 66
4052 Basel, Switzerland

This is a reprint of articles from the Special Issue published online in the open access journal *Viruses* (ISSN 1999-4915) (available at: www.mdpi.com/journal/viruses/special_issues/11MAW3E5S1).

For citation purposes, cite each article independently as indicated on the article page online and as indicated below:

Lastname, A.A.; Lastname, B.B. Article Title. *Journal Name* **Year**, *Volume Number*, Page Range.

ISBN 978-3-0365-9233-6 (Hbk)
ISBN 978-3-0365-9232-9 (PDF)
doi.org/10.3390/books978-3-0365-9232-9

Contents

About the Editors

Pietro Hiram Guzzi

Pietro Hiram Guzzi's research focuses on developing novel methods for efficient analysis of biological networks derived from large (big) biomedical data generated from high-throughput experiments. The ultimate goal of his research is to support the development of novel drugs in a translational and precision medicine scenario. From a computational point of view, his research interests include, but are not limited to, network generation and analysis (including comparison and integration) by integrating both clinical and geographical data. From a biological point of view, his research focuses on gene and protein function analysis, as well as identification of the interplay among molecules related to disease insurgence and progression. More recently, he is involved in the analysis of diseases through the integration of heterogeneous data into a single model based on networks and the development of novel methods for brain data analysis. His research efforts have resulted in 2 books, 3 book chapters, 38 peer-reviewed publications in top-class journals, and more than 40 papers presented in bioinformatics conferences.

Marianna Milano

Marianna Milano is an assistant professor and a senior research scientist in the field of omics and biological network data analysis at the University Magna Graecia, Catanzaro, Italy. Her research interests focus on the development of innovative algorithms for the analysis of clinical and omic data through the application of biological knowledge formalized in ontologies; the extraction of knowledge from biological and biomedical data; the use of formal knowledge representation tools in the field of computational biology; and the development of algorithms for the analysis of biological and biomedical networks through the application of graph theory. She is a Member of BITS (Bioinformatics Italian Society). In 2022, she organized and chaired the 1st workshop on "Foundations of Network Analysis" in conjunction with the 2022 IEEE International Conference on Bioinformatics and Biomedicine (IEEE BIBM 2022). In 2023, she organized and chaired the 1st thematic track on "Network Models and Analysis: From Foundations to Complex Systems (NMA) in conjunction with the 2023 International Conference on Computational Science (ICCS 2023).

Marianna Milano currently serves as a reviewer for several scientific journals as well as for national and international conferences. She is a chair and program committee member of several national and international conferences, and has managed many Special Issues.

Jayanta Kumar Das

Jayanta Kumar Das received a B. Tech degree in Computer Science and Engineering and an M. Tech degree in Information Technology in 2012 and 2014, respectively, from GCECT. He pursued his Ph.D. working in the Applied Statics Unit, Indian Statistical Institute, during 2015–2019 and obtained his degree in Computer Science and Engineering from the University of Calcutta in 2019. Currently, he is a post-doctoral fellow at the School of Medicine, Johns Hopkins University, MD, USA. His areas of research include computational bioinformatics and mathematical genomics, machine learning, integral value transformation (IVT), discrete dynamics, data structure and graph theory, and algorithm design, sequence, and pattern analysis.

Preface

In the face of unprecedented global challenges brought forth by diseases like COVID-19, influenza, Varicella virus, and HPV, the scientific community finds itself at the forefront of a battle against ever-evolving pathogens. The advent of effective vaccines has undoubtedly marked a significant triumph in the realm of medicine, saving countless lives and preventing the spread of debilitating illnesses. Yet, this momentous achievement does not signal the end of our journey but rather a new beginning—a clarion call for continuous exploration and innovation.

This reprint emerges as a beacon guiding researchers, scientists, and medical professionals through uncharted territories. In this volume, we delve deep into the intricate world of immunization, where the fusion of bioinformatics and network medicine reshapes our understanding of vaccines' efficacy, safety, and immunogenicity.

The papers presented in this reprint serve as testaments to the power of interdisciplinary collaboration. By synergizing the analytical prowess of bioinformatics with the holistic approach of network medicine, we uncover novel avenues for advancing vaccine research. The critical importance of this synergy lies not only in enhancing the quality of existing vaccines but also in propelling the development of groundbreaking ones.

One of the central themes explored within these pages revolves around the pressing need to sustain vaccine-induced protection against variant strains and breakthrough infections. The urgency of this matter cannot be overstated. Thus, the contributors to this volume have meticulously studied existing vaccines and probed into the creation of prioritization campaigns. By understanding the complex interplay between pathogens, immune responses, and human networks, we pave the way for more targeted, efficient, and adaptable immunization strategies.

Moreover, this reprint is a testament to our commitment to inclusivity. By examining age- and sex-based differences in vaccine responses, we aim to ensure that our advancements in medicine benefit diverse populations equitably. In doing so, we recognize that the quest for comprehensive healthcare extends beyond scientific innovation—it encompasses societal responsibility and compassion.

As the editors of this volume, we are profoundly inspired by the dedication and brilliance of the contributors who have embarked on this intellectual odyssey. Their tireless efforts have illuminated the path ahead, reminding us that our pursuit of knowledge is boundless. We extend our deepest gratitude to these trailblazers and to you, our readers, for joining us on this expedition.

Pietro Hiram Guzzi, Marianna Milano, and Jayanta Kumar Das
Editors

viruses

MDPI

Article

Expression of Toll-like Receptors on the Immune Cells in Patients with Common Variable Immune Deficiency after Different Schemes of Influenza Vaccination

Aristitsa Mikhailovna Kostinova [1,2,*], Elena Alexandrovna Latysheva [2,3], Nelly Kimovna Akhmatova [4], Anna Egorovna Vlasenko [5], Svetlana Anatolyevna Skhodova [4], Ekaterina Alexandrovna Khromova [4], Andrey Viktorovich Linok [1], Arseniy Alexandrovich Poddubikov [1], Tatyana Vasilievna Latysheva [2] and Mikhail Petrovich Kostinov [1,4]

1 Federal State Autonomous Educational Institution of Higher Education I.M. Sechenov First Moscow State Medical University of the Ministry of Health of the Russian Federation (Sechenov University), Trubetskaya Str. 8/2, 119991 Moscow, Russia; arseniypoddibikov@gmail.com (A.A.P.); monolit.96@mail.ru (M.P.K.)
2 National Research Center—Institute of Immunology Federal Medical-Biological Agency of Russia, Kashirskoe Shosse, 24, 115478 Moscow, Russia; tvlat@mail.ru (T.V.L.)
3 Pirogov Russian National Research Medical University, Ostrovitianov Str. 1, 117997 Moscow, Russia
4 Russian Federal State Budgetary Scientific Institution «I.I. Mechnikov Research Institute of Vaccines and Sera», Malyi Kazenniy Pereulok, 5a, 105064 Moscow, Russia; kate.khromova@mail.ru (E.A.K.)
5 Federal State Budgetary Educational Institution of Higher Education "Samara State Medical University" of the Ministry of Healthcare of the Russian Federation, Chapaevskaya Street, 89, 443099 Samara, Russia
* Correspondence: aristica_kostino@mail.ru; Tel.: +7-9166226839

Abstract: Background: for the first time, the effect of one and two doses of adjuvanted influenza vaccines on toll-like receptors (TLRs) in patients with common variable immunodeficiency (CVID) was studied and compared (primary vaccination with one vs. two doses, primary vs. repeated vaccination). Materials and methods: Six patients received one dose of quadrivalent adjuvanted influenza vaccine during the 2018–2019 and 2019–2020 influenza seasons, and nine patients with CVID received two doses of trivalent inactivated influenza vaccine during 2019–2020. Expression of TLRs was measured by flow cytometry. Results: The expression of toll-like receptors in patients with CVID was noted both with repeated (annual) administration of the influenza vaccine and in most cases was accompanied by an increase in the proportion of granulocytes (TLR3 and TLR9), lymphocytes (TLR3 and TLR8), and monocytes (TLR3 and TLR9). When carried out for the first time as a simultaneous vaccination with two doses it was accompanied by an increase in the proportion of granulocytes, lymphocytes expressing TLR9, and on monocytes—TLR3 and TLR9. Conclusion: in CVID patients, the use of adjuvanted vaccines is promising, and research on the influence of the innate immunity and more effective regimens should be continued.

Keywords: quadrivalent adjuvanted influenza vaccines; toll-like receptors; influenza; CVID; azoximer bromide

Citation: Kostinova, A.M.; Latysheva, E.A.; Akhmatova, N.K.; Vlasenko, A.E.; Skhodova, S.A.; Khromova, E.A.; Linok, A.V.; Poddubikov, A.A.; Latysheva, T.V.; Kostinov, M.P. Expression of Toll-like Receptors on the Immune Cells in Patients with Common Variable Immune Deficiency after Different Schemes of Influenza Vaccination. *Viruses* **2023**, *15*, 2091. https://doi.org/10.3390/v15102091

Academic Editors: Pietro Hiram Guzzi, Marianna Milano and Jayanta Kumar Das

Received: 14 September 2023
Revised: 6 October 2023
Accepted: 11 October 2023
Published: 14 October 2023

1. Introduction

Inborn errors of immunity (IEI; primary immunodeficiencies (PIDs)) encompass a heterogeneous group of orphan diseases (prevalence 1 to 5 in 1000), which are based on approximately 500 currently known genetic defects of the immune system [1]. The most common clinical manifestation of IEI is an increased risk of recurrent and potentially life-threatening infections due to defects of immunological defense mechanisms and its critical pathway. Immunodeficiencies affecting humoral immunity (antibody synthesis) engage more than 50% of all diseases in the structure of IEI. The main treatment for patients with impaired antibody synthesis is lifelong replacement therapy with intravenous immunoglobulin (IVIG), containing a wide variety of donor class G immunoglobulins that reduce susceptibility to infectious agents. However, this treatment is not expected to protect recipients from all currently circulating infections as antibodies to rare or highly variable infectious agents are often absent in donor plasma [2].

Nowadays many patients with IEI survive into adulthood, and regular and effective therapy allows them to have a high quality of life that does not differ from healthy people: they work and study, have hobbies, make friends, and have families. But at the same time, this active lifestyle in the community leads to frequent contact with respiratory viruses, and the high susceptibility of IEI patients contributes to exacerbations of chronic sinopulmonary diseases. That is why we can say that the prevention of infections is particularly acute.

Vaccination is recognized in the whole world as the most effective means of protection against infectious diseases. Most countries have already implemented vaccination programs for the population against respiratory infections that help to reduce the risk of contamination and morbidity, both through the direct protection of each vaccinated individual and through the development of herd immunity [3]. However, vaccination coverage among immunocompromised people still remains below the WHO recommendations, even in developed countries [4]. This is facilitated by many socially significant factors, such as the influence of the media, existing misconceptions about the effectiveness and safety of vaccination, low awareness of medical workers, and the doctor–patient relationship [5]. Additionally, certain features of the pathogens themselves, for example, high mutational variability, contributes to the risk of pandemics even nowadays.

Over the past 100 years, four influenza pandemics have occurred on the planet, with the 1918 pandemic, caused by the influenza A/H1N1 virus, being the most devastating as it claimed the lives of more than 40 million people [6]. The last influenza pandemic "engulfed" the world in 2009–2010. Therefore, vaccination against the influenza virus still remains actual and relevant.

Immunization is the only way to form protection against seasonal influenza, not only in healthy people but also in patients with IEI. In order to reduce morbidity and mortality from vaccine-preventable infections in immunocompromised patients, for the first time in 2013 in the guidelines of the American Society of Infectious Diseases and in the world in 2015, patients with IEI were recommended to undergo annual vaccination against the influenza virus regardless of the type of the immune defect [7,8]. However, currently available data on the formation of post-vaccination immunity in this cohort of patients are limited and in some cases contradictory. It is not only due to different vaccination schemes, the use of different vaccines, but also because of the various examined parameters of the immune system. Considering that in patients with IEI there is a need to increase the effectiveness of vaccines, adjuvant vaccines which enhance a specific immune response are preferable [9,10].

However, no study has been conducted comparing both immunogenicity of different influenza vaccination regimens in patients with impaired humoral immunity and the clinical significance of immunization on the course of the underlying disease of patients for the purposes of practical health care, and not only from the standpoint of scientific research. Approximately 10–20% of patients from the most common group of patients with a defect in the humoral immunity—common variable immune deficiency (CVID)—have a residual response to vaccination against protein antigens and, to a lesser extent, against polysaccharide antigens. However, post-vaccination immunity in patients with CVID has not been sufficiently studied, which is due to the different use of various vaccine preparations, immunization regimens, determined markers, research methods, and the lack of clear standards for many indicators of innate immunity. Moreover, the effect of quadrivalent adjuvant influenza vaccines on the parameters of innate and adaptive immunity has not been studied.

Innate immunity is the first line of the immune system defense, both in evolutionary terms and in terms of time response, which develops in the first hours and days after exposure to a pathogen. The speed of the response is realized through the involvement of already existing stereotypical mechanisms for recognizing specific molecular structures of pathogens. It is the recognition of molecular structures of microbial origin that is the key component of the immune response that initiates inflammation [11]. This response is mediated by a special family of receptors that recognize the most common molecular patterns (PAMPs—pathogen associated molecular patterns) of microorganisms (viruses, bacteria, parasites, etc.) and are called PRRs (pattern recognition receptors) [12]. PRRs are genetically stable. Functionally, they can be divided into two groups: endocytic and signaling. Endocytic PRRs (mannose receptors and scavenger receptors) have been known in immunology for a long time and provide phagocytosis processes with subsequent delivery of the pathogen inside the phagosome to lysosomes, providing the start of an adaptive immune response. Among signaling PRRs are three families of great importance: toll-like (TLR), nod-like (NLR), and rig-like receptors (RLR). TLRs were the first PRRs to be identified, and that is why they are the most studied.

Fifteen TLRs have been described in mammals and humans. They are located on the membrane, in endosomes, or in the cytoplasm of cells that serve as the first line of defense (neutrophils, macrophages, dendritic, endothelial, and epithelial cells of the skin and mucous membranes) from pathogens.

TLR1, TLR2, TLR4, TLR5, and TLR6 are expressed on the extracellular surface and recognize microbial wall components including lipopolysaccharides, lipopeptides, and flagellin. TLR3, TLR7, TLR8, and TLR9 are mostly located in endosomal compartments; this arrangement allows these receptors to recognize the DNA and RNA breakdown products of viral and bacterial origin. TLRs are expressed in a variety of cell types, including monocytes, phagocytes, dendritic cells, and B cell subpopulations. TLR1, TLR2, TLR6, TLR7, TLR9, and TLR10 are expressed on human B cells.

While B cells are traditionally considered to play a key role in adaptive immunity due to their ability to produce antibodies, activation of innate immune receptors also expressed on B cells provides a co-stimulatory effect that promotes both their function and survival [13] and the coordination of innate and adaptive immune signals, resulting in a wide range of cellular responses. Memory B cells are usually formed in germinal centers in response to T-dependent or T-independent antigens. Like other antigen-presenting cells, B cells express various TLRs [14,15]—conservative membrane proteins which provide alternative ways to activate B cells [16].

One of the most powerful stimulators of B cell activation and maturation are endosomal TLRs in connection with single-stranded RNAs or various synthetic agonists (TLR7) and unmethylated CpG motifs in microbial DNA (CpG DNA) (TLR9) [17]. Binding of TLR9 by CpG DNA has been shown to activate normal B cells, increase the expression of costimulatory molecules, trigger the secretion of IL-6 and IL-10, and mediate T independent isotype switching and production of specific antibodies independent of B cell receptor (BCR) binding [18–22]. Cross-linking of the B cell receptor (BCR) leads to rapid activation of their expression. Naive B cells express low levels of TLR while memory B cells constitutively express TLR7, TLR8, and TLR9 at higher levels [23–26]. These differences between naive and B memory cells are connected with different adaptive functions: memory B cells express higher levels of TLR and have a more ability to differentiate into plasma cells via TLR stimulation compared to naive B cells [27]. On the contrary, it is suggested that prolonged PAMP activation has undesirable consequences for the organism since the TLR signaling pathway is controlled by a variety of feedback mechanisms [28].

Both in vivo and in vitro experiments have shown that switching of B cells to IgG isotypes requires the simultaneous presence of at least two signals along with BCR involvement: TLR activation and the involvement of CD40 and/or IFN-alpha [29]. These observations have led to studies suggesting that TLR activation may provide a long-lasting stimulus important for maintaining memory B cell proliferation and differentiation into mature antibody-secreting cells, which are initially induced by BCR and by T cells [30,31]. Binding and activation of TLR7 and TLR9 can serve as a signal for the start of B cell differentiation after antigen stimulation via BCR. However, there are data reporting defects in TLR7 and TLR9 on B cells from patients with CVID [17]. Since TLR activation seems to be an essential mechanism for the activation and subsequent survival of memory B cells [32], a study of TLR9 defects in patients with CVID revealed that their B cells were not activated by the CpG ODN ligand even upon costimulation of BCR, as well as by secretion of IL-6 and IL-10; thus, there was no TLR activation, low proliferation of B cells, absence of their maturation, isotype switching, and production of IgG and IgA [17].

Thus, the exact role of TLRs in B cell biology is not clear: are they necessary for the development of some normal humoral immunity pathways, or does stimulation of TLRs serve more as an adjuvant to existing functions? It is believed that TLR signaling pathways may provide secondary stimuli for B cell development, but other molecular mechanisms may compensate for defective signaling through these innate receptors [33]. A more detailed understanding of the innate immune status upon ligand-activated TLRs is needed to identify specific defects in innate immune responses in CVID patients that may explain the likely variability in clinical symptoms.

2. Materials and Methods

In an open-label, single-center, non-randomized, prospective, cohort, controlled study the effect of influenza tetravalent inactivated subunit adjuvanted vaccine on the expression of endosomal toll-like receptors on immunocompetent cells in healthy volunteers and patients with CVID using different influenza vaccination schemes were investigated.

2.1. Participants

In 2018, from the registry of the Institute of Immunology that includes all adult patients with IEI from the whole Russian Federation, 297 outpatient records of patients with IEI were analyzed, and 203 patients were selected with a diagnosis of CVID, which was established in accordance with the criteria of the European Society on Immunodeficiency Disorders (ESID). However, due to the comorbidity of patients and in accordance with strict criteria for inclusion in the study, as well as the need to relieve exacerbations when patients were hospitalized, the annual change in the strain composition of influenza vaccines, and the limited timing of influenza vaccination (autumn–winter period), immunization was administered to only 15 patients with CVID.

2.2. Inclusion Criteria

- Confirmed diagnosis of CVID in accordance with diagnostic criteria established by the European Society for Immunodeficiency Diseases and the American Academy of Allergy, Asthma and Immunology for the diagnosis and treatment of IEI.
- IVIG therapy no later than 28 days before vaccination and no earlier than 21 days after it; that is, there was a break between two subsequent administrations of immunoglobulins of at least 7 weeks.
- Signed informed consent.
- Excluded causes of secondary hypogammaglobulinemia.
- No symptoms of flu or flu-like illness in the past six months.
- No glucocorticosteroid or other immunosuppressive therapy admission at the time of the study and 3 months before the start.
- The absence of symptoms of protein-losing enteropathy as well as any suspicion of oncological or lymphoproliferative disease in patients with CVID at the time of the study.
- No level of specific antiviral antibodies in protective titers (>1:40) in pre-vaccination blood samples.
- Individuals with cognitive or behavioral impairments, psychiatric disorders, or alcohol abusers that could preclude participation in the study were excluded.
- Vaccination against any other infections within 1.5–2 months prior to enrollment in the study.

- All contraindications for use in accordance with the instructions (a history of allergy to egg whites or any component of the study vaccine, symptoms of acute infection at the time of vaccination, pregnancy, etc.) were followed.

2.3. Vaccination Schemes

Thus, in the 2018–2019 influenza season at the Department of Immunopathology in the National Research Center—Institute of Immunology Federal Medical-Biological Agency of Russia and in the laboratory of the Mechanisms of Immune Regulation in Mechnikov Research Institute of Vaccines and Sera of the Ministry of Health of the Russian Federation in Moscow, 6 patients with a diagnosis of CVID were involved in the study. They received a single dose (0.5 mL) of a quadrivalent subunit adjuvant vaccine (aQIV) against the influenza virus "Grippol Quadrivalent" (NPO Petrovax Pharm LLC, Moscow, Russia). Exactly one year later 5 out of these 6 patients were re-vaccinated with the trivalent subunit adjuvant influenza virus vaccine Grippol Plus (NPO Petrovax Pharm LLC, Russia) from the Grippol family of vaccines. The mean age of patients with CVID was 36.6 ± 2.03 years. In the next 2019–2020 influenza season 9 more patients with CVID were immunized simultaneously with a double dose (2 × 0.5 mL) of the trivalent subunit adjuvant influenza virus vaccine (aTIV) Grippol Plus.

Thirty-two healthy volunteers aged 19 to 48 years (32.56 ± 1.67 years) in the 2018–2019 season were also vaccinated with a single dose of aQIV according to the same schemes as patients with CVID and served as a control group to determine the working concentrations for immunological parameters. All participants had not been vaccinated against influenza in the previous two seasons (2016–2017, 2017–2018), nor were there any confirmed cases of influenza reported.

2.4. Used Vaccines

Grippol Quadrivalent (NPO Petrovax Pharm LLC, Moscow, Russia) and Grippol Plus (NPO Petrovax Pharm LLC, Moscow, Russia) vaccines were available for immunization of people aged 18 to 60 years during the 2018–2019 and 2019–2020 flu season, respectively, and included strains that met the 2018–2019 WHO recommendations for the northern hemisphere for quadrivalent vaccines (A/Michigan/45/2015 (H1N1)pdm09-like virus; A/Singapore/INFIMH-16-0019/2016 (H3N2)-like virus; B/Colorado/06/2017-like virus (B/Victoria/2/87lineage); B/Phuket/3073/2013-like virus (B/Yamagata/16/88 lineage)), and in 2019–2020 for trivalent vaccines (A/Brisbane/02/2018 (H1N1)pdm09-similar virus; A/Kansas/14/2017 (H3N2)-like virus; B/Colorado/06/2017-like virus (lineage B/Victoria/2/87)). The vaccines contained 5 μg of hemagglutinin of each included influenza strain and 500 μg of azoximer bromide (without preservative). The key benefit of the vaccine is a three-fold decreased amount of hemagglutinin protein as compared to traditional technologies due to the use of Polyoxidonium (azoximer bromide)—a water-soluble, high-molecular immune system adjuvant that enhances the immune response to vaccination.

2.5. Methods

The expression of TLRs on granulocytes (CD45+CD15+), lymphocytes (CD45+CD19+), and monocytes (CD45+CD14+) in the peripheral blood in patients with CVID and healthy patients before vaccination and 24 ± 3 days after was studied in vitro by flow cytometry using mAbs to TLR3-PE, TLR8-FITC, and TLR9-PE (e-Bioscience, San Diego, CA, USA) on the Cytomix FC 500 flow cytometer (Beckman Coulter, Indianapolis, IN, USA) according to the method described in the manufacturer's instructions.

The study was conducted according to the Russian Federation National Standard Protocol ГОСТР 52379-2005 Good Clinical Practice and International GCP standards [34]. The study was based on the ethical principles and recommendations of the WHO and the Russian Ministry of Health.

2.6. Statistics

Checking the normality of the distribution of signs was carried out by the Shapiro–Wilk normality test. As a result, significant deviations from the normal distribution were revealed in the distribution of signs. Analysis of the dynamics of signs and comparison between study groups was carried out using the construction of a robust linear mixed effects model (RLMEM) [35]. The statistical significance of the model coefficients was determined using Satterthwaite's degrees of freedom approximation [36]. A posteriori comparisons (between groups at control points and between control points for each group) were carried out by constructing the corresponding contrasts based on the calculated model using the emmeans package [37].

When comparing the formation of post-vaccination immunity after the primary vaccination with one and two doses, the time after the introduction of the vaccine preparation (24 ± 3 days) and one or two doses of the vaccine were determined as fixed factors, individual patients were set as accidental factors. When comparing the formation of post-vaccination immunity after primary and after re-vaccination with a single dose of the drug, the time after the introduction of the vaccine drug (24 ± 3 days), and the stage of vaccination (primary/revaccination), individual patients were determined as fixed factors (the associated nature of the samples was taken into account both in time and stages of vaccination) were given as random factors. Comparison of delta cell number changes between study groups was performed using the Mann–Whitney test for unrelated samples (one and two doses of vaccine) and using the paired Wilcoxon test for related samples (primary and revaccination with one dose).

Differences were considered statistically significant at $p \leq 0.05$ and insignificant at $p \geq 0.10$; in intermediate cases ($0.05 < p < 0.1$), trends towards differences were discussed [38]. Calculations and graphical constructions were made using the GraphPad Prism program (v.9.3.0 licensed GPS-1963924) and the statistical environment R (v.3.6, GNU GPL2 license).

3. Results

No one in the control nor the CVID groups came down with influenza virus infection within the 24 days post-vaccination in the 2018–2019 and 2019–2020 seasons.

3.1. Comparing the Content of TLR3, TLR8, and TLR9 in Patients with CVID after the Primary Vaccination and a Year Later with One Dose of the Quadrivalent Vaccine

3.1.1. Granulocytes

The proportion of granulocytes expressing TLR3 and TLR9 after the primary vaccination decreased statistically significantly—from 93.8 (91.3–95.0)% to 89.5 (81.9–91.1)% ($p = 0.03$) and from 93.6 (92.4–96.9)% to 86.0 (81.0–91.4)% ($p = 0.001$), respectively. A year after the primary vaccination, these indicators remained statistically significantly lower than the initial values: 87.0 (69.4–89.0)% for TLR3 ($p < 0.001$) and 87.1 (80.0–88.3)% for TLR9 ($p = 0.001$). After the second vaccination one year after the primary, there was a statistically significant increase in the percentage of granulocytes expressing TLR3 and TLR9, in relation to the values before the second vaccination—up to 92.2 (84.4–93.9)% for TLR3 ($p = 0.02$) and 94.2 (86.5–98.0)% for TLR9 ($p = 0.001$). Thus, as a result of the primary vaccination and repeated vaccination one year after, the proportion of granulocytes expressing TLR3 (Figure 1a) and TLR9 (Figure 1b) did not change statistically significantly relative to the initial level.

(a)

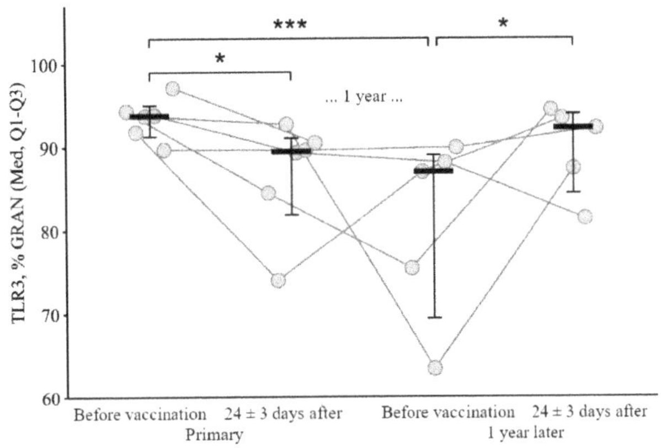

(b)

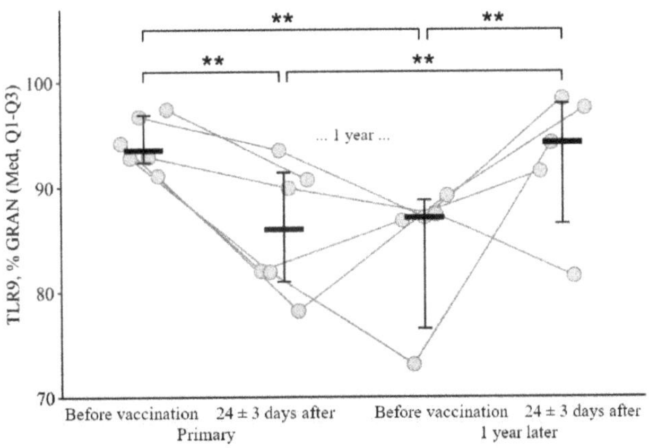

Figure 1. Percentage of granulocytes expressing TLR3 and TLR9 in patients with CVID at different stages of vaccination with a single vaccine dose (individual values as well as the median and interquartile range are given). *p*-value for comparing TLRs expression at different stages of vaccination (before vs. 24 ± 3 days after immunization with one vaccine dose during primary vs. a year later vaccination). * Statistically significant differences between different stages of vaccination at $p \leq 0.05$ level. ** statistically significant differences between different stages of vaccination at $p < 0.01$ level. *** statistically significant differences between different stages of vaccination at $p < 0.001$. A robust linear mixed effects model was used for calculations.

However, it is worth noting the changes in these indicators after primary and second vaccination: a decrease after the primary vaccination (5.6 [17.8; 0.1]% for TLR3 and 8.8 [12.9; 3]% for TLR9) and growth after the second vaccination (+6.4 [6.7; +24]% for TLR3 and +8.3 [5.9; +21.1]% for TLR9), $p = 0.04$ and $p = 0.02$, respectively), compared with changes after the primary vaccination (Figure 2a,b).

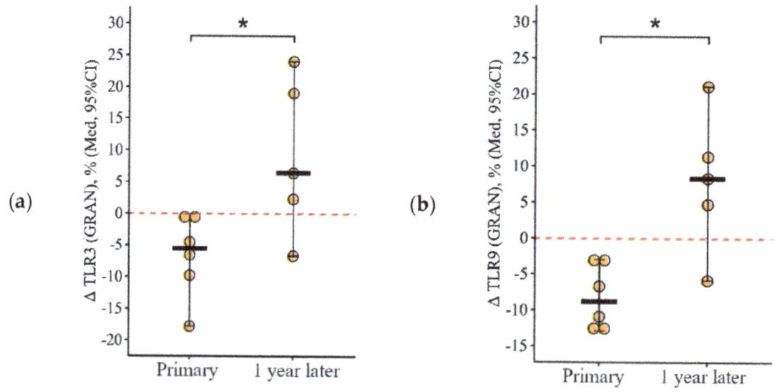

Figure 2. Delta (Δ—difference between the parameters before and 24 ± 3 days after vaccination) of the percentage of granulocytes expressing TLR3 and TLR9 in patients with CVID vaccinated with a single vaccine dose (individual values as well as the median and its 95% confidence interval are given). *p*-value for comparing delta of the percentage of granulocytes expressing TLR3 and TLR9 during primary and a year later vaccination. *—statistically significant differences between changes after primary and second vaccination at $p \leq 0.05$ level. Mann–Whitney test was used.

The percentage of granulocytes expressing TLR8 (data not shown) did not show statistically significant changes during the primary vaccination and repeated vaccination one year after and was 93.7 (85.2–96.0)% initially and 89.4 (88.2–93.0)% 24 ± 3 days after repeated vaccination.

3.1.2. Lymphocytes

The percentage of B-lymphocytes expressing TLR3 and TLR8 shows similar dynamics during immunization: no changes after the primary vaccination ($p = 0.51$ for TLR3 and $p = 0.41$ for TLR8) and an increase after the second vaccination (Figure 3a,b). The proportion of lymphocytes expressing TLR3 increased by 2.7 [1.04; +12.7]% in relation to the level before the second vaccination (from 0.60 (0.25–1.85)%, $p < 0.001$) and by 2.6 [0; +13.9]% in relation to the initial level (from 0.25 (0.09–1.73)%, $p < 0.001$). The proportion of lymphocytes expressing TLR8 increased by 1.8 [9.6; +10.2]% comparatively to the level before the second vaccination (from 0.48 (0.30−5.66)%, $p = 0.08$) and by 1.9 [1.4; +8.6]% comparatively to the initial level (from 0.25 (0.10−1.85)%, $p = 0.02$).

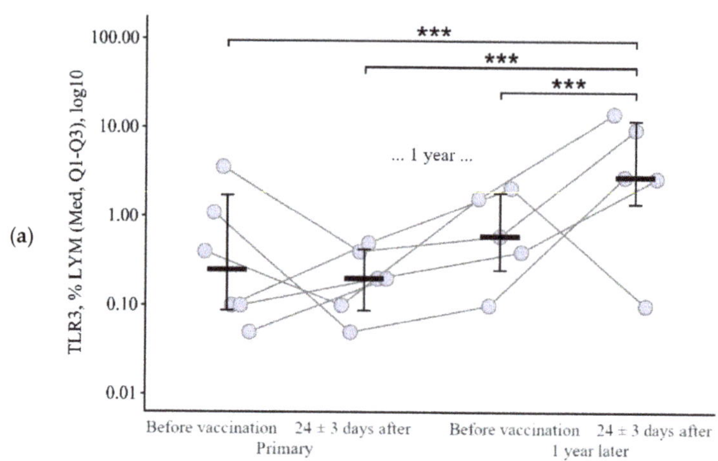

Figure 3. *Cont.*

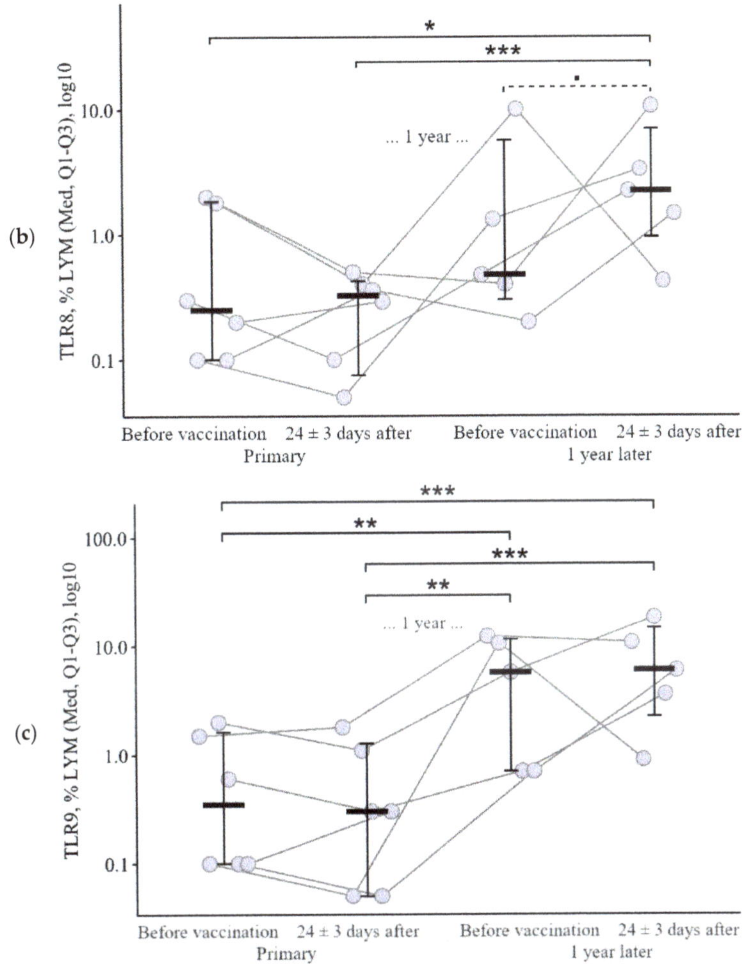

Figure 3. Percentage of lymphocytes expressing TLR3, TLR8, and TLR9 in patients with CVID at different stages of vaccination with a single vaccine dose (individual values as well as the median and interquartile range are shown). *p*-value for comparing TLRs expression at different stages of vaccination (before vs. 24 ± 3 days after immunization with one vaccine dose during primary vs. a year later vaccination). ▪ Differences between different stages of vaccination at the $p < 0.10$ trend level. * Statistically significant differences between different stages of vaccination at $p \leq 0.05$ level. ** Statistically significant differences between different stages of vaccination at $p < 0.01$ level. *** Statistically significant differences between different stages of vaccination at $p < 0.001$. A robust linear mixed effects model was used for calculations.

Thus, changes in the proportion of lymphocytes expressing TLR3 and TLR8 after primary and after the second vaccination differ ($p = 0.05$ for TLR3 and $p = 0.08$ for TLR8): there is no change after the primary vaccination and an increase after the second vaccination, one year later (Figure 4a,b).

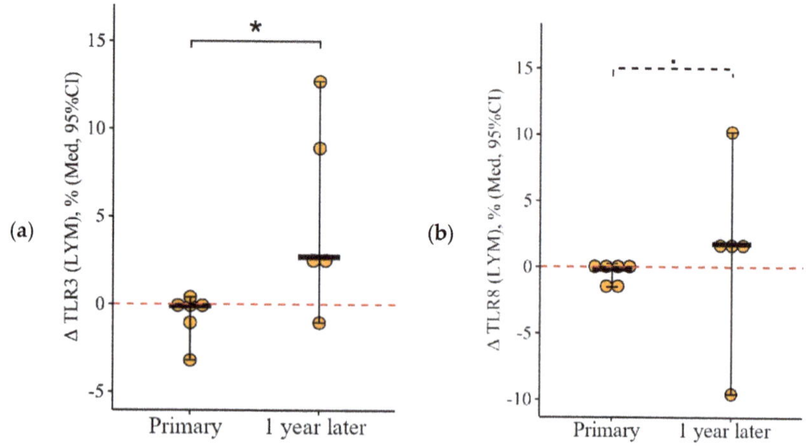

Figure 4. Delta (Δ—difference between the parameters before and 24 ± 3 days after vaccination) of the percentage of lymphocytes expressing TLR3 and TLR8 in patients with CVID vaccinated with a single vaccine dose, at various stages of vaccination (individual values are given, median and its 95% confidence interval). *p*-value for comparing delta of the percentage of lymphocytes expressing TLR3 and TLR8 during primary and a year later vaccination. ▪—differences between changes after primary and second vaccination one year later at the $p < 0.10$ trend level. *—statistically significant differences between changes after primary and second vaccination one year later at $p \leq 0.05$ level. Mann–Whitney test was used.

The percentage of lymphocytes expressing TLR9 (data not shown) shows slightly different dynamics—no changes after the primary vaccination, but a statistically significant increase after a year (by +5.9 [+0.8; +16.4]% to 5.70 (0.70–11.45)%, $p = 0.01$ compared to initial) (Figure 3c). After the second vaccination one year later, the proportion of lymphocytes expressing TLR9 did not change statistically significantly ($p = 0.15$ relative to the level before revaccination) and remained higher than the initial level ($p < 0.001$).

3.1.3. Monocytes

The proportion of monocytes expressing TLR3 and TLR9 after the primary vaccination decreased statistically significantly: by 25.1 [71.3; +0.4]% from 53.1 (22.8–75.6)% to 7.3 (2.7–44.0)% for TLR3 ($p = 0.01$) and by 23.2 [66.7; 5.5]% from 66.1 (35.3–80.8)% to 6.3 (1.6–73.0)% for TLR9 ($p < 0.001$). One year later, the proportion of monocytes expressing TLR3 and TLR9 increased slightly (up to 22.1 (5.1–35.7)% for TLR3 and up to 32.9 (22.6–44.1)% for TLR9) but still remained below the initial levels ($p = 0.05$ and $p = 0.04$, respectively). A statistically significant increase was observed after the second vaccination (relative to the level before the second vaccination): by +20.9 [1.6; +62.4]% to 45.4 (28.7 62.8)% ($p = 0.04$) for TLR3 and by +12.6 [8.6; +30.2]% to 50.2 (33.2 64.6)% for TLR9 ($p = 0.05$). As a result of the increase after the second vaccination, the percentage of monocytes expressing TLR3 and TLR9 became comparable to the initial level ($p = 0.89$ for TLR3 and $p = 0.40$ for TLR9) (Figure 5a,b).

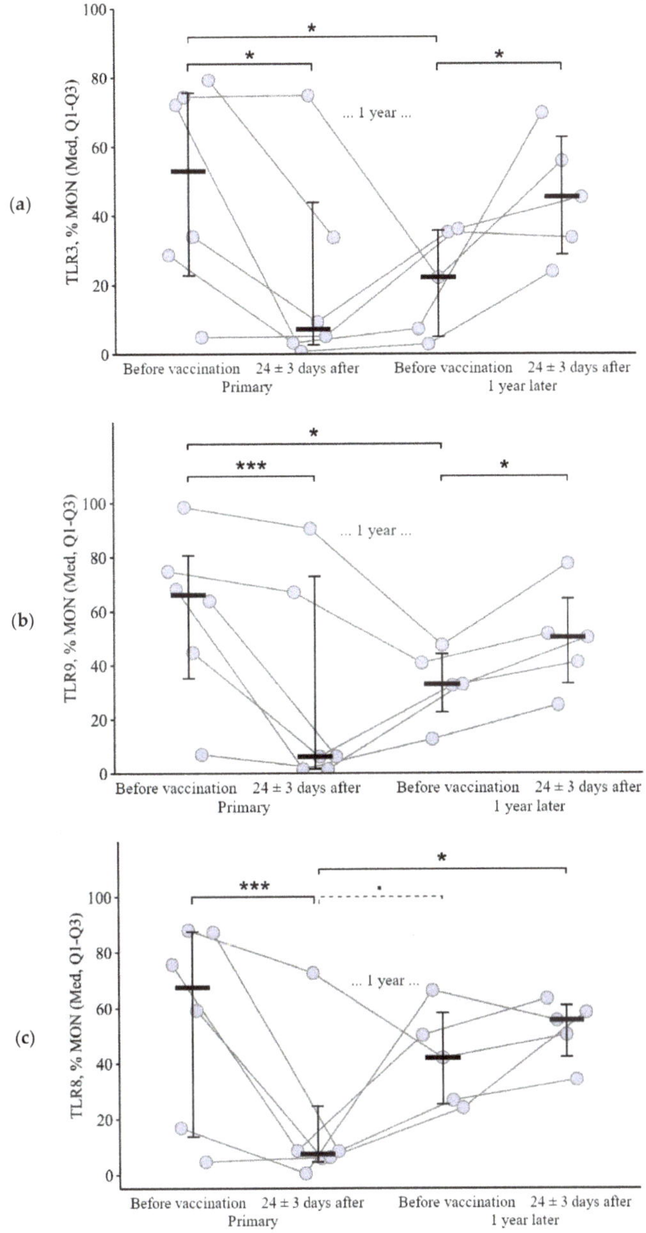

Figure 5. Percentage of monocytes expressing TLR3, TLR8, and TLR9 in patients with CVID at different stages of vaccination with a single vaccine dose (individual values as well as the median and interquartile range are shown). *P*-value for comparing TLRs expression at different stages of vaccination (before vs. 24 ± 3 days after immunization with one vaccine dose during primary vs. a year later vaccination). ▪ Differences between different stages of vaccination at the $p < 0.10$ trend level. * Statistically significant differences between different stages of vaccination at $p \leq 0.05$ level. *** Statistically significant differences between different stages of vaccination at $p < 0.001$. A robust linear mixed effects model was used for calculations.

The proportion of monocytes expressing TLR8 also decreased 24 ± 3 days after the primary vaccination: by 34.9 [78.7; +1.6]% from 67.5 (13.9–87.4)% to 7.4 (4.6–24.7)%, $p < 0.001$. But a year after the primary vaccination (before the second vaccination), there was an increase to 41.9 (25.4–58.2)% ($p = 0.07$ relative to the level after the primary vaccination). A total of 24 ± 3 days after the second vaccination, the proportion of monocytes expressing TLT9 was 55.6 (42.3–60.9)%; there were no significant changes compared to the level before the second vaccination ($p = 0.82$); it remained comparable to the initial level ($p = 0.82$) and was statistically significantly higher than the value after the primary vaccination ($p = 0.02$) (Figure 5c).

The percentage of lymphocytes expressing TLR9 (data not shown) shows slightly different dynamics—no changes after the primary vaccination, but a statistically significant increase after a year (by +5.9 [+0.8; +16.4]% to 5.70 (0.70–11.45)%, $p = 0.01$ compared to initial) (Figure 3c). After the second vaccination one year later, the proportion of lymphocytes expressing TLR9 did not change statistically significantly ($p = 0.15$ relative to the level before revaccination) and remained higher than the initial level ($p < 0.001$).

It is worth noting the statistically significant multidirectional nature of changes after the primary and after the second vaccination for all examined TLRs on monocytes ($p = 0.02$ for TLR3, $p = 0.01$ for TLR8, and $p = 0.006$ for TLR9): a decrease after the primary vaccination and increase (or no change, as in the case of TLR8) after the second vaccination (Figure 6a–c).

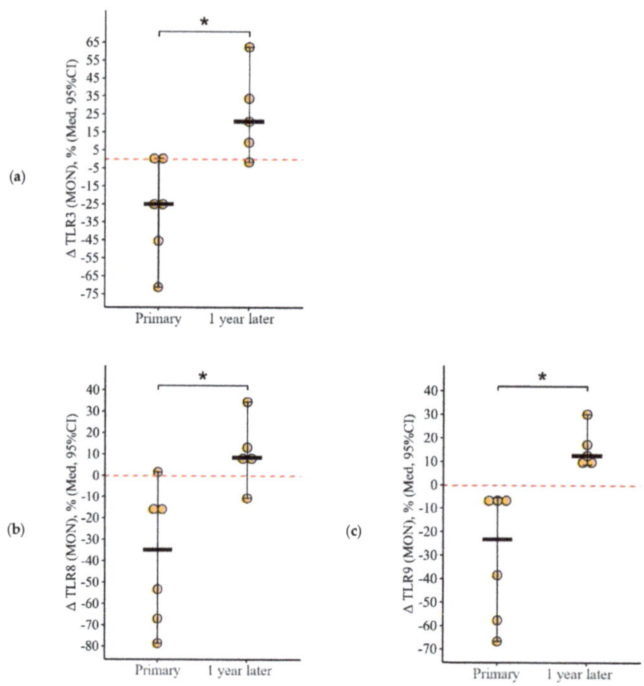

Figure 6. Delta (Δ—difference between the parameters before and 24 ± 3 days after vaccination) of the percentage of lymphocytes expressing TLR3, TLR8, and TLR9 in patients with CVID vaccinated with a single vaccine dose at various stages of vaccination (individual values are given as well as the median and its 95% confidence interval). p-value for comparing delta of the percentage of monocytes expressing TLR3, TLR8, and TLR9 during primary and a year later vaccination. ▪differences between changes after primary and second vaccination one year later at the $p \leq 0.10$ trend level. *—statistically significant differences between changes after primary and second vaccination one year later at $p \leq 0.05$ level. Mann–Whitney test was used.

3.2. *Comparing the Content of TLR3, TLR8, and TLR9 in Patients with CVID after Administration of One Dose and Simultaneously Two Doses of the Adjuvanted Influenza Vaccine*

3.2.1. Granulocytes

After vaccination with one dose, there was a statistically significant decrease in the proportion of granulocytes expressing TLR9, from 93.6 (92.4–96.9)% to 86.0 (81.0–91.4)% (p = 0.05) and at a borderline level of significance expressing TLR 3, from 93.8 (91.3–95.0)% to 89.5 (81.9–91.1)% (p = 0.08) (Figure 7a,b). While after vaccination with two doses, the proportion of granulocytes expressing TLR9 was statistically significantly higher than after vaccination with one dose—96.7 (90.4–98.1)% vs. 86.0 (81. 0–91.4)% (p = 0.02). The delta change in the proportion of granulocytes (relative to the initial level) expressing TLR3 and TLR9 was 5.6 [9.8; 1.0]% and 8.8 [12.2; 3.2]% with one dose of the vaccine and +3.5 [2.9; +19.4]% and +4.5 [0.7; +26.5]% after vaccination with two doses (p = 0.11 for TLR3 and p = 0.03 for TLR9 compared with one dose) (Figure 8a,b).

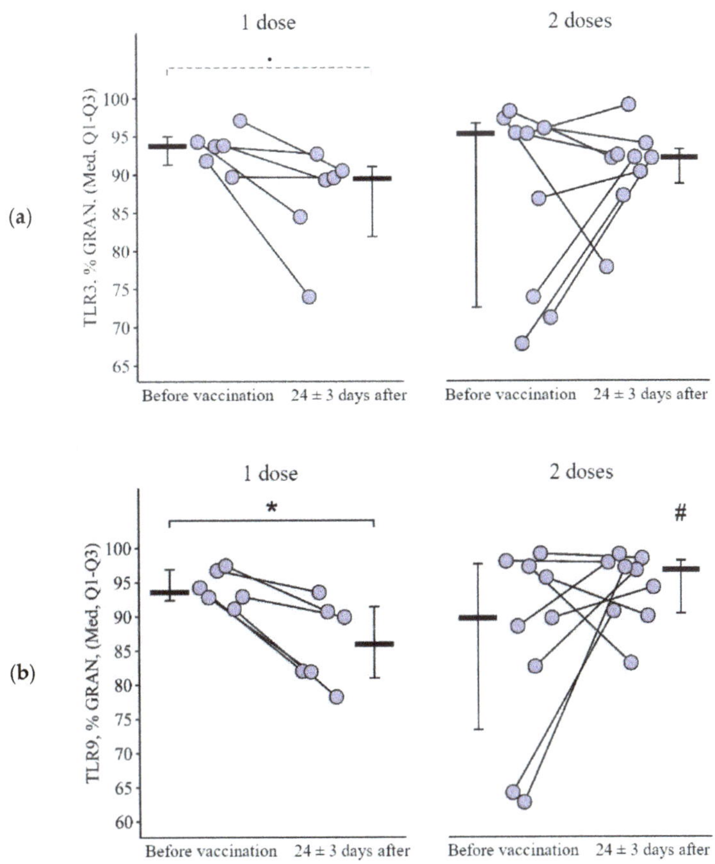

Figure 7. Percentage of granulocytes expressing TLR3 and TLR9 in patients with CVID after administration of single and double doses of adjuvanted influenza vaccine initially and 24 ± 3 days after immunization (individual values as well as the medians and interquartile range are given). p-value for comparing TLRs expression at different stages of vaccination (before vs. 24 ± 3 days after immunization with one vs. two vaccine doses. #—statistically significant differences between vaccination with one and two doses at the $p \leq 0.05$ level. •—differences before and after vaccination at the $p < 0.10$ trend level. *—statistically significant differences before and after vaccination at $p \leq 0.05$ level. A robust linear mixed effects model was used for calculations.

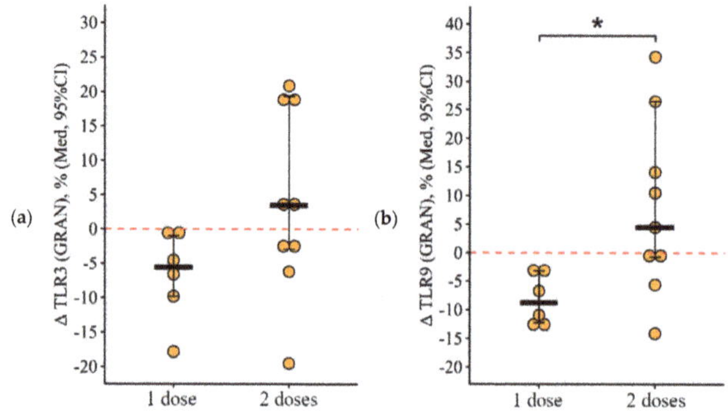

Figure 8. Delta (Δ—difference between the parameters before and 24 ± 3 days after vaccination) of the percentage of granulocytes expressing TLR3 and TLR9 in patients with CVID vaccinated with single and double doses (individual values as well as the median and its 95% confidence interval). *p*-value for comparing delta of the percentage of granulocytes expressing TLR3 and TLR9 after one vs. two vaccine doses administration. *—statistically significant differences between vaccination with one and two doses at the $p \leq 0.05$ level. Mann–Whitney test was used.

The percentage of granulocytes expressing TLR8 (data not shown) did not change significantly before or after vaccination and did not depend on the number of doses; it showed comparable dynamics for one and two doses. The medians were (before and after vaccination, respectively) 93.7 (85.2–96.0)% and 88.9 (85.7–91.0)% after one dose, 94.5 (82.7–99.1)% and 89.8 (84.8–97.9)% after two doses of the vaccine.

3.2.2. Lymphocytes

After vaccination with two doses, there was a statistically significant increase in the proportion of lymphocytes expressing TLR9, from 2.70 (0.40−6.60)% to 15.40 (1.90−26.75)% (p = 0.02). No such changes were observed after vaccination with one dose. A total of 24 ± 3 days after vaccination with two doses, the proportion of lymphocytes expressing TLR9 was higher than after vaccination with one dose: 15.40 (1.90–26.75)% vs. 0.30 (0.05–1.28)% (p < 0.001) (Figure 9).

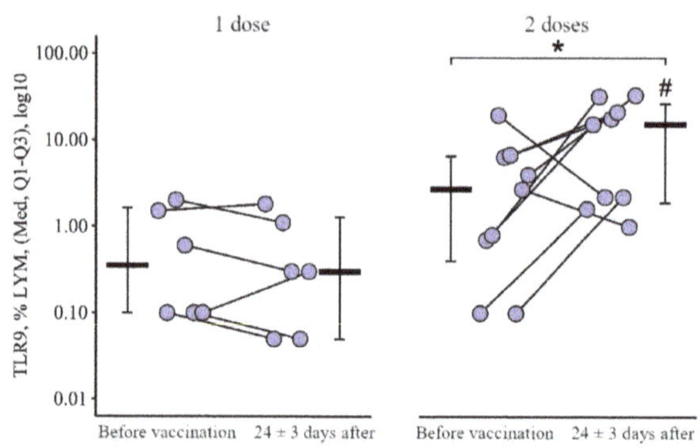

Figure 9. Percentage of lymphocytes expressing TLR9 in patients with CVID after administration of single and double doses of adjuvanted influenza vaccine initially and 24 ± 3 days after immunization (individual values as well as the medians and interquartile range are given). *—statistically significant differences before and after vaccination at $p \leq 0.05$ level. #—statistically significant differences between vaccination with one and two doses at the $p \leq 0.05$ level. A robust linear mixed effects model was used for calculations.

The change in the proportion of lymphocytes expressing TLR9 as a result of vaccination with two doses was +11 [+1.5; +29.4]%, and as a result of vaccination with one dose it was 0.1 [0.3; +0.3]%, in which the differences were statistically significant ($p = 0.05$) (Figure 10).

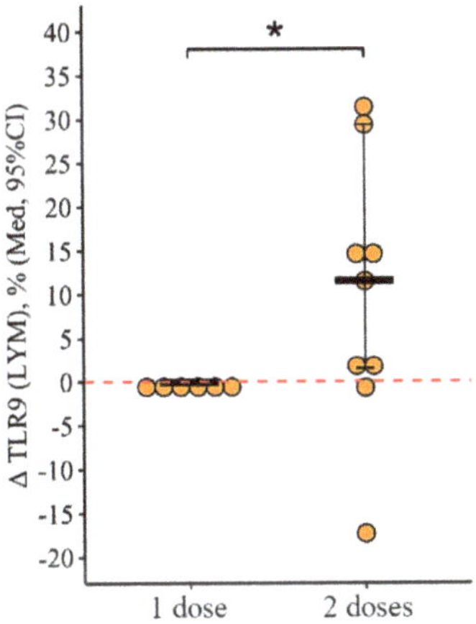

Figure 10. Delta (Δ—difference between the parameters before and 24 ± 3 days after vaccination) of the percentage of lymphocytes expressing TLR9 in patients with CVID vaccinated with single and double doses (individual values as well as the median and its 95% confidence interval). *—statistically significant differences between vaccination with one and two doses at the $p \leq 0.05$ level. Mann–Whitney test was used.

The percentage of lymphocytes expressing TLR3 and TLR8 did not change statistically significantly during vaccination, was independent of dose, and showed comparable dynamics for one and two doses.

3.2.3. Monocytes

As a result of vaccination with two doses, the initial level of monocytes expressing TLR3 and TLR9 increased by +10.7 [2.7; +29.2]% from 16.2 (9.3–27.8)% to 31.2 (17.3–39.1)% ($p = 0.05$) and by 21.4 [+7.0; +51.3]% from 38.4 (16.0–42.6)% to 49.2 (30.8–73.0)% ($p = 0.04$), respectively, while after vaccination with one dose the decrease in these parameters was 25.1 [45.6; +0.2]% from 53.1 (22.8–75.6)% to 7.3 (2.7–44.0)% ($p < 0.001$ for TLR3) and by 23.3 [57.6; 7.8]% from 66.1 (35.3–80.8)% to 6.3 (1.6–73.0)% ($p = 0.01$ for TLR9) (Figure 11a,b). The rate of change in the proportion of monocytes expressing TLR3 and TLR9 as a result of vaccination between the one- and two-dose groups differed statistically significantly ($p = 0.01$ in both cases) (Figure 12a,b).

As a result of vaccination with both one and two doses, the proportion of monocytes expressing TLR8 decreased statistically significantly—by 34.9 [67.0; 15.4]% from 67.5 (13.9–87.4)% to 7.4 (4.6–24.7)% ($p < 0.001$) in the single-dose group and by 39.0 [58.3; 10.3]% from 59.4 (26.2–62.2)% to 7.2 (1.5–23.1)% in the two-dose group (Figure 12c).

The intensity of the reduction (delta of the percentage) was not statistically significantly different between the one-dose and two-dose groups ($p = 0.96$).

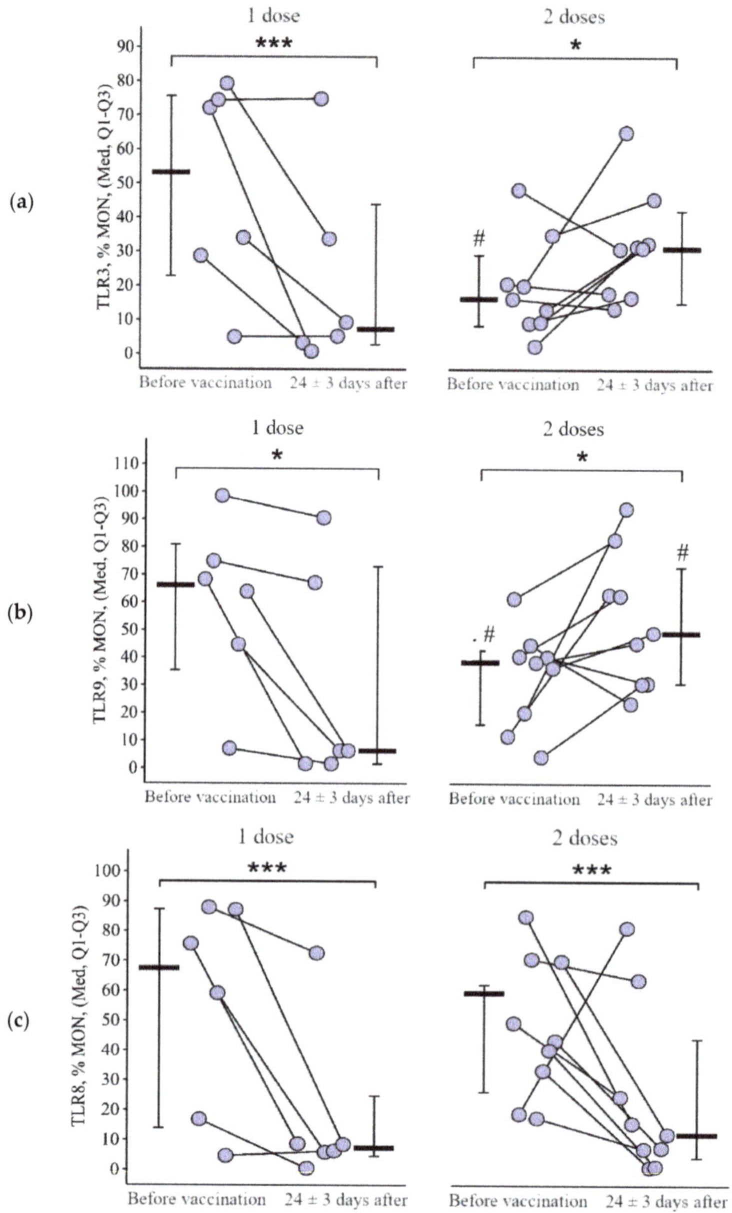

Figure 11. Percentage of monocytes expressing TLR3, TLR8, and TLR9 in patients with CVID after administration of single and double doses of adjuvanted influenza vaccine initially and 24 ± 3 days after immunization (individual values as well as the medians and interquartile range are given). *p*-value for comparing TLRs expression at different stages of vaccination (before vs. 24 ± 3 days after immunization with one vs. two vaccine doses). #—statistically significant differences between vaccination with one and two doses at the $p \leq 0.05$ level. *—statistically significant differences before and after vaccination at $p \leq 0.05$ level. ***—statistically significant differences before and after vaccination at $p < 0.001$ level. A robust linear mixed effects model was used for calculations.

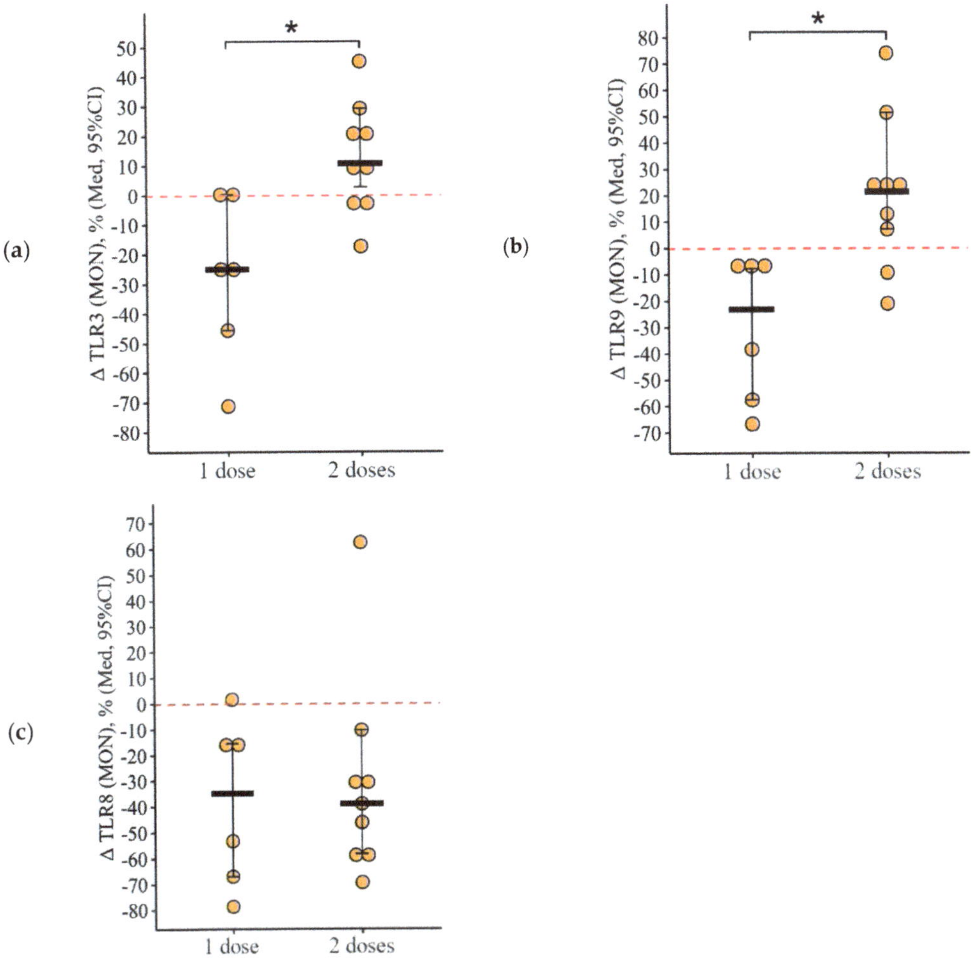

Figure 12. Delta (Δ—difference between the parameters before and 24 ± 3 days after vaccination) of the percentage of monocytes expressing TLR3, TLR8, and TLR9 in patients with CVID vaccinated with single and double doses (individual values as well as the median and its 95% confidence interval). *P*-value for comparing delta of the percentage of monocytes expressing TLR3, TLR8, and TLR9 after one vs. two vaccine doses administration. *—statistically significant differences between vaccination with one and two doses at the $p \leq 0.05$ level. Mann–Whitney test was used.

To summarize, after vaccination with one dose there was a decrease in most indicators (the proportion of granulocytes expressing TLR3 and TLR9 and the proportion of monocytes expressing TLR3, TLR8, and TLR9), while after vaccination with two doses, on the contrary, an increase in some indicators was observed (the proportion of lymphocytes expressing TLR9, and the proportion of monocytes expressing TLR3 and TLR9).

3.3. *Comparing the Content of TLR3, TLR8, and TLR9 in Healthy Volunteers after Vaccination with One Dose of the Quadrivalent Vaccine*

On the granulocytes there were no dynamics in the level of TLR expression: TLR3 was 91.65 (88.13–96.63)% and after vaccination it became 93.35 (90.7–95.53)% ($p = 0.57$); TLR9 was 92.7 (87.3–95.3)% and after vaccination it became 91.55 (88.65–95.3)% ($p = 0.67$);

and TLR8 was 86.5 (79.9–92.4)% before and became 85.03 (80.3–86.5)% (p = 0.27) after vaccination (Figure 13).

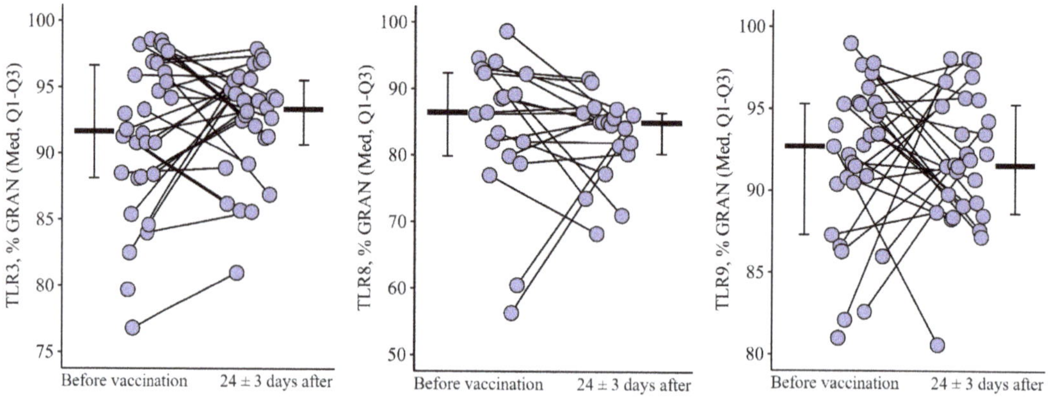

Figure 13. Percentage of granulocytes expressing TLR3, TLR8, and TLR9 in healthy volunteers after vaccination with one dose of the quadrivalent vaccine (individual values as well as the median and interquartile range are shown). Wilcoxon test for related samples was used.

On lymphocytes there was a statistically significant increase: TLR3—from 0.10 (0–0.2)% up to 0.15 (0.1–0.225)% (p = 0.03), TLR8—from 0.1 (0–0.2)% up to 0.2 (0.1–0.3)% (p = 0.05), and TLR9—from 0.1 (0–0.1)% up to 0.15 (0.1–0.3)% (p = 0.02) (Figure 14).

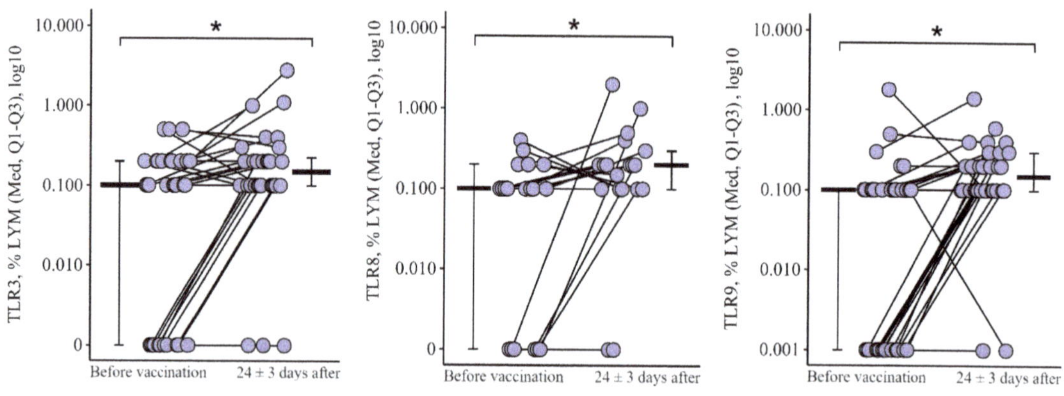

Figure 14. Percentage of lymphocytes expressing TLR3, TLR8, and TLR9 in healthy volunteers after vaccination with one dose of the quadrivalent vaccine (individual values as well as the median and interquartile range are shown). *—$p \leq 0.05$. Wilcoxon test for related samples was used.

On monocytes there is also a statistically significant increase: TLR3—from 8.6 (4.7–23.7)% to 23.6 (12.5–32.15)% (p = 0.03), TLR8—from 6.2 (2.8–17)% up to 15.25 (8.3–25.3)% (p = 0.05), and TLR9—from 12.45 (4.15–24.98)% up to 19.25 (9.7–43.98)% (p = 0.04) (Figure 15).

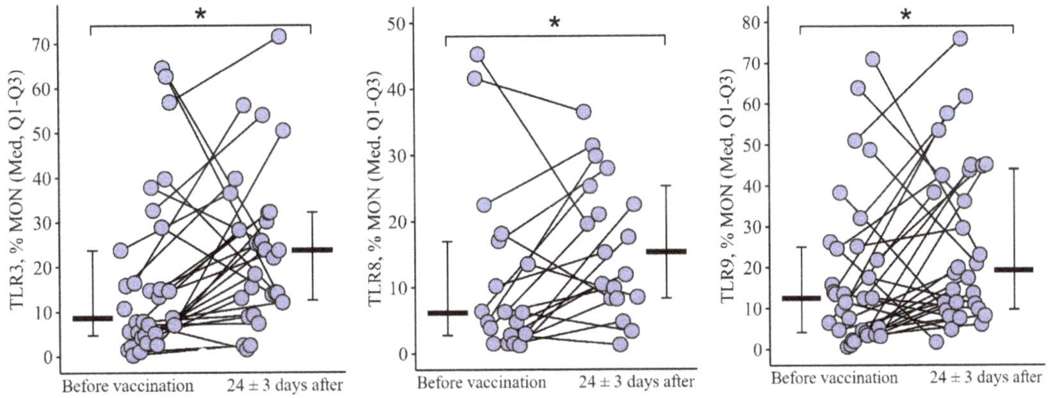

Figure 15. Percentage of monocytes expressing TLR3, TLR8, and TLR9 in healthy volunteers after vaccination with one dose of the quadrivalent vaccine (individual values as well as the median and interquartile range are shown). *—$p \leq 0.05$. Wilcoxon test for related samples was used.

4. Discussion

Immunodeficiencies affecting humoral immunity represent a heterogeneous group of diseases, of which CVID is the most common. Common variable immune deficiency (CVID) is a primary immunodeficiency, a rare disease, diverse in clinical manifestations (phenotypes), and characterized by hypogammaglobulinemia [9]. Despite the fact that variants of monogenic forms of CVID have already been described, the mode of inheritance is mainly polygenic [9,39]. Clinical manifestations of CVID include primarily recurrent sinopulmonary infections, but patients also have an increased susceptibility to developing cancer, autoimmune, or inflammatory diseases. Autoimmune disorders are diagnosed in approximately 20–25% of patients with CVID [40].

The main laboratory sign of CVID is the inability of B cells to produce antibodies, and the main therapy is immunoglobulin replacement therapy to fight infections. Decades ago, Hermans et al. [41] were the first to describe the absence of vaccine-specific antibody synthesis in a cohort of patients with CVID. Subsequently, inadequate response to vaccination became one of the key diagnostic criteria.

At the moment, it is not completely known whether vaccination with inactivated vaccines against viral infections can have a sufficient preventive effect in patients with CVID, since large cohort studies have not been conducted due to the rare occurrence of this pathology in the population, as well as the delay in diagnosis for decades [42].

However, data obtained in recent years have led to a revision of the position on vaccination of patients with IEI. Currently, vaccination against respiratory infections is recommended for all patients by the same scheme as in the healthy population [7]. There is currently no reliable data on whether the immune response in patients with CVID will be sufficient (protective), and therefore the study of the feasibility of vaccination specifically against viral infections becomes even more relevant.

Our study compared the effectiveness of influenza vaccination in patients with CVID with one dose of quadrivalent adjuvanted and two doses of trivalent adjuvanted influenza vaccine without IVIG immunotherapy for 7 weeks. An immune response was expected due to the presence of an adjuvant in the vaccines that activates innate immune factors, as well as the expanded spectrum of antigens in the vaccine, in contrast to the study using the monovaccine Pandemrix [43].

Toll-like receptors (TLRs) play an important role in the formation of the post-vaccination immune response. TLRs are pattern recognition receptors (PRRs) that play a key role in the elimination of microbial agents. They initiate a series of signal cascades, constituting the primary line of defense against microorganisms through the recruitment of phagocytes or activation of dendritic cells [13,14,44]. Additionally, these signals trigger the maturation of dendritic cells, thereby orchestrating the secondary immune response known as acquired immunity. These are conserved membrane proteins that provide alternative modes of B cell activation. TLRs are expressed in a variety of cell types, including monocytes, phagocytic cells, dendritic cells, and B cell subsets. TLRs have established roles in the physiological regulation of pro-inflammatory cytokine production, essential for mounting immune responses against bacterial, fungal, and viral infections [45]. For example, TLR3 plays a pivotal role in cross-priming naive CD8 T cells, which subsequently differentiate into the cytotoxic T cells crucial for eliminating virus-infected cells [46,47], and the stimulation of TLR2, TLR4, and TLR9 results in respiratory burst and altered expression of adhesion molecules [48,49].

Dendritic cell activation is primarily associated with TLR2, TLR3, TLR4, TLR7, and TLR9. TLRs significantly contribute to the activation of antigen-presenting cells, not only by inducing pro-inflammatory cytokine production but also by enhancing the expression of various co-stimulatory molecules necessary for effective antibody recognition [50,51]. Furthermore, TLRs govern the dendritic cell maturation and antigen-presenting function [52] that is very important and plays a crucial role in susceptibility of patients with CVID toward infections. Studies have reported that influenza vaccines activate innate effectors—comprising both myeloid and lymphoid lineages of dendritic cells, which constitute the first line of defense against infections [53].

Besides its beneficial effects on a host's innate immune responses, the influenza vaccine also includes enhancement of phagocytosis and its anti-toxic effects, such as the reduction in free radicals [54].

In 2022 there was an analysis conducted of the safety and immunogenicity profile of an azoximer bromide polymer-adjuvanted subunit influenza vaccine, which included trials performed between 1993 and 2016 and comprised 11,736 participants aged between 6 months and 99 years [55]. It showed that Grippol family vaccines induced antibody production in both children and adults up to 60 years at levels similar to vaccines with the standard amount of HA. In another study of the influence of adjuvanted vaccines with azoximer bromide on the effectors of inborn immunity it was shown that all evaluated (split, subunit, and adjuvanted) influenza vaccines elicited a statistically significant ($p < 0.05$) increase in the counts of granulocytes expressing TLR2, TLR6, TLR8, and TLR9 in peripheral blood mononuclear cell (PBMC) cultures when compared to unstimulated cells [56]. But unlike "classic" vaccines, adjuvanted vaccine showed high induction potential on TLR9- and TLR8-expressing cells, compared to the subunit vaccine ($p = 0.012$ and $p < 0.001$, respectively) and split vaccine ($p = 0.003$ and $p < 0.001$, respectively) possibly because of the co-stimulating effect of the adjuvant in the adjuvanted vaccine.

We studied expressions of TLR3, TLR8, and TLR 9, which like TLR7, are mainly located in the endosomal compartment; this location allows these receptors to recognize DNA and RNA breakdown products of viral and bacterial origin. TLR9 is expressed on human B cells. Activation of TLR3 on peripheral blood mononuclear cells, which include lymphocytes (T, B, and NK cells), monocytes and dendritic cells, as well as fibroblasts in CVID, leads to normal production of IFN-α and IFN-β, potentially providing adequate protection against viruses.

Our study also obtained data on the activation of TLR3 expressed on granulocytes and lymphocytes after repeated administration of one dose of adjuvanted influenza vaccine (as is recommended according to present guidelines—annually) and on monocytes upon administration of two doses simultaneously and, conversely, the absence of TLR3 activation upon the first vaccination with one dose. Consequently, annual vaccination against influenza using the same adjuvanted vaccine contributes to temporary nonspecific prevention of respiratory infections [57–63]. Simultaneous administration of two doses of vaccines also contributes to TLR3 expression which is observed only on monocytes, and, probably, the protective effect against other respiratory infections may be less pronounced.

One of the most powerful stimulators of B cell activation and maturation are endosomal toll-like receptors, the agonists of which are single-stranded RNA or various synthetic agonists (TLR7) and unmethylated CpG motifs in microbial DNA (CpG DNA) (TLR9), respectively [64]. Binding of TLR9 by CpG DNA has been shown to activate normal B cells, enhance the expression of co-stimulatory molecules, trigger the secretion of IL-6 and IL-10, and mediate T independent isotype switching and the production of specific antibodies independent of B cell receptor (BCR) binding [18–22]. Our study also found significant expression of TLR9, both one year after repeated vaccination with one dose and with the simultaneous administration of two doses of adjuvanted influenza vaccines, which was not observed after the first year of vaccination of patients with CVID with one dose. The role of TLR8, according to the results of our study, is less significant in the mechanisms of formation of post-vaccination immunity in patients with CVID, although one cannot fail to note the increase in its expression on lymphocytes a year after repeated vaccination with one dose of the study cohort of patients.

Moving on to the analysis of the data on TLR receptors in patients with CVID, it is necessary to indicate the results of studying similar indicators in healthy people, who were vaccinated as a control group and showed a statistically significant increase in the expression of all toll-like receptors, types 3, 8, and 9, on all studied immunocytes, except for granulocytes. While in patients with CVID there is an increase in the expression of intracellular receptors TLR3 and TLR9 also on granulocytes. On lymphocytes the dynamics of the expression of TLR3 and TLR8 is similar to that in healthy participants and is partially observed at the trend level after repeated vaccination a year later as well as with TLR9 after administration of two doses. On monocytes from patients with CVID, regardless of the immunization regimen, completely comparable results were obtained on the expression of all intracellular receptors, which probably indicates the activation of innate immune parameters and nonspecific protection against pathogens and an enhanced immune response.

Author Contributions: A.M.K.: investigation and writing—original draft; E.A.L.: conceptualization and writing—review and editing; N.K.A.: methodology, resources, data curation, and writing—review and editing; A.E.V.: visualization and formal analysis; S.A.S.: investigation; E.A.K.: investigation; A.V.L.: investigation; A.A.P.: visualization; T.V.L.: writing—review and editing; M.P.K.: methodology, resources, and project administration. All authors have read and agreed to the published version of the manuscript.

Funding: This research received no external funding.

Institutional Review Board Statement: The study was conducted in accordance with the Declaration of Helsinki, and approved by the Institutional Ethics Committee of National Research Center—Institute of Immunology Federal, Medical-Biological Agency of Russia (protocol code №11-1, date of approval 12.11.2018 and protocol code №7, date of approval 08.07.2019). This study was based on the ethical principles and recommendations of the WHO and the Russian Ministry of Health. All patients signed an informed consent form for participation in this study before the beginning of the research.

Informed Consent Statement: Informed consent was obtained from all subjects involved in the study.

Data Availability Statement: Not applicable.

Conflicts of Interest: The authors declare no competing interests. The authors declare that the research was conducted in the absence of any commercial or financial relationships that could be construed as a potential conflict of interest.

Abbreviations

CVID	common variable immune deficiency
IEI	inborn errors of immunity
IVIG	intravenous *immunoglobulin*
PAMPs	pathogen associated molecular patterns
PBMC	peripheral blood mononuclear cell
PID	*primary immune deficiency*
PRRs	pattern recognition receptors
aQIV	adjuvant quadrivalent inactivated vaccine
aTIV	adjuvant trivalent inactivated vaccine
TLRs	toll-like receptors

References

1. Tangye, S.G.; Al-Herz, W.; Bousfiha, A.; Chatila, T.; Cunningham-Rundles, C.; Etzioni, A.; Franco, J.L.; Holland, S.M.; Klein, C.; Morio, T.; et al. Human Inborn Errors of Immunity: 2022 Update on the Classification from the International Union of Immunological Societies Expert Committee. *J. Clin. Immunol.* **2022**, *42*, 1473–1507. [CrossRef] [PubMed]
2. Wasserman, R.L. Recombinant human hyaluronidase-facilitated subcutaneous immunoglobulin infusion in primary immunodeficiency diseases. *Immunotherapy* **2017**, *9*, 1035–1050. [CrossRef] [PubMed]
3. Eichner, M.; Schwehm, M.; Eichner, L.; Gerlier, L. Direct and indirect effects of influenza vaccination. *BMC Infect. Dis.* **2017**, *17*, 308. [CrossRef]
4. World Health Organization. *Global $ 2019–2030*; World Health Organization: Geneva, Switzerland, 2019; 34p.
5. Dubé, È.; Farrands, A.; Lemaitre, T.; Boulianne, N.; Sauvageau, C.; Boucher, F.D.; Tapiero, B.; Quach, C.; Ouakki, M.; Gosselin, V.; et al. Overview of knowledge, attitudes, beliefs, vaccine hesitancy and vaccine acceptance among mothers of infants in Quebec, Canada. *Hum. Vaccines Immunother.* **2019**, *15*, 113–120. [CrossRef] [PubMed]
6. Krammer, F.; Smith, G.J.D.; Fouchier, R.A.M.; Peiris, M.; Kedzierska, K.; Doherty, P.C.; Palese, P.; Shaw, M.L.; Treanor, J.; Webster, R.G.; et al. Influenza. *Nat. Rev. Dis. Primers* **2018**, *4*, 4. [CrossRef] [PubMed]
7. Eibl, M.M.; Wolf, H.M. Vaccination in patients with primary immune deficiency, secondary immune deficiency and autoimmunity with immune regulatory abnormalities. *Immunotherapy* **2015**, *7*, 1273–1292. [CrossRef] [PubMed]
8. Rubin, L.G.; Levin, M.J.; Ljungman, P.; Davies, E.G.; Avery, R.; Tomblyn, M.; Bousvaros, A.; Dhanireddy, S.; Sung, L.; Keyserling, H.; et al. 2013 IDSA Clinical Practice Guideline for Vaccination of the Immunocompromised Host. *Clin. Infect. Dis.* **2013**, *58*, e44–e100. [CrossRef]
9. Bonilla, F.A.; Barlan, I.; Chapel, H.; Costa-Carvalho, B.T.; Cunningham-Rundles, C.; de la Morena, M.T.; Espinosa-Rosales, F.J.; Hammarström, L.; Nonoyama, S.; Quinti, I.; et al. International Consensus Document (ICON): Common Variable Immunodeficiency Disorders. *J. Allergy Clin. Immunol. Pr.* **2015**, *4*, 38–59. [CrossRef]
10. Sobh, A.; Bonilla, F.A. Vaccination in Primary Immunodeficiency Disorders. *J. Allergy Clin. Immunol. Pr.* **2016**, *4*, 1066–1075. [CrossRef]
11. Janeway CA, Medzhitov, R. Introduction: The role of innate immunity in the adaptive immune response. *Semin Immunol.* **1998**, *10*, 349–350. [CrossRef]
12. Dzhitov, R.; Janeway, C. Innate Immunity. *N. Engl. J. Med.* **2000**, *343*, 338–344. [CrossRef] [PubMed]
13. Booth, J.; Wilson, H.; Jimbo, S.; Mutwiri, G. Modulation of B cell responses by Toll-like receptors. *Cell Tissue Res.* **2010**, *343*, 131–140. [CrossRef]
14. Bernasconi, N.L.; Onai, N.; Lanzavecchia, A.; Cotta, C.V.; Zhang, Z.; Kim, H.-G.; Klug, C.A. A role for Toll-like receptors in acquired immunity: Up-regulation of TLR9 by BCR triggering in naive B cells and constitutive expression in memory B cells. *Blood* **2003**, *101*, 4500–4504. [CrossRef] [PubMed]
15. Bourke, E.; Bosisio, D.; Golay, J.; Polentarutti, N.; Mantovani, A. The toll-like receptor repertoire of human B lymphocytes: Inducible and selective expression of TLR9 and TLR10 in normal and transformed cells. *Blood* **2003**, *102*, 956–963. [CrossRef] [PubMed]
16. Trinchieri, G.; Sher, A. Cooperation of Toll-like receptor signals in innate immune defence. *Nat. Rev. Immunol.* **2007**, *7*, 179–190. [CrossRef]
17. Yu, J.E.; Knight, A.K.; Radigan, L.; Marron, T.U.; Zhang, L.; Sanchez-Ramón, S.; Cunningham-Rundles, C. Toll-like receptor 7 and 9 defects in common variable immunodeficiency. *J. Allergy Clin. Immunol.* **2009**, *124*, 349–356e3. [CrossRef]
18. Hannoun, C. The evolving history of influenza viruses and influenza vaccines. *Expert Rev. Vaccines* **2013**, *12*, 1085–1094. [CrossRef]
19. Recommended composition of influenza virus vaccines for use in the 2018–2019 northern hemisphere influenza sea-son. *REH* **2018**, *93*, 133–141.
20. Ambrose, C.S.; Levin, M.J. The rationale for quadrivalent influenza vaccines. *Hum. Vaccines Immunother.* **2012**, *8*, 81–88. [CrossRef]
21. Caini, S.; Huang, Q.S.; Ciblak, M.A.; Kusznierz, G.; Owen, R.; Wangchuk, S.; Henriques, C.M.P.; Njouom, R.; Fasce, R.A.; Yu, H.; et al. Epidemiological and virological characteristics of influenza B: Results of the Global Influenza B Study. *Influ. Other Respir. Viruses* **2015**, *9*, 3–12. [CrossRef]

22. Kharit, S.M.; Рудакоba, A.; Yckob, A.; Konovalova, L.N.; Lobzin, Y.V. The averted costs due to influenza vaccination with trivalent and quadrivalent vaccines. *J. Infectology* **2017**, *9*, 17–22. [CrossRef]
23. Eichner, M.; Schwehm, M.; Hain, J.; Uphoff, H.; Salzberger, B.; Knuf, M.; Schmidt-Ott, R. 4Flu-an individual based simulation tool to study the effects of quadrivalent vaccination on seasonal influenza in Germany. *BMC Infect. Dis.* **2014**, *14*, 365. [CrossRef] [PubMed]
24. Crépey, P.; De Boer, P.T.; Postma, M.J.; Pitman, R. Retrospective public health impact of a quadrivalent influenza vaccine in the United States. *Influenza Other Respir. Viruses* **2015**, *9* (Suppl. S1), 39–46. [CrossRef] [PubMed]
25. Moa, A.M.; Chughtai, A.A.; Muscatello, D.J.; Turner, R.M.; MacIntyre, C.R. Immunogenicity and safety of inactivated quadrivalent influenza vaccine in adults: A systematic review and meta-analysis of randomised controlled trials. *Vaccine* **2016**, *34*, 4092–4102. [CrossRef] [PubMed]
26. Guidance. Summary of Data to Support the Choice of Influenza Vaccination for Adults in Primary Care. Public Health England. Published 29 January 2018. Available online: https://www.gov.uk/government/publications/flu-vaccination-supporting-data-for-adult-vaccines/summary-of-data-to-support-the-choice-of-influenza-vaccination-for-adults-in-primary-care (accessed on 21 March 2022).
27. Crampton, S.P.; Voynova, E.; Bolland, S. Innate pathways to B-cell activation and tolerance. *Ann. N. Y. Acad. Sci.* **2010**, *1183*, 58–68. [CrossRef]
28. De Waure, C.; Boccalini, S.; Bonanni, P.; Amicizia, D.; Poscia, A.; Bechini, A.; Barbieri, M.; Capri, S.; Specchia, M.L.; Di Pietro, M.L.; et al. Adjuvanted influenza vaccine for the Italian elderly in the 2018/19 season: An updated health technology assessment. *Eur. J. Public Health.* **2019**, *29*, 900–905. [CrossRef]
29. Heer, A.K.; Shamshiev, A.; Donda, A.; Uematsu, S.; Akira, S.; Kopf, M.; Marsland, B.J. TLR Signaling Fine-Tunes Anti-Influenza B Cell Responses without Regulating Effector T Cell Responses. *J. Immunol.* **2007**, *178*, 2182–2191. [CrossRef]
30. Lanzavecchia, A.; Bernasconi, N.; Traggiai, E.; Ruprecht, C.R.; Corti, D.; Sallusto, F. Understanding and making use of human memory B cells. *Immunol. Rev.* **2006**, *211*, 303–309. [CrossRef]
31. Ruprecht, C.R.; Lanzavecchia, A. Toll-like receptor stimulation as a third signal required for activation of human naive B cells. *Eur. J. Immunol.* **2006**, *36*, 810–816. [CrossRef]
32. Bernasconi, N.L.; Traggiai, E.; Lanzavecchia, A. Maintenance of Serological Memory by Polyclonal Activation of Human Memory B Cells. *Science* **2002**, *298*, 2199–2202. [CrossRef]
33. Pasare, C.; Medzhitov, R. Control of B-cell responses by Toll-like receptors. *Nature* **2005**, *438*, 364–368. [CrossRef] [PubMed]
34. Good Clinical Practice. Russian Federation National Standard, GOSTR 52379-2005/Moscow. 2005. Available online: http://www.medtran.ru/rus/trials/gost/52379-2005.html (accessed on 1 April 2006).
35. Koller, M. robustlmm: An *R* Package for Robust Estimation of Linear Mixed-Effects Models. *J. Stat. Softw.* **2016**, *75*, 1–24. [CrossRef]
36. Luke, S.G. Evaluating significance in linear mixed-effects models in R. *Behav. Res. Methods* **2016**, *49*, 1494–1502. [CrossRef]
37. Lenth Russell, V. Package 'emmeans': Estimated Marginal Means, aka Least Squares Means, Version 1.8.8.; CRAN. 2023. Available online: https://cran.r-project.org/web/packages/emmeans/emmeans.pdf (accessed on 1 April 2006).
38. Nohrin, D.Y. *Laboratory Workshop on Biostatistics*; Publishing House of Chelyabinsk State University: Chelyabinsk, Russia, 2018; 289p.
39. Kienzler, A.; Hargreaves, C.E.; Patel, S.Y. The role of genomics in common variable immunodeficiency disorders. *Clin. Exp. Immunol.* **2017**, *188*, 326–332. [CrossRef] [PubMed]
40. Boileau, J.; Mouillot, G.; Gérard, L.; Carmagnat, M.; Rabian, C.; Oksenhendler, E.; Pasquali, J.-L.; Korganow, A.-S. Autoimmunity in common variable immunodeficiency: Correlation with lymphocyte phenotype in the French DEFI study. *J. Autoimmun.* **2011**, *36*, 25–32. [CrossRef]
41. Hermans, P.E.; Diaz-Buxo, J.A.; Stobo, J.D. Idiopathic late-onset immunoglobulin deficiency: Clinical observations in 50 patients. *Am. J. Med.* **1976**, *61*, 221–237. [CrossRef]
42. Hong, R.; Agrawal, S.; Gollapudi, S.; Gupta, S. Impaired Pneumovax-23-Induced Monocyte-Derived Cytokine Production in Patients with Common Variable Immunodeficiency. *J. Clin. Immunol.* **2010**, *30*, 435–441. [CrossRef]
43. Gardulf, A.; Abolhassani, H.; Gustafson, R.; Eriksson, L.E.; Hammarström, L. Predictive markers for humoral influenza vaccine response in patients with common variable immunodeficiency. *J. Allergy Clin. Immunol.* **2018**, *142*, 1922–1931.e2. [CrossRef]
44. Cunningham-Rundles, C.; Radigan, L.; Knight, A.K.; Zhang, L.; Bauer, L.; Nakazawa, A. TLR9 Activation Is Defective in Common Variable Immune Deficiency. *J. Immunol.* **2006**, *176*, 1978–1987. [CrossRef]
45. Engel, A.L.; E Holt, G.; Lu, H. The pharmacokinetics of Toll-like receptor agonists and the impact on the immune system. *Expert Rev. Clin. Pharmacol.* **2011**, *4*, 275–289. [CrossRef]
46. Schulz, O.; Diebold, S.S.; Chen, M.; Näslund, T.I.; Nolte, M.A.; Alexopoulou, L.; Azuma, Y.-T.; Flavell, R.A.; Liljestrom, P.; Reis E Sousa, C. Toll-like receptor 3 promotes cross-priming to virus-infected cells. *Nature* **2005**, *433*, 887–892. [CrossRef] [PubMed]
47. Azuma, M.; Ebihara, T.; Oshiumi, H.; Matsumoto, M.; Seya, T. Cross-priming for antitumor CTL induced by soluble Ag + polyI:C depends on the TICAM-1 pathway in mouse $CD11c^{+}/CD8\alpha^{+}$dendritic cells. *OncoImmunology* **2012**, *1*, 581–592. [CrossRef] [PubMed]
48. Yoshimura, A.; Ohishi, H.M.M.; Aki, D.; Hanada, T. Regulation of TLR signaling and inflammation by SOCS family proteins. *J. Leukoc. Biol.* **2004**, *75*, 422–427. [CrossRef] [PubMed]

49. Hyang, L.T.; Paredes, C.J.; Papoutsakis, E.T.; Miller, W.M. Gene expression analysis illuminates the transcriptional programs un-derlying the functional activity of ex vivo-expanded granulocytes. *Physiol. Genom.* **2007**, *31*, 114–125. [CrossRef] [PubMed]
50. Kawai, T.; Akira, S. The role of pattern-recognition receptors in innate immunity: Update on Toll-like receptors. *Nat. Immunol.* **2010**, *11*, 373–384. [CrossRef]
51. Kumagai, Y.; Takeuchi, O.; Kato, H.; Kumar, H.; Matsui, K.; Morii, E.; Aozasa, K.; Kawai, T.; Akira, S. Alveolar macrophages are the primary interferon-alpha producer in pulmonary infec-tion with RNA viruses. *Immunity* **2007**, *27*, 240–252. [CrossRef]
52. Manicassamy, S.; Pulendran, B. Modulation of adaptive immunity with Toll-like receptors. *Semin. Immunol.* **2009**, *21*, 185–193. [CrossRef]
53. Khromova, E.A.; Akhmatova, E.A.; Skhodova, S.A.; Semochkin, I.A.; Khomenkov, V.G.; Akhmatova, N.K.; Kostinov, M.P. Effect of in-fluenza vaccines on subpopulations of blood dendritic cells. *J. Microbiol. Epidemiol. Immunobiol.* **2016**, *5*, 23–28. (In Russian) [CrossRef]
54. Committee for Proprietary Medicinal Products (CPMP). *CPMP: Note for Guidance on Preclinical Pharmacological and Toxico-Logical Testing of Vaccines*; CPMP/SWP/565/95 adopted December 97; CPMP: Amsterdam, The Netherlands, 1997.
55. Kompier, R.; Neels, P.; Beyer, W.; Hardman, T.; Lioznov, D.; Kharit, S.; Kostinov, M. Analysis of the safety and immunogenicity profile of an azoximer bromide poly-mer-adjuvanted subunit influenza vaccine. *F1000Research* **2022**, *11*, 259. [CrossRef]
56. Kostinov, M.P.; Akhmatova, N.K.; Khromova, E.A.; Skhodova, S.A.; Stolpnikova, V.N.; Cherdantsev, A.P.; Vlasenko, A.E. The Impact of Adjuvanted and Non-Adjuvanted Influenza Vaccines on the Innate and Adaptive Immunity Effectors. *Infect. Dis.* **2018**, *1*, 83–109.
57. Kostinov, M.P.; Chuchalin, A.D. *Guidance for Clinical Immunology in Respiratory Medicine*, 2nd ed.; MDV Group: Moscow, Russia, 2018; p. 304. ISBN 978-5-906748-08-9. (In Russian)
58. Pružinec, P.; Chirun, N.; Sveikata, A. The safety profile of Polyoxidonium in daily practice: Results from postauthorization safety study in Slovakia. *Immunotherapy* **2018**, *10*, 131–137. [CrossRef]
59. Pinegin, B.V.; Ilyina, N.I.; Latysheva, T.V. *Polyoxidonium in Clinical Practice*; Karaulov, A.V., Ed.; GEOTAR–Media: Moscow, Russia, 2008; p. 136. (In Russian)
60. Karaulov, A.V.; Bykov, A.S.; Volkova, N.V. Review of Grippol Family Vaccine Studies and Modern Adjuvant Development. *Epidemiol. Vaccinal Prev.* **2019**, *18*, 101–119. [CrossRef]
61. Alexia, C.; Cren, M.; Louis-Plence, P.; Vo, D.-N.; El Ahmadi, Y.; Dufourcq-Lopez, E.; Lu, Z.-Y.; Hernandez, J.; Shamilov, F.; Chernysheva, O.; et al. Polyoxidonium® Activates Cytotoxic Lymphocyte Responses Through Dendritic Cell Maturation: Clinical Effects in Breast Cancer. *Front. Immunol.* **2019**, *10*, 2693. [CrossRef] [PubMed]
62. Kostinov, M.P.; Akhmatova, N.K.; Khromova, E.A.; Skhodova, S.A.; Stolpnikova, V.N.; Cherdantsev, A.P.; Vlasenko, A.E. The impact of adjuvanted and non-adjuvanted influenza vaccines on in vitro lymphocyte immunophenotype. *Russ. J. Infect. Immun.* **2023**, *13*, 430–438. [CrossRef]
63. Kostinov, M.P.; Cherdantsev, A.P.; Shmitko, A.D.; Akhmatova, N.K.; Kostinova, A.M.; Praulova, D.A.; Polyschuk, V.B.; Protasov, A.D.; Zhestkov, A.V.; Tezikov, Y.V. The efficacy of immunoadjuvant-containing influenza vaccines in pregnancy. In *Vaccines*; Afrin, F., Hemeg, H., Ozbak, H., Eds.; 2017; Section 3, Chapter 4; pp. 67–93. [CrossRef]
64. Salem, M.L.; El-Hennawy, D. The possible beneficial adjuvant effect of influenza vaccine to minimize the severity of COVID-19. *Med. Hypotheses* **2020**, *140*, 109752. [CrossRef] [PubMed]

viruses

Article

Assessment of Peste des Petits Ruminants Antibodies in Vaccinated Pregnant Ewes of Kazakh Breed Fine-Fleeced and Determination of the Decreasing Trend of Maternal Immunity in Their Lambs

Zhanat Amanova *, Sholpan Turyskeldy, Zhanat Kondybaeva, Zhanna Sametova, Abdurakhman Usembai, Aslan Kerimbayev and Yerbol Bulatov

Research Institute for Biological Safety Problems, Gvardeiskiy 080409, Kazakhstan; smankizi@biosafety.kz (S.T.); zh.kondybaeva@biosafety.kz (Z.K.); zh.sametova@biosafety.kz (Z.S.); a.ussenbay@biosafety.kz (A.U.); a.kerimbayev@biosafety.kz (A.K.); ye.bulatov@biosafety.kz (Y.B.)

* Correspondence: zh.amanova@biosafety.kz; Tel.: +7-(701)-415-63-79

Abstract: In this article, we first assessed peste des petits ruminants (PPR) antibodies in vaccinated pregnant ewes of Kazakh breed fine-fleeced immunized with the PPR vaccine and the duration of maternal immunity in their lambs. Ewes in the last trimester of pregnancy and gestation were immunized with a vaccine from the Nigeria 75/1 strain of the PPR virus (PPRV) produced by the Research Institute of Biological Safety Problems (RIBSP), Kazakhstan. Serum samples from lambs born from vaccinated and unvaccinated ewes were collected a week after birth and at intervals of 7 days for 18 weeks after birth. Serum samples collected from lambs were tested for PPR antibodies using competitive ELISA and virus neutralization test (VNT). Maternal antibodies (MAs) in lambs born from vaccinated ewes were detected for up to 18 weeks, with a tendency to decrease starting at week 14, and by the end of the experiment receded below the protective level (<1:8). In the blood serum of a 14-week-old lamb with MAs (1:8), post vaccination with a field dose (10^3 TCID50) of the vaccine against PPR, the titers of protective antibodies against PPRV increased to 1:16 on day 14 post vaccination, and the lamb was protected from infection with the field PPRV. A lamb of the same age with MAs in the 1:8 titer was 100% protected from infection with the field PPRV. Therefore, it is recommended that lambs of the Kazakh fine-wool breed be immunized from the age of 14 weeks or older to avoid a period of susceptibility.

Keywords: peste des petits ruminants; ewes; lambs; vaccination; passive immunity

Citation: Amanova, Z.; Turyskeldy, S.; Kondybaeva, Z.; Sametova, Z.; Usembai, A.; Kerimbayev, A.; Bulatov, Y. Assessment of Peste des Petits Ruminants Antibodies in Vaccinated Pregnant Ewes of Kazakh Breed Fine-Fleeced and Determination of the Decreasing Trend of Maternal Immunity in Their Lambs. *Viruses* **2023**, *15*, 2054. https://doi.org/10.3390/v15102054

Academic Editors: Pietro Hiram Guzzi, Marianna Milano and Jayanta Kumar Das

Received: 31 August 2023
Revised: 29 September 2023
Accepted: 3 October 2023
Published: 6 October 2023

1. Introduction

Peste des petits ruminants (PPR) is a highly contagious, infectious viral disease of small ruminant species that is caused by the peste des petits ruminants virus (PPRV), the prototype member of the *Morbillivirus* genus in the Paramyxoviridae family [1]. The disease is currently endemic in most of Africa, the Middle East, South Asia and China [2]. Despite strict control measures including statutory regulations along with the availability of vaccines and diagnostics, PPR remains a constant threat [3]. Currently, PPR is not an endemic disease for the Republic of Kazakhstan. However, it should be noted that the results obtained by Lundervold et al. in 1997–1998, when conducting a study on the detection of antibodies in sheep, goats and cattle in Central Kazakhstan, showed that PPRV could circulate in the country unnoticed [4]. Five years later, in 2003, an outbreak of PPR occurred in the Turkestan region in small ruminant livestock [4,5]. Since then, until the end of 2014, the OIE has not received official reports of cases of infection with the PPRV from Kazakhstan. However, more than 10 years later, PPR caused three outbreaks in individual farms at the end of 2014 in Southern Kazakhstan (Zhambyl region). Based on partial N gene sequencing, the identified strains showed high similarity to PPRV strains from China from 2013/2014, suggesting PPR transboundary spread between the two countries. The three outbreaks did not have any obvious epidemiological linkages, suggesting that PPRV may have been persistently present at a subclinical level despite vaccination efforts [6,7]. These data are of the greatest concern because Kazakhstan is home to the world's three largest populations of *Saiga* antelope, which are susceptible to the PPR virus. However, a serological study conducted in Kazakhstan in the period from 2012 to 2014 did not reveal *Saiga* antelopes seropositive for PPR [6].

To date, due to the outbreak of PPR in Mongolia, the regions of Eastern Kazakhstan are at risk of PPR infection.

To achieve PPR eradication, which is targeted for 2030, a PPR Global Control and Eradication Strategy (PPR GCES) was developed, based on a progressive reduction in PPR incidence and spread through targeted vaccination [8].

Within the territory of the Republic of Kazakhstan, annual routine vaccination of small ruminants against PPR is carried out in border regions in order to prevent the introduction of PPR from neighboring countries that are disadvantaged by this disease. In Kazakhstan, live attenuated vaccines are mainly used for the prevention of PPR, which create 1-year immunity against PPR.

To increase the immunogenicity of the PPR vaccine, we developed a live vaccine based on the Nigeria/75/1 strain. The live vaccine from the Nigeria/75/1 strain is the most widely used PPR vaccine approved by the OIE [9]. It has been reported that the Nigeria 75/1 strain of the PPRV causes persistent immunity in once-immunized animals for up to 3 years [10].

There are few reports on the duration of persistence of maternal antibodies in lambs/kids born from ewes vaccinated against PPR [11]. Maternal antibodies transmitted to lambs from vaccinated ewes through colostrum protect newborn lambs from infection for a certain period. However, maternal antibodies can negatively affect the effectiveness of vaccination with live vaccines. Therefore, before starting vaccination, it is recommended to check the immunological status of lambs.

To avoid negative consequences when immunizing young lambs and to determine the optimal age of lambs for vaccination against PPR, we studied the persistence of maternal antibodies in lambs born from ewes of Kazakh breed fine-fleeced vaccinated against PPR.

2. Materials and Methods

2.1. Animals

In the southern regions of Kazakhstan, sheep mating occurs more often in the months of October and November, and lambing occurs most often in March and April. In this regard, in September, we purchased sheep of Kazakh breed fine-fleeced of either sex (3 males and 15 females aged 12–12.5 months). All animals were dewormed by oral administration of albendazole. Food and water were available in unlimited quantities throughout the experiment. The animals were labeled and kept in isolation in the RIBSP quarantine zone for 30 days. Body temperature was regularly measured in the animals, and blood serum was collected to determine the presence of specific antibodies to PPRV. Detection of antibodies to PPRV in sheep sera was performed using a virus neutralization test (VNT) [12]. The animals were not found to have specific antibodies to the PPRV, and they had not previously been vaccinated against this disease.

Ewes in the third trimester of pregnancy were used for the research. Pregnancy periods in ewes were determined using ultrasonography. Estrus synchronization in ewes was not performed; therefore, as soon as the lambs were born, they were included in experimental studies. Newborn lambs were kept together with their mothers so that they could suck colostrum freely.

Experimental studies were conducted in compliance with international and national ethical standards. The protocol was approved by the Ethics Committee for Animal Experimentation at the RIBSP (Permission number: 2908/22).

2.2. Vaccination

Out of 15 females, 10 became pregnant. Ewes (n = 6) in the 3rd trimester of pregnancy were immunized with PPR vaccine (Nigeria 75/1 strain) produced by the RIBSP, Kazakhstan. Ewes were immunized subcutaneously with a single field dose of the vaccine (1.0×10^3 TCID 50/mL). Four healthy ewes in the 3rd trimester of pregnancy were used as control animals (unvaccinated). All of the vaccinated and unvaccinated ewes were monitored daily for clinical signs of PPR, and rectal temperature was measured for 21 days post vaccination (dpv).

2.3. Blood Sample Collection

Blood samples of vaccinated and control ewes taken on days 0, 7, 14 and 21 were tested for the presence of antibodies in blood sera. Blood samples were taken from newborn lambs at the age of 1 week with an interval of 7 days for 18 weeks. The obtained blood sera of experimental animals were inactivated at a temperature of 56 °C for 30 min and placed in freezers with a storage temperature of −20 °C.

2.4. Serology Tests

2.4.1. VNT

PPRV antibodies were detected using VNT [12] and ELISA [13]. The titer of virus neutralizing antibodies (VNA) was calculated by the Reed and Muench method [14]. The viral neutralizing activity of the serum was expressed in the neutralization index, which is the difference in the logarithms of the titer of the virus in the presence of specific and normal serum.

The highest serum dilution was taken as the antibody titer, which was able to suppress the activity of the virus injected at the specified dose in 50% of the infected culture.

VNT was carried out in two repetitions, and the average value of the two tests was used when analyzing the results of the study.

2.4.2. Competitive ELISA (c-ELISA)

Blood sera of experimental animals were additionally examined for the presence of antibodies in c-ELISA. As an additional test for detecting anti-PPRV antibodies to nucleoprotein (NP) in the blood sera of experimental animals, a c-ELISA kit (ID Screen®PPR Competition (PPRC-4P), ID.vet, Montpellier, France) was used [13]. The ELISA test was performed according to the manufacturer's instructions. The results of the ELISA analysis were read using a spectrophotometer at a wavelength of 450 nm. The test results were considered reliable if the average value of the ODc− > 0.7 and the ratio of the average values of the ODc+/ODc− < 0.3. The S/N percentage (S/N%) value was calculated for each sample using the formula sample OD/ODc− ×100%. Samples showing a ratio of S/N% = 50% were considered positive, whereas S/N > 60% were considered negative. Samples with a ratio of 50% < S/N% ≤ 60% were considered doubtful.

2.5. Molecular Test

Real-Time RT-PCR

Total RNA was isolated from the collected blood samples and swabs using a kit (ID Gene™ Mag Fast Extraction Kit (IDvet, Grabels, France). Aliquots of RNA were stored in a freezer at a temperature of −70 °C until tested.

All types of samples were analyzed by real-time RT-PCR to detect the presence of viral nucleic acid with the RT-qPCR kit (ID Gene Tempeste des Petits Ruminants Duplex, IDvet genetics, Grabels, France), using the Applied Biosystems 7500 system (Carlsbad, CA, USA).

2.6. Evaluation of the Protective Effectiveness of Passive Immunity

When the first two newborn lambs reached the age of 14 weeks, they were used for research to assess the effectiveness of passive immunity. Before starting the study, the lambs' blood sera were tested for the presence of maternal antibodies in c-ELISA. The titer of maternal antibodies circulating in the blood sera of these lambs was 1:8. One of them was subcutaneously immunized with a vaccine at a dose of 1.0×10^3 TCID50/mL. On 0, 7, and 14 dpv, blood samples were collected to determine the titer of serum antibodies.

The second 14-week-old lamb was used to assess protective passive immunity.

As unvaccinated animals, 2 lambs of 14 weeks of age were used, born first from unvaccinated ewes.

2.7. Challenge Study

On 14 dpv, 2 experimental lambs (one unvaccinated and one vaccinated) born with maternal antibodies and 2 unvaccinated lambs were inoculated with the control strain of PPRV (Kentau-7) [15] subcutaneously into the subscapular region in a volume of 1.0 mL ($10^{5.0}$ TCID50). After control infection, experimental and unvaccinated animals were observed for 14 days with daily measurement of body temperature, sampling of swabs (nasal, ocular, oral and rectal) and blood, as well as the identification of clinical signs of PPR. Clinical signs of PPR detected in all animals were evaluated in points [16,17]. During the challenge study, severely affected lambs were withdrawn from the study and humanely euthanized.

2.8. Data Analysis

During the statistical analysis, GraphPad Prism software (version 8.0.1.) was used. Differences between antibody titers and between temperature and clinical parameters of experimental and control animals after immunization and infection with a control virus were determined using bilateral ANOVA tests. The value of $p \leq 0.05$ indicated that there was a significant difference between the data obtained. The difference in efficiency between the groups was compared using a one-sided Fisher exact criterion for two proportions at an alpha level of <0.05.

3. Results

3.1. Adverse Reaction Monitoring

Post vaccination, the physiological state of pregnant ewes was within normal limits. The pregnancies of the ewes proceeded without any complications and as a result, healthy lambs were born. On the 2nd dpv, two ewes had a local reaction in the form of swellings with a diameter of 0.6 cm^2, which disappeared within 4–5 days. During the observation period (21 dpv), the body temperature of the ewes fluctuated in the range of 38.5–39.8 °C (Figure 1).

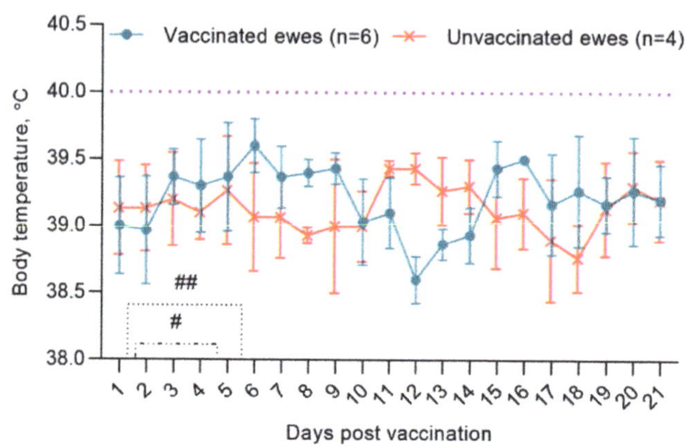

Figure 1. Dynamics of changes in rectal temperature of vaccinated and unvaccinated ewes. (#), (##) Duration of local reactions that occurred in vaccinated ewes post vaccination.

3.2. Post-Vaccination Titers of Neutralizing Antibodies to the PPRV in Pregnant Ewes

Before vaccination (day 0), all pregnant ewes were seronegative for PPRV antibodies. In the first week post immunization, the average titers of neutralizing antibodies (NAs) increased to 1.6 log2. On 14 dpv, the level of NAs to the PPRV in pregnant ewes was 5.6 log2, and the indicated antibody titer increased to 7.2 log2 on the 21st day post vaccination (Figure 2). Unvaccinated ewes were seronegative for the PPRV (Figure 2).

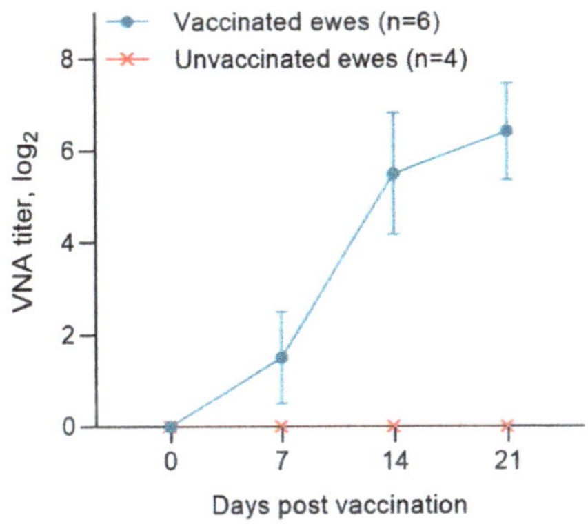

Figure 2. Accumulation of antibodies in ewes of the experimental and unvaccinated groups post immunization with the PPR vaccine.

3.3. Post-Vaccination Titers of Antibodies to the PPRV in ELISA in Pregnant Ewes

All samples of blood sera of pregnant ewes were negative before immunization with the vaccine against PPRV and had a ratio of S/N > 150%. Within 7 dpv, the S/N ratio in pregnant ewes was 87%. Protective titers of antibodies to the PPRV were found in 43% of animals. In 32% of animals, the titer of the produced antibodies exceeded the recommended threshold of S/N values (60%) for positive samples. In the remaining 25% of pregnant ewes, the results of c-ELISA were doubtful (50–60%). On 14 dpv, judging by the antibody titers, 92% of pregnant ewes were protected from PPRV, and the S/N ratio was 46%. The lowest S/N% value in vaccinated ewes (18%) was recorded on 21 dpv (Figure 3).

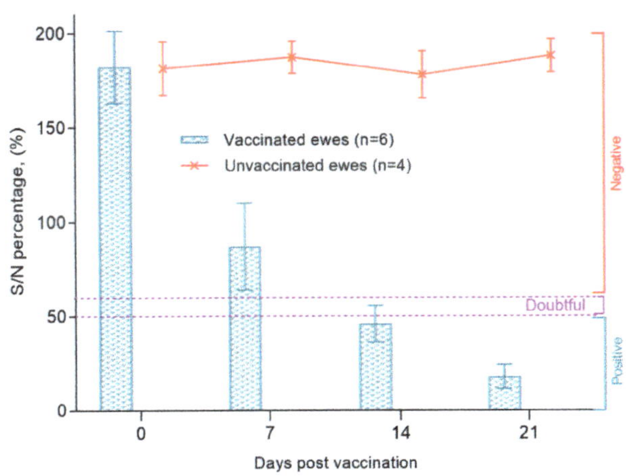

Figure 3. Antibody responses in ewes after vaccination against peste des petits ruminant virus, as measured by ELISA S/N ratio (%).

3.4. Titers of Maternal Antibodies in Lambs Born from Vaccinated Ewes

A total of 10 lambs were born from vaccinated ewes at different times. In all lambs (n = 6) born from vaccinated pregnant ewes, the level of MAs produced was in the range of 5.5 log2 and 7.2 log2, and provided 100% humoral protection of the offspring from the PPRV up to the 14th week. Starting from the 14th week, there was a gradual decrease in MAs titers below the protective level (<1:8), and by the end of the 18th week, the MA titers in lambs were in the range from 1.0 to 2.25 log2 (Figure 4). Unvaccinated lambs (n = 4) were seronegative for the PPRV (Figure 4).

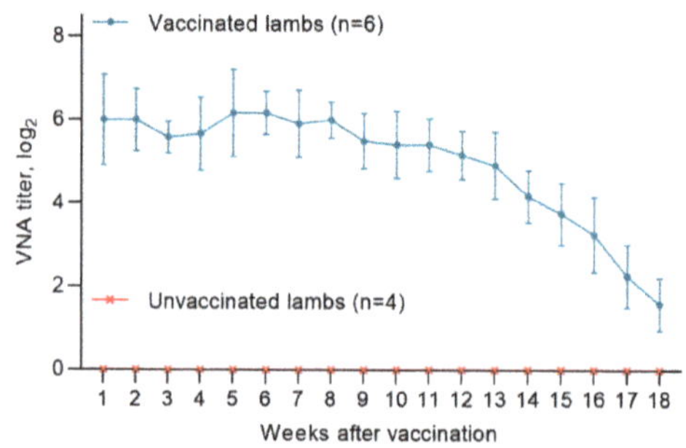

Figure 4. Dynamics of changes in the level of maternal antibodies in offspring born from vaccinated ewes.

3.5. Titers of Antibodies to the PPRV in ELISA in Lambs Born from Vaccinated Ewes

Blood serum samples were collected from all lambs a week after birth to assess the level of MAs. In all collected samples of lambs' blood serum, the S/N ratio was lower than <50%, which indicated a high level of passive immunity in newborn lambs.

The level of MAs reproduced in kids persisted until the 14th week, while there was no significant difference between antibody titers ($p > 0.05$). At the same time, the average value of S/N was in the range of 20–55% (Figure 5). The blood sera of unvaccinated lambs were negative, since all samples had a ratio of S/N > 150% (Figure 5). From the 14th week, there was a gradual increase in the S/N ratio, which increased to 150% at the end of the experiment (week 18). Accordingly, by the end of the experiment, the level of produced antibodies in all lambs exceeded >60%, which indicated a decrease in MA titers in lambs below the protective level. The serum samples of the unvaccinated lambs were negative, since all samples had an S/N ratio >150% (Figure 5).

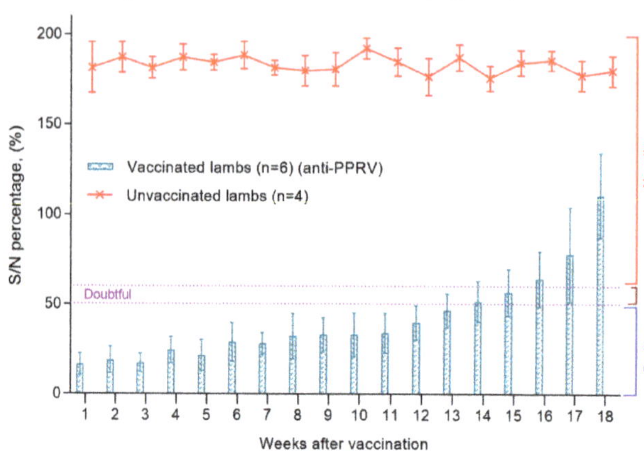

Figure 5. Assessment of the level of MAs reproduced in vaccinated and unvaccinated groups of lambs after birth from vaccinated ewes.

3.6. Post-Vaccination Titer of Neutralizing Antibodies to the PPRV in a Lamb Born with MAs

Before vaccination (day 0), the titer of NAs in the vaccinated lamb was 4.0 log2. In the first week post vaccination, the titer of NAs increased to 4.7 log2. At 2 weeks post vaccination, the level of NAs was 5.4 log2. PPRV antibodies were not found in the blood of the unvaccinated lambs.

3.7. Assessment of Viral Genomic Load in Blood and Swabs in Lambs

PPRV genome was not detected in blood samples and smears of lambs born with MAs (Figure 6). Similar results were obtained when testing all types of samples collected from 14-week-old lambs (one vaccinated and one unvaccinated), used to assess the protective effectiveness of passive immunity (Figure 6).

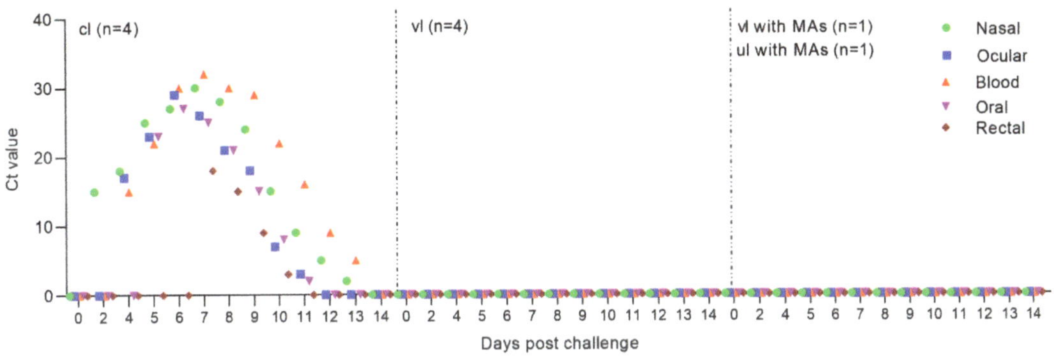

Figure 6. Quantitative determination of viral nucleic acid in blood samples and swabs of vaccinated and unvaccinated lambs post challenge. PPRV-specific RNA was measured by real-time reverse transcription PCR (RT-qPCR), and the amount of viral RNA is expressed as a Ct, a value that increases as the amount of viral RNA increases. Abbreviations: cl, challenged lambs; vl, vaccinated lambs.; vl with MAs, vaccinated lamb born with MAs; ul with MAs, unvaccinated lamb born with MAs.

In contrast, according to the results of RT-qPCR analysis, all types of samples taken from challenged lambs turned out to be positive (Figure 6). In challenged lambs, PPRV genome was detected from 2–3 days in nasal swabs, from 4–5 days in blood and ocular swabs, from 5–6 days in oral swabs and from 6–7 days in rectal swabs.

3.8. Evaluation of the Resistance of Experimental and Unvaccinated Lambs against Inoculation with the Field PPRV

Vaccinated at 14 weeks of age, the lamb was not ill with the manifestation of clinical signs of the PPRV. However, within 2 days (on the 3rd–4th day), he had an increase in body temperature to 40.0 °C, which then normalized (Figure 7); at the same time, the level of clinical indicators increased to 1 point (Figure 8). Also, the lamb had pink spot of irregular rounded shape with a diameter of 0.2 cm^2 at the injection site of the vaccine, which resolved within 4 dpc. However, the results of the RT-qPCR analysis were negative.

An unvaccinated 14-week-old lamb born with MAs post challenge with the control strain had an increase in rectal temperature to 40.2 °C on the 3rd dpc, which normalized on the 6th day of the control test; at this stage (Figure 7), the lamb's clinical score reached 1 (Figure 8). Also, a pink swelling of irregular rounded shape with a diameter of 0.3 cm^2 was observed in the lamb at the injection site of the vaccine, which resolved within 6 dpc. All types of samples and swabs collected from an unvaccinated lamb as a result of PCR analysis were negative.

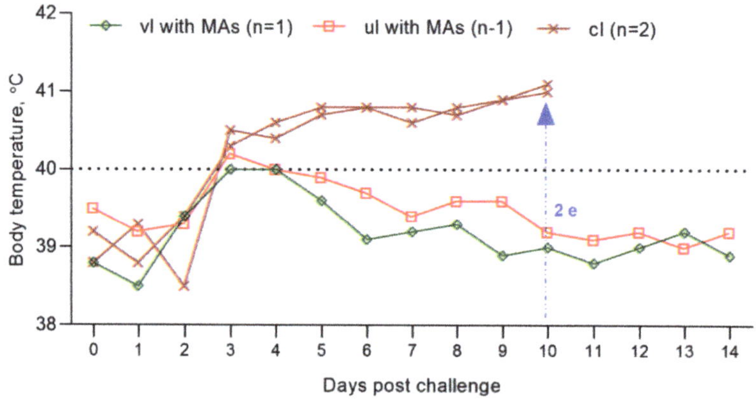

Figure 7. Dynamics of rectal temperature changes in lambs of the experimental (14-week-old lambs born with MAs (one unvaccinated and one vaccinated)) and unvaccinated groups after inoculation with the field strain PPRV. The graph shows the temperature reactions of one vaccinated, one unvaccinated (14-week-old lambs born with MAs) and two unvaccinated lambs post challenge. (2 e) On the 10th dpc, two unvaccinated lambs were humanely euthanized. Abbreviations: vl with MAs, vaccinated lamb born with MAs; ul with MAs, unvaccinated lamb born with MAs; cl, challenged lambs.

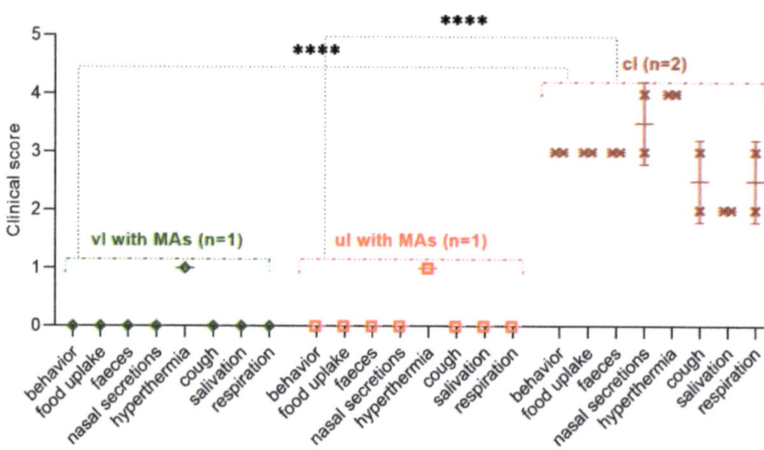

Figure 8. Manifestation of clinical signs in experimental (14-week-old lambs born with MAs (one unvaccinated and one vaccinated)) and unvaccinated groups of lambs after challenge with the strain Kentau-7. The graph shows the clinical scores of one vaccinated, one unvaccinated (14-week-old lambs born with MAs) and two unvaccinated lambs post challenge. Abbreviations: vl with MAs, vaccinated lamb born with MAs; ul with MAs, unvaccinated lamb born with MAs; cl, challenged lambs. ****, The difference between total clinical scores in experimental and challenged lambs.

On 3 dpc, challenged lambs had an increase in rectal temperature, which reached up to 40.8 °C on 5 dpc (Figure 7). Pyrexia in infected lambs lasted for 8 days. On the 10th day of the control trial, pyrexia was observed in lambs equal to a clinical score of 4 (41.1 °C) (Figures 7 and 8).

In addition, the challenged lambs had clinical signs characteristic of PPRV (liquid transparent discharge from the eyes and nose), and a pink spot with a diameter of 2.0 cm^2 was observed at the site of vaccine administration. On the 5th–6th dpc in intact lambs, nasal discharge became thick, purulent and yellow-greenish in color. In both challenged lambs, the gums became hyperemic. On the 7th–8th dpc wheezing was heard in the breasts of the lambs, while they had no appetite and were apathetic and sluggish, and liquid pus accumulated in the corners of the eyes (acute conjunctivitis). In addition, both lambs had loose stools, estimated at 3 clinical points (Figure 8). At 10 dpc, due to the deterioration in the general physiological condition of both challenged lambs, a decision was made on immediate humane euthanasia. The total clinical score in intact lambs before euthanasia was 41 points (Figure 8). After euthanasia, both lambs were subjected to necropsy (Figure 9).

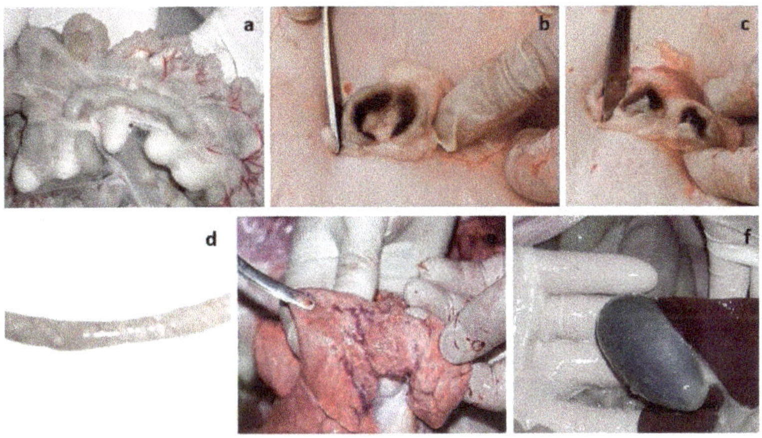

Figure 9. Pathoanatomic examination of euthanized lambs. (**a–c**) Enlarged lymph nodes. (**d**) pathological changes in the intestinal walls of euthanized lambs. (**e**) Hemorrhages under the serous membrane of the lungs. (**f**) A filled gallbladder.

3.9. Necropsy of Euthanized Lambs

During the necropsy and examination of the internal organs of the euthanized lambs, changes were found in the lymph nodes (mesenteric, pre-scapular, patellar), small intestine and lungs. The lymph nodes were enlarged (Figure 9a–c). Catarrhal inflammation and hemorrhages were observed in the mucous membrane of the small intestine (Figure 9d). Multiple dark red hemorrhages were observed under the pleura of the lungs (Figure 9e). The gallbladder was filled (Figure 9f).

4. Discussion

It is known that proper vaccination of ewes before mating is of great importance for the immune system of the offspring, since vaccines stimulate the production of MAs, which then pass into colostrum and provide newborn lambs with additional passive protection.

Conversely, vaccination of newborn lambs causes an insignificant immune response, because their immune system is not yet fully developed and cannot produce antibodies until the third or fourth week of life. In addition, it is known that vaccines administered to lambs up to 2 weeks of age bind MAs, and the young body remains defenseless.

Therefore, determining the optimal age for vaccination of young animals, which will come at a time when the level of MAs will significantly decrease and the young body will be able to develop its own protective adaptive immunity in response to the introduction of the vaccine with the formation of immunological memory, is of great importance. It should be noted that the formation of maternal immunity in offspring largely depends on the breed of the animal and the effectiveness of the strain used in the manufacture of a vaccine applicable for the immunization of pregnant animals.

Previously, in Kazakhstan, a live attenuated vaccine from the G45-MK strain was used to prevent PPR, which protected immunized animals from PPR for 1 year. Due to the short duration of immunity in vaccinated animals after immunization with a vaccine against PPR from the G-45MK strain, RIBSP employees decided to use a highly immunogenic strain (Nigeria 75/1) as the main agent for the development of a vaccine against PPR.

Although the Nigeria 75/1 strain is the main agent recommended by the OIE for the development of a vaccine against PPR, there is currently limited information on the optimal age at which lambs should be vaccinated with this vaccine.

These studies were devoted to assessing the duration of maternal immunity in lambs born from pregnant ewes of Kazakh breed fine-fleeced, immunized with a vaccine from the attenuated Nigeria 75/1 strain.

The post-vaccination results for pregnant ewes obtained by us proved the effectiveness of the vaccine used for sheep of the Kazakh breed fine-fleeced, since immunized pregnant ewes had protective antibody titers formed on day 14 (SN $\geq$ 1:8) to the PPRV (Figures 2 and 3). In previous studies, the authors reported that protective antibodies to the PPRV were formed in Kano brown goats a week post vaccination with a similar vaccine [18]. This difference in the periods for formation of protective antibody titers is probably due to differences in the types of experimental animals.

Lambs born from sheep of Kazakh breed fine-fleeced immunized with a vaccine from the Nigeria 75/1 strain after receiving a sufficient amount of colostrum had protective levels of maternal antibodies against PPR for 14 weeks. From week 14 to the end of the experiment (18 weeks), a decrease in antibody activity was observed, and at week 18, all lambs had titers <1:8 (Figures 4 and 5). In this connection, lambs of the Kazakh breed fine-fleeced are proposed to be immunized at the age of 14 weeks or older, when the maternal immunity is sufficiently weakened and the lamb can develop its own immune response to the introduction of the vaccine.

Similar results were reported in a previously published work by Bodo et al. (2006), where the authors recommend immunizing lambs born from Djallonké ewes in the interval from 11 to 14 weeks after birth [19].

Markus et al. (2019) and Abdollahi et al. (2023), who conducted similar studies, recommend immunizing kids born from Kano brown and Saanen goats vaccinated with the Nigeria 75/1 strain at the age of 10 weeks [18,20], while Olushola S. Olaolu et al. (2021) suggest vaccinating Yankasa lambs at the age of 9 weeks [21]. In these studies, differences in the age at which kids should be vaccinated may be related to the types and breeds of experimental animals used in the studies. It has previously been proven that the same PPR vaccine can cause different immune responses in different breeds of goats [22].

It is known that for reliable protection of animals from PPR, the level of VNA in the blood serum of animals should be $\geq$1:8 [23,24]. Since the lambs were born at different dates, two lambs of 14 weeks of age with a serum antibody titer of 1:8, born first from vaccinated ewes, and two lambs of the same age, born first from unvaccinated ewes, were used to evaluate the effectiveness of passive immunity in lambs. At the same time, in the blood serum of the vaccinated lamb, the antibody titer increased above 1:16 for 14 days, and the lamb was 100% protected from the control virus (Figure 7). An unvaccinated lamb of 14 weeks of age, born with maternal antibodies, was also fully protected during active infection with the control PPRV (Figure 7).

In a similar study conducted by Balamurugan et al. (2012) in a lamb immunized with the Sungri vaccine strain, protective titers of antibodies to the PPRV began to increase from the 21st day post vaccination (21 dpv) [11]. It is possible that the difference in the period of the formation of immunity in kids in these similar studies is due to the difference in the vaccine strains used in the two live vaccines, and also, the possible influence of the type and breed of experimental animals on the obtained research results is not excluded.

The total clinical score calculated by us during the control trial using the field strain PPRV was for the experimental and control groups of lambs 2 and 41 (Figure 8), respectively.

At the same time, the lambs of the experimental group had a two-day increase in body temperature and pink spots on the site of the introduction of the vaccine, which disappeared by themselves within 4 dpc (Figure 7).

However, the results of RT-qPCR analysis turned out to be negative when examining blood samples and swabs collected from an experimental group of lambs after vaccination and control infection (Figure 6).

The unvaccinated lambs showed pronounced clinical signs of PPRV after the control test. In addition, all swabs and blood samples collected from the unvaccinated group of lambs were positive when examined using RT-qPCR (Figure 6). Due to the deterioration in their general condition, both lambs of the unvaccinated group were euthanized using an injectable drug belonging to the barbiturate group (Figure 7).

The results of this study once again prove the safety and effectiveness of the vaccine from the Nigeria 75/1 strain for pregnant ewes, since the vaccine did not affect the course of pregnancy in ewes of the Kazakh breed fine-fleeced, while all ewes were protected from PPR from 2 weeks post vaccination. The maternal immunity formed in lambs born from vaccinated sheep of the Kazakh breed fine-fleeced persisted until 14 weeks after birth, with slight fluctuations in antibody titers ($p > 0.05$). Although a minimum number of lambs were used to determine the effectiveness of the vaccine, the results obtained indicate that lambs born from a fine-fleeced Kazakh breed should be vaccinated from the age of 14 weeks or older.

It is important to note that, despite the limited number of animals in our study, the results offer opportunities for careful design, meticulous data analysis, and collaboration. By implementing these strategies and viewing this limitation as a stepping stone for future research, we can work toward eliminating this constraint and gaining more robust insights in the future.

5. Conclusions

Analyzing the results we obtained in this study, we came to the conclusion that during routine vaccination against PPR, lambs born from sheep of Kazakh breed fine-fleeced should be immunized from the age of 14 weeks or older to avoid a period of susceptibility in lambs to the PPRV. Since the same vaccine against PPR can cause different immune reactions in different breeds of sheep and goats, it is necessary to conduct further studies on other breeds of sheep and goats living in Kazakhstan in order to determine the appropriate period of immunization of small cattle with a vaccine from the Nigeria 75/1 strain.

Author Contributions: Conceptualization, Z.A.; Data curation, Z.A. and Y.B.; Formal analysis, Z.A.; Funding acquisition, A.K.; Investigation Z.K., Z.S., S.T. and A.U.; Methodology, Z.A., Z.K. and S.T.; Project administration, A.K.; Resources, Z.S. and A.U.; Supervision, Y.B.; Validation, Z.A.; Visualization, Z.A., Z.K., Z.S., S.T. and A.U.; Writing—original draft, Z.A.; Writing—review and editing, Y.B. and Z.A. All authors have read and agreed to the published version of the manuscript.

Funding: This work was supported by the Ministry of Agriculture of the Republic of Kazakhstan: for No04/8-21-29 "Program-targeted financing of scientific research and activities" 2021–2023, carried out within the framework of the scientific and technical program "Biological safety of the Republic of Kazakhstan: threat assessment, scientific and technical basis for their prevention and elimination" for 2021–2023.

Institutional Review Board Statement: The study was conducted according to the guidelines of the Declaration of Helsinki, and approved by the Ethics Committee for Animal Experimentation at the RIBSP, RK (Permission number: 2908/22).

Informed Consent Statement: Not applicable.

Data Availability Statement: Not applicable.

Acknowledgments: We thank Ye. Zh. Alimbayev for his help during the animal experiments. We also thank the computer equipment maintenance operator for technical assistance.

Conflicts of Interest: The authors declare no conflict of interest. The funder M.A., R.K. had no role in the development of the study; in the collection, analysis and interpretation of data; in the writing of the report; and in deciding whether to submit the article for publication.

References

1. Mantip, S.E.; Shamaki, D.; Farougou, S. Peste des petits ruminants in Africa: Meta-analysis of the virus isolation in molecular epidemiology studies. *Onderstepoort J. Vet. Res.* **2019**, *86*, 1–15. [CrossRef] [PubMed]
2. Kumar, N.; Maherchandani, S.; Kashyap, S.K.; Singh, S.V.; Sharma, S.; Chaubey, K.K.; Ly, H. Peste Des Petits Ruminants Virus Infection of Small Ruminants: A Comprehensive Review. *Viruses* **2014**, *6*, 2287–2327. [CrossRef]
3. Balamurugan, V.; Kumar, K.V.; Dheeraj, R.; Kurli, R.; Suresh, K.P.; Govindaraj, G.; Shome, B.R.; Roy, P. Temporal and Spatial Epidemiological Analysis of Peste Des Petits Ruminants Outbreaks from the Past 25 Years in Sheep and Goats and Its Control in India. *Viruses* **2021**, *13*, 480. [CrossRef] [PubMed]
4. Lundervold, M.; Milner-Gulland, E.; O'Callaghan, C.; Hamblin, C.; Corteyn, A.; Macmillan, A. A Serological Survey of Ruminant Livestock in Kazakhstan During Post-Soviet Transitions in Farming and Disease Control. *Acta Vet. Scand.* **2004**, *45*, 211–224. [CrossRef]
5. Abduraimov, Y.O.; Yershebulov, Z.D.; Zhugunisov, K.D.; Ye, A.B.; Nurgaziyev, R.Z.; Ye, D.K. The study of immunobiological properties of the vaccine against ovine rinderpest. *Bull ASAU* **2016**, *2*, 121–125. (In Russian)
6. Legnardi, M.; Raizman, E.; Beltran-Alcrudo, D.; Cinardi, G.; Robinson, T.; Falzon, L.C.; Djomgang, H.K.; Okori, E.; Parida, S.; Njeumi, F.; et al. Peste des Petits Ruminants in Central and Eastern Asia/West Eurasia: Epidemiological Situation and Status of Control and Eradication Activities after the First Phase of the PPR Global Eradication Programme (2017–2021). *Animals* **2022**, *12*, 2030. [CrossRef]
7. Kock, R.A.; Orynbayev, M.B.; Sultankulova, K.T.; Strochkov, V.M.; Omarova, Z.D.; Shalgynbayev, E.K.; Rametov, N.M.; Sansyzbay, A.R.; Parida, S. Detection and Genetic Characterization of Lineage IV Peste Des Petits Ruminant Virus in Kazakhstan. *Transbound. Emerg. Dis.* **2015**, *62*, 470–479. [CrossRef]
8. Benfield, C.T.O.; Legnardi, M.; Mayen, F.; Almajali, A.; Cinardi, G.; Wisser, D.; Chaka, H.; Njeumi, F. Peste Des Petits Ruminants in the Middle East: Epidemiological Situation and Status of Control and Eradication Activities after the First Phase of the PPR Global Eradication Program (2017–2021). *Animals* **2023**, *13*, 1196. [CrossRef] [PubMed]
9. Shatar, M.; Khanui, B.; Purevtseren, D.; Khishgee, B.; Loitsch, A.; Unger, H.; Settypalli, T.B.K.; Cattoli, G.; Damdinjav, B.; Dundon, W.G. First genetic characterization of peste des petits ruminants virus from Mongolia. *Arch. Virol.* **2017**, *162*, 3157–3160. [CrossRef] [PubMed]
10. Amanova, Z.; Zhugunissov, K.; Barakbayev, K.; Kondybaeva, Z.; Sametova, Z.; Shayakhmetov, Y.; Kaissenov, D.; Dzhekebekov, K.; Zhunushov, A.; Abduraimov, Y.; et al. Duration of Protective Immunity in Sheep Vaccinated with a Combined Vaccine against Peste des Petits Ruminants and Sheep Pox. *Vaccines* **2021**, *9*, 912. [CrossRef]
11. Balamurugan, V.; Sen, A.; Venkatesan, G.; Rajak, K.K.; Bhanuprakash, V.; Singh, R.K. Study on passive immunity: Time of vaccination in kids born to goats vaccinated against Peste des petits ruminants. *Virol. Sin.* **2012**, *27*, 228–233. [CrossRef]
12. OIE. Peste des petits ruminants (infection with peste des petits ruminant's virus). In *Testerial Manual 2018*; OIE: Paris, France, 2018; Chapter 3.7.9; pp. 1–16.
13. Libeau, G.; Préhaud, C.; Lancelot, R.; Colas, F.; Guerre, L.; Bishop, D.H.; Diallo, A. Development of a competitive ELISA for detecting antibodies to the Peste des Petits Ruminants virus using a recombinant nucleoprotein. *Res. Vet. Sci.* **1995**, *58*, 50–55. [CrossRef] [PubMed]
14. Reed, L.J.; Muench, H. A simple method of estimating fifty per cent endpoints. *Am. J. Epidemiol.* **1938**, *27*, 493–497. [CrossRef]
15. Mamadaliev, S.M.; Koshemetov, Z.K.; Nurabaev, S.S.; Orynbayev, M.B.; Kasenov, M.M.; Bulatov, Y.A.; Mambetaliev, M. Kentau-7 PK-4-05D Strain of Peste des Petits Ruminants Virus, Suitable for Manufacturing Diagnostic Drugs and Monitoring the Immunogenicity of Vaccine Strains. Patent #20025, 29 January 2007. (In Kazakhstan).
16. Bamouh, Z.; Fakri, F.; Jazouli, M.; Safini, N.; Tadlaoui, K.O.; Elharrak, M. Peste des petits ruminants pathogenesis on experimental infected goats by the Moroccan 2015 isolate. *BMC Vet. Res.* **2019**, *15*, 452. [CrossRef]
17. Hamdi, J.; Bamouh, Z.; Jazouli, M.; Alhyane, M.; Safini, N.; Omari Tadlaoui, K.; Fassi Fihri, O.; El Harrak, M. Experimental infection of indigenous North African goats with goatpox virus. *Acta Vet. Scand.* **2021**, *63*, 9. [CrossRef]
18. Markus, T.P.; Adamu, J.; Kazeem, H.M.; Olaolu, O.S.; Woma, T.Y. Assessment of Peste des petits ruminants antibodies in vaccinated pregnant Kano brown does from Nigeria and subsequent maternal immunity in their kids. *Small Rumin. Res.* **2019**, *174*, 53–56. [CrossRef]
19. Bodjo, S.C.; Couacy-Hymann, E.; Koffi, M.Y.; Danho, T. Assessment of the duration of maternal antibodies specific to the homologous peste des petits ruminant vaccine "Nigeria 75/1" in Djallonké lambs. *Biokemistri* **2006**, *18*, 99–103. [CrossRef]
20. Abdollahi, M.; Lotfi, M.; Lotfollahzadeh, S.; Adibi, M.; Kamalzadeh, M.; Firuzyar, S. Determining the decreasing trend of maternal immunity against small ruminant morbillivirus in goat kids. *Vet. Med. Sci.* **2023**, *9*, 1818–1823. [CrossRef] [PubMed]
21. Olaolu, O.; Kazeem, H.; Adamu, J.; Markus, T.; Woma, T. Assessment of Peste Des Petits Ruminants Antibodies in Vaccinated Yankasa Pregnant Ewes from Nigeria and the Duration of Maternal Immunity in Their Lambs. *Vaccine Res.* **2021**, *8*, 47–51. [CrossRef]
22. Begum, S.S.; Mahato, G.; Sharma, P.; Hussain, M.; Saleque, A. Assessment of immune response to a lyophilized peste-des-petitsruminants virus vaccine in three different breeds of goats. *Vet. World* **2016**, *9*, 568–571. [CrossRef]

23. Rossitter, P.B.; Jessett, D.M.; Taylor, W.P. Microneutralisation systems for use with different strains of peste des petits ruminants virus and rinderpest virus. *Trop. Anim. Health Prod.* **1985**, *17*, 75–81. [CrossRef] [PubMed]
24. Saravanan, P.; Sen, A.; Balamurugan, V.; Rajak, K.; Bhanuprakash, V.; Palaniswami, K.; Nachimuthu, K.; Thangavelu, A.; Dhinakarraj, G.; Hegde, R.; et al. Comparative efficacy of peste des petits ruminants (PPR) vaccines. *Biologicals* **2010**, *38*, 479–485. [CrossRef] [PubMed]

viruses

MDPI

Article

Stochastic Interventional Vaccine Efficacy and Principal Surrogate Analyses of Antibody Markers as Correlates of Protection against Symptomatic COVID-19 in the COVE mRNA-1273 Trial

Ying Huang [1,2,3], Nima S. Hejazi [1,4], Bryan Blette [5,†], Lindsay N. Carpp [1], David Benkeser [6], David C. Montefiori [7], Adrian B. McDermott [8,‡], Youyi Fong [1,2], Holly E. Janes [1,2], Weiping Deng [9], Honghong Zhou [9], Christopher R. Houchens [10], Karen Martins [10], Lakshmi Jayashankar [10], Britta Flach [8], Bob C. Lin [8], Sarah O'Connell [8], Charlene McDanal [7], Amanda Eaton [7], Marcella Sarzotti-Kelsoe [7], Yiwen Lu [1], Chenchen Yu [1], Avi Kenny [3], Marco Carone [3], Chuong Huynh [10], Jacqueline Miller [9], Hana M. El Sahly [11], Lindsey R. Baden [12], Lisa A. Jackson [13], Thomas B. Campbell [14], Jesse Clark [15], Michele P. Andrasik [1], James G. Kublin [1], Lawrence Corey [1,16], Kathleen M. Neuzil [17], Rolando Pajon [9], Dean Follmann [18], Ruben O. Donis [10], Richard A. Koup [8], Peter B. Gilbert [1,2,3,*], on behalf of the Immune Assays [§], Moderna, Inc. [§], Coronavirus Vaccine Prevention Network (CoVPN)/Coronavirus Efficacy (COVE) [§] and United States Government (USG)/CoVPN Biostatistics Teams [§]

1 Vaccine and Infectious Disease Division, Fred Hutchinson Cancer Center, Seattle, WA 98109, USA; yhuang@fredhutch.org (Y.H.); nhejazi@hsph.harvard.edu (N.S.H.); lcarpp@fredhutch.org (L.N.C.); yfong@fredhutch.org (Y.F.); hjanes@fredhutch.org (H.E.J.); ylu2@scharp.org (Y.L.); cyu@scharp.org (C.Y.); mandrasik@fredhutch.org (M.P.A.); jkublin@fredhutch.org (J.G.K.); lcorey@fredhutch.org (L.C.)
2 Public Health Sciences Division, Fred Hutchinson Cancer Center, Seattle, WA 98109, USA
3 Department of Biostatistics, University of Washington, Seattle, WA 98195, USA; avikenny@uw.edu (A.K.); mcarone@uw.edu (M.C.)
4 Department of Biostatistics, T.H. Chan School of Public Health, Harvard University, Boston, MA 02115, USA
5 Department of Biostatistics, Epidemiology and Informatics, University of Pennsylvania, Philadelphia, PA 19104, USA; bryan.blette@vumc.org
6 Department of Biostatistics and Bioinformatics, Rollins School of Public Health, Emory University, Atlanta, GA 30322, USA; benkeser@emory.edu
7 Department of Surgery, Duke Human Vaccine Institute, Duke University Medical Center, Durham, NC 27710, USA; monte@duke.edu (D.C.M.); charlene.mcdanal@duke.edu (C.M.); amanda.eaton@duke.edu (A.E.); msarzott@duke.edu (M.S.-K.)
8 Vaccine Research Center, National Institute of Allergy and Infectious Diseases, National Institutes of Health, Bethesda, MD 20892, USA; britta_f@yahoo.com (B.F.); bob.lin@nih.gov (B.C.L.); rkoup@mail.nih.gov (R.A.K.)
9 Moderna, Inc., Cambridge, MA 02139, USA; weiping.deng@modernatx.com (W.D.); honghong.zhou@modernatx.com (H.Z.); jacqueline.miller@modernatx.com (J.M.); rolando.pajon@modernatx.com (R.P.)
10 Biomedical Advanced Research and Development Authority, Washington, DC 20201, USA; christopher.houchens@hhs.gov (C.R.H.); kalaren.martins@hhs.gov (K.M.); lakshmi.jayashankar@hhs.gov (L.J.); chuong.huynh@hhs.gov (C.H.); ruben.donis@hhs.gov (R.O.D.)
11 Department of Molecular Virology and Microbiology, Baylor College of Medicine, Houston, TX 77030, USA; hana.elsahly@bcm.edu
12 Brigham and Women's Hospital, Boston, MA 02115, USA; lbaden@bwh.harvard.edu
13 Kaiser Permanente Washington Health Research Institute, Seattle, WA 98101, USA; lisa.a.jackson@kp.org
14 Division of Infectious Diseases, University of Colorado Anschutz Medical Campus, Aurora, CO 80045, USA; thomas.campbell@ucdenver.edu
15 Department of Medicine, Division of Infectious Diseases, David Geffen School of Medicine at UCLA, Los Angeles, CA 90095, USA; jlclark@mednet.ucla.edu
16 Department of Laboratory Medicine and Pathology, University of Washington, Seattle, WA 98195, USA
17 Center for Vaccine Development and Global Health, University of Maryland School of Medicine, Baltimore, MD 21201, USA; kneuzil@som.umaryland.edu
18 Biostatistics Research Branch, National Institute of Allergy and Infectious Diseases, National Institutes of Health, Bethesda, MD 20892, USA; dfollmann@niaid.nih.gov
* Correspondence: pgilbert@fredhutch.org
† Current address: Department of Biostatistics, Vanderbilt University Medical Center, Nashville, TN 37203, USA.
‡ Current address: Sanofi Vaccines R&D, 1541 Avenue Marcel Mérieux, 69280 Marcy l'Etoile, France.

Citation: Huang, Y.; Hejazi, N.S.; Blette, B.; Carpp, L.N.; Benkeser, D.; Montefiori, D.C.; McDermott, A.B.; Fong, Y.; Janes, H.E.; Deng, W.; et al. Stochastic Interventional Vaccine Efficacy and Principal Surrogate Analyses of Antibody Markers as Correlates of Protection against Symptomatic COVID-19 in the COVE mRNA-1273 Trial. *Viruses* **2023**, *15*, 2029. https://doi.org/10.3390/v15102029

Academic Editor: Koki Taniguchi

Received: 14 July 2023
Revised: 25 September 2023
Accepted: 28 September 2023
Published: 29 September 2023

§ The Members of the Immune Assays; Moderna, Inc.; CoVPN/COVE; and USG/CoVPN Biostatistics Teams are listed in the Supplementary Material.

Abstract: The COVE trial randomized participants to receive two doses of mRNA-1273 vaccine or placebo on Days 1 and 29 (D1, D29). Anti-SARS-CoV-2 Spike IgG binding antibodies (bAbs), anti-receptor binding domain IgG bAbs, 50% inhibitory dilution neutralizing antibody (nAb) titers, and 80% inhibitory dilution nAb titers were measured at D29 and D57. We assessed these markers as correlates of protection (CoPs) against COVID-19 using stochastic interventional vaccine efficacy (SVE) analysis and principal surrogate (PS) analysis, frameworks not used in our previous COVE immune correlates analyses. By SVE analysis, hypothetical shifts of the D57 Spike IgG distribution from a geometric mean concentration (GMC) of 2737 binding antibody units (BAU)/mL (estimated vaccine efficacy (VE): 92.9% (95% CI: 91.7%, 93.9%)) to 274 BAU/mL or to 27,368 BAU/mL resulted in an overall estimated VE of 84.2% (79.0%, 88.1%) and 97.6% (97.4%, 97.7%), respectively. By binary marker PS analysis of Low and High subgroups (cut-point: 2094 BAU/mL), the ignorance interval (IGI) and estimated uncertainty interval (EUI) for VE were [85%, 90%] and (78%, 93%) for Low compared to [95%, 96%] and (92%, 97%) for High. By continuous marker PS analysis, the IGI and 95% EUI for VE at the 2.5th percentile (519.4 BAU/mL) vs. at the 97.5th percentile (9262.9 BAU/mL) of D57 Spike IgG concentration were [92.6%, 93.4%] and (89.2%, 95.7%) vs. [94.3%, 94.6%] and (89.7%, 97.0%). Results were similar for other D29 and D57 markers. Thus, the SVE and PS analyses additionally support all four markers at both time points as CoPs.

Keywords: binding antibody assay; immune correlates of protection; modified treatment policy; neutralizing antibody assay; principal stratification; principal surrogate; SARS-CoV-2; stochastic intervention; stochastic interventional vaccine efficacy

1. Introduction

In the coronavirus efficacy (COVE) phase 3 clinical trial of the mRNA-1273 COVID-19 vaccine, participants were randomized 1:1 to receive mRNA-1273 vaccine (n = 15,209 assigned) or placebo (n = 15,206 assigned), administered on Day 1 (D1) and Day 29 (D29) [1,2]. Estimated vaccine efficacy (VE) in baseline negative per-protocol participants against the primary endpoint of virologically confirmed, symptomatic COVID-19 (hereafter, "COVID-19") starting ≥14 days post-D29 through a median follow-up of 5.3 months, corresponding to completion of the blinded phase, was 93.2% (95% confidence interval (CI), 91.0% to 94.8%) [2]. Vaccine safety was also assessed, with no safety concerns identified during the trial [1,2]. As part of the United States government (USG)-coordinated effort to identify a correlate of protection (CoP) for COVID-19 vaccines [3], we developed a "master protocol" Statistical Analysis Plan (SAP) for harmonizing immune correlates analyses across all of the USG/COVID-19 Response Team phase 3 COVID-19 vaccine trials [4]. As obtaining evidence from multiple analysis frameworks is typically needed to establish an immunologic biomarker for applications such as regulatory decisions or immunobridging, the SAP laid out multiple correlate of risk (CoR) and CoP objectives, each of which addresses a different question.

In Gilbert et al. [5], we reported the COVE trial results for some of these immune correlate objectives, for the immune markers IgG binding antibodies (bAbs) against the SARS-CoV-2 Spike protein (Spike IgG), IgG bAbs against the Spike receptor binding domain (RBD IgG), 50% inhibitory dilution pseudovirus neutralizing antibody titer (nAb-ID50), and 80% inhibitory dilution pseudovirus neutralizing antibody titer (nAb-ID80), measured on D29 and on D57 in all vaccine recipient breakthrough cases and a randomly sampled immunogenicity subcohort. The IgG markers were measured against the original index strain and nAb-ID50 against the B.1/B.1.2 lineage (NC_045512.2), which is the index strain except with the D614G mutation. All four D57 antibody markers correlated inversely with COVID-19 and impacted mRNA-1273 VE against COVID-19 through ~4 months post-D29 [5]. Similar results were obtained for the D29 antibody markers. In additional analyses of COVE, nAb-ID50 was the strongest independent correlate of risk as determined by machine learning analyses that evaluated multivariable correlates of risk (CoRs) [6]. However, these studies did not report on all the immune correlate objectives outlined in the master protocol SAP. In particular, the assessment of CoPs was based on the controlled VE framework [7] and the natural direct and indirect effects mediation framework [8]. The former framework considers a hypothetical intervention that assigns all participants to receive a vaccine and to have a specific immune marker value, estimates COVID-19 risk under this intervention, and then estimates VE by contrasting this risk with COVID-19 risk under the hypothetical intervention that assigns all participants to receive placebo. The latter framework estimates the proportion of VE that is causally mediated through the immune marker, defined by comparing the natural direct effect to the overall VE.

Here, we report the results from additional CoP analyses of the COVE trial, completing the suite of pre-specified blinded-phase immune correlates analyses of the D29 and D57 antibody markers and COVID-19. Specifically, we evaluated each of the four markers, Spike IgG, RBD IgG, nAb-ID50, and nAb-ID80, measured at D29 or D57, as a CoP against COVID-19 defined using two additional causal inference frameworks for CoP assessment as specified in the master protocol SAP: the stochastic interventional vaccine efficacy (SVE) framework and principal surrogate (PS) framework (within the principal stratification framework of causal inference [9]). The SVE approach, like the controlled VE approach, is based on a hypothetical intervention to modify the immune marker but considers the more plausible stochastic intervention of shifting each vaccine recipient's immune marker value by a fixed amount relative to their observed marker value instead of the static intervention of deterministically setting the marker level to the same fixed value for all vaccine recipients; thus, the SVE approach defines a contrast relative to the observed marker values, which may plausibly arise in future hypothetical scenarios. The PS approach, in contrast, does not intervene on the immune marker and instead estimates how VE varies across subgroups defined by the value of the immune marker of vaccine recipients (which is a counterfactual variable for participants in the placebo arm). Table 1 summarizes the four statistical frameworks for assessing a correlate of protection from a vaccine efficacy trial.

Table 1. Four statistical frameworks for assessing an immune marker measured at a post-vaccination time point as an immune correlate of protection (CoP) against a clinical outcome from a vaccine efficacy trial, all of which were applied to the COVE trial.

Statistical Framework for Assessing a CoP	Objective of the CoP Analysis Applied to an Immune Marker in COVE
Controlled vaccine efficacy (VE) [7]	To assess the vaccine efficacy that would be observed under a hypothetical intervention that assigns all participants to the vaccine arm and to a specific value of the marker, as opposed to assigning all participants to placebo *
Mediation of VE [8]	To assess the proportion of the overall VE against COVID-19 that is mediated through the marker, through assessment of the natural direct effect (NDE) of vaccine assignment on COVID-19 (NDE = the component of VE that remains after setting (deactivating) the marker to the level it would have if assigned to the placebo arm)
Stochastic interventional VE [10]	To assess how overall VE would change under user-specified shifts of marker values of vaccine recipients from their observed values
Principal surrogate VE [11]	To assess how VE varies over subgroups defined by the marker value if assigned to the vaccine arm

* This objective/definition attains in studies where all placebo recipients have the same value of the immune marker. This occurs in COVE, as baseline negative placebo recipients all have antibody levels to SARS-CoV-2 antigens that are below the assay detection limit. For VE trials where placebo recipients have variability in the immune marker (i.e., studies that enroll individuals previously infected with the pathogen), the objective/definition must be updated: to assess the VE that would be observed under a hypothetical intervention that assigns all participants to the vaccine arm and to a specific value of the marker, as opposed to assigning all participants to placebo and to a specific value of the marker.

2. Materials and Methods

2.1. COVE Trial and Study Endpoint

The COVE trial (NCT04470427), conducted in the United States, enrolled 30,420 adults age 18 and over at appreciable risk of SARS-CoV-2 infection and/or high risk of severe COVID-19 disease and randomly assigned them in a 1:1 ratio to receive either vaccine or placebo [1,2]. The study endpoint used in the correlates analysis, which we refer to as "COVID-19", is the first occurrence of acute symptomatic COVID-19 with virologically-confirmed SARS-CoV-2 infection [1,2]. Virological confirmation refers to a positive SARS-CoV-2 reverse-transcriptase–polymerase-chain-reaction assay of a nasopharyngeal swab, nasal, or saliva sample. As in Gilbert et al. [5], COVID-19 endpoints beginning seven days post-D29 or -D57, depending on whether D29 or D57 markers were assessed, respectively, through completion of the blinded phase of follow-up were included in the correlates analyses. The calendar dates of this timeframe were 27 July 2020 to 26 March 2021.

2.2. Ethics Statement

All study participants provided written informed consent before enrollment and the protocol and consent forms were approved by the central institutional review board.

2.3. Case-Cohort Sets Included in the Correlates Analyses

A case-cohort sampling design [12] detailed in Gilbert et al. [5] was used to sample participants randomly for measurement of antibody markers on D1, D29, and D57. In all vaccine recipients with a breakthrough COVID-19 endpoint, antibody markers were also measured on the same days. As defined in previous studies [1,2], correlates analyses were conducted in baseline negative (no immunologic or virologic evidence of prior COVID-19 at enrollment) per-protocol (received both doses without major protocol violations) participants. The analysis cohort included a randomly sampled immunogenicity subcohort of size 1010 from the vaccine arm plus all vaccine cases starting 7 days after D29 (for D29 correlates analysis) or starting 7 days after D57 (for D57 correlates analysis). The analysis cohort also included all baseline negative per-protocol placebo recipients, without making use of any antibody data because all immune marker values are constant at "zero" (below assay detection or quantitation limits). See Figure S2 in Gilbert et al. [5] for a schematic of participant flow from enrollment through to the analysis. The numbers of vaccine arm cases and non-cases with measured antibody marker data (for each of the four antibody markers) included in the D29 correlates analyses and in the D57 correlates analyses are provided in Table S3. There are 1005 vaccine arm non-cases and 46 (36) vaccine arm cases with D29 (D57) antibody marker data and, hence, they were included for D29 (D57) correlates analyses. All baseline negative per-protocol placebo recipients were included in the analysis (13,221 non-cases and 751 (659) cases).

2.4. Pseudovirus Neutralizing Antibody Assay

Serum nAb activity against SARS-CoV-2 was measured in a validated assay utilizing lentiviral vector pseudotyped with full-length Spike of the D614G strain NC_045512.2 [13]. Assay readouts were calibrated to the World Health Organization 20/136 anti-SARS-CoV-2 immunoglobulin International Standard [14] and are expressed in international units (IU50/mL and IU80/mL for nAb-ID50 and nAb-ID80, respectively). The arithmetic mean calibration factors used to convert assay readouts to international units are provided in Table S4. Table S5 gives the assay limits, with limit of detection (LOD) 2.42 IU50/mL for nAb-ID50 and 15.02 IU80/mL for nAb-ID80. Values below the LOD were assigned the value of LOD/2.

2.5. Binding Antibody Assay

Serum IgG bAbs against Spike and RBD were measured using a validated solid-phase electrochemiluminescence S-binding IgG immunoassay [5]. Assay readouts were converted to binding antibody units per ml (BAU/mL) using the World Health Organization 20/136 anti-SARS-CoV-2 immunoglobulin International Standard [14]. Table S5 gives the assay limits, with LOD = 0.3076 BAU/mL for Spike and LOD = 1.593648 for RBD. Values below the LOD were assigned the value of LOD/2.

2.6. Stochastic Interventional VE

For each antibody marker, measured levels were hypothetically shifted along a grid, (−1.0, −0.8, −0.6, −0.4, −0.2, 0, 0.2, 0.4, 0.6, 0.8, 1.0), on the $\log_{10}$ scale such that −1.0 represents a 10-fold decrease in geometric mean and 1.0 represents a 10-fold increase in geometric mean. For each shift, the average risk of COVID-19 in per-protocol baseline negative vaccine recipients was estimated via the method of Hejazi et al. [10]. Downward (negative) shifts that would result in more than 10% of participants having counterfactual values of the marker below the assay's LOD were omitted. The zero-shift value corresponds to the observed $\log_{10}$ geometric mean marker level. As described in the SAP, these average risk estimates can be translated to the VE scale by also estimating the average risk of per-protocol baseline negative placebo recipients, which are the results that are presented. The analyses were implemented with the txshift and sl3 packages [15–17] for the R language and environment for statistical computing [18,19].

2.7. Binary Principal Surrogate Evaluation

For each antibody marker, the method of Gilbert et al. [20] was used to estimate VE for two principal strata, defined by the immune marker in vaccine recipients being above vs. below the median marker value, with parameters of interest denoted VE(1) = VE(High) and VE(0) = VE(Low). That is, let Y(1) and Y(0) be potential outcomes indicating whether COVID-19 occurs during follow-up if assigned vaccine or placebo, respectively. Let S(1) indicate the marker value at the time point of interest (D29 or D57) if assigned vaccine and let sc indicate a specified cut-point value for S(1) (i.e., median value). Then, the causal estimands of interest are VE(1), VE(0), and RR ratio = (1 − VE(0))/(1 − VE(1)), with

$$VE(1) = 1 - P(Y(1) = 1 \mid S(1) > sc)/P(Y(0) = 1 \mid S(1) > sc) \text{ and}$$

$$VE(0) = 1 - P(Y(1) = 1 \mid S(1) \leq sc)/P(Y(0) = 1 \mid S(1) \leq sc),$$

where all the probabilities also condition on not experiencing the COVID-19 endpoint by the marker time point of interest (D29 or D57) under both randomization assignments [20]. The relative VE parameter (relative risk ratio = RR ratio) is the degree to which the vaccine confers greater risk reduction for the High subgroup compared to the Low subgroup. The Supplementary Text lists the assumptions needed for the method, which include No Early Harm (NEH) [20], i.e., there are no individuals who would have a COVID-19 outcome before the marker was measured under assignment to vaccine, but not under assignment to placebo. This method relies on user specification of three sensitivity parameters, β_2, β_3, and β_4, to construct IGI and EUI bounds. These parameters reflect different types and degrees of post-randomization selection bias, all with a log-odds ratio scale, with details in the Supplementary Text and in Gilbert et al. [20]. For data analysis, first, each sensitivity parameter was set to zero, such that the VE parameters were point-identified, and point estimates and 95% CIs were calculated. Then, each of the sensitivity parameters were set to vary from log(0.75) to −log(0.75) (medium robustness) and from log(0.5) to −log(0.5) (high robustness), such that the estimands were partially identifiable, and IGIs and 95% EUIs were calculated.

Table S6 provides the median cut-points for each of the D29 and D57 antibody markers. The method was implemented using the psbinary R package [21].

2.8. Continuous Marker Principal Surrogate Evaluation

For each antibody marker measured on a continuous scale, the methods of Huang, Zhuang, and Gilbert [22] were used to estimate the VE curve, i.e., the curve of VE for the subgroup of vaccine recipients with immune markers at each specific level s, i.e.,

VE(*s*) = 1 − P(Y(1) = 1 | S(1) = *s*)/P(Y(0) = 1 | S(1) = *s*). Specifically, under the NEH assumption, a participant disease-free (i.e., COVID-19 endpoint-free) at the time of marker measurement can belong to either one of the two strata: the "equal-always-at-risk" stratum (where the participant will be disease-free at the time of marker measurement if assigned placebo) or the "early-protected" stratum (where the participant will have experienced the disease outcome by the time of marker measurement if assigned placebo). A model for the mixing probability of the two strata among all vaccine recipients disease-free at the time of marker measurement is assumed with a sensitivity parameter β that equals the log-odds ratio of remaining early-at-risk if assigned placebo for early-at-risk vaccine recipients with Y(1) = 1 relative to Y(1) = 0. Sensitivity analyses were conducted for estimating the VE curve for each marker, with the sensitivity parameter β varying from $-\log(4)$ to 0. For comparison, a VE curve estimator was also produced under the EECR assumption, which, in addition to NEH, also assumes that no individuals would have COVID-19 before the marker was measured under placebo assignment but not under vaccine assignment.

3. Results

3.1. Stochastic Interventional VE Analysis Supports Each of the Four Antibody Markers as a Correlate of Protection

For a given immune marker measured at a post-vaccination time point, SVE analysis [10] posits hypothetical individual-level shifts of the marker's observed values by different, user-specified magnitudes [10]. The output of SVE analysis is that of the overall VE estimates had the vaccine hypothetically elicited marker values increased (or decreased) as specified by different shift magnitudes. Applying this framework to COVE, estimated mRNA-1273 VE against COVID-19 increased as each D57 marker was hypothetically increased, and vice versa for hypothetical decreases (Figure 1). For example, the observed geometric mean D57 Spike IgG concentration in vaccine recipients was 2737 BAU/mL and the actual cumulative VE from 7 to 100 days post-D57 was 92.9% (95% CI: 91.4%, 93.9%). Overall estimated VE decreased to 84.2% (95% CI: 79.0%, 88.1%) when shifting D57 Spike IgG marker values down one $\log_{10}$ (to a geometric mean of 274 BAU/mL), and increased to 97.6% (97.4%, 97.7%) when shifting D57 Spike IgG marker values up one $\log_{10}$ (to a geometric mean of 27,368 BAU/mL) (Figure 1A). Results were similar for each of the other three D57 markers.

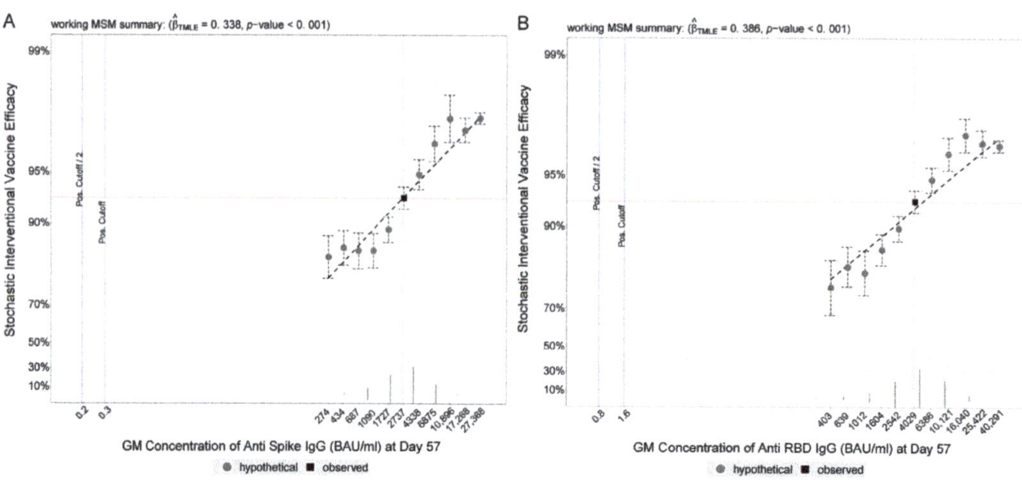

Figure 1. *Cont.*

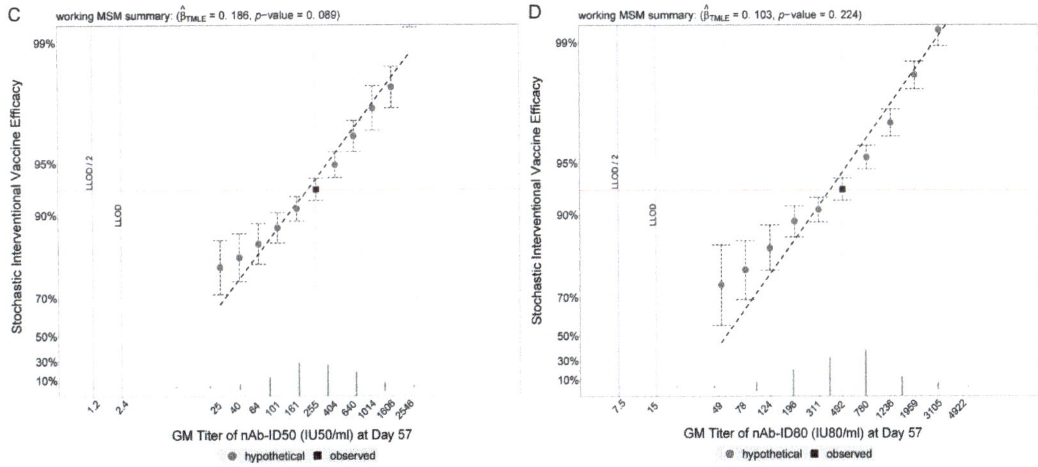

Figure 1. Stochastic interventional vaccine efficacy (SVE) estimates against COVID-19 with hypothetical shifts in geometric mean D57 antibody marker level. SVE, with 95% confidence intervals, for D57 (**A**) Spike IgG, (**B**) RBD IgG, (**C**) nAb-ID50, or (**D**) nAb-ID80, estimated using the method of Hejazi et al. [10]. The y-axis plots the estimated vaccine efficacy (VE) for a vaccine that elicits hypothetical D57 geometric mean value indicated on the x-axis. The vertical red line corresponds to the geometric mean concentration or titer in the COVE study population (baseline negative per-protocol vaccine recipients in the immunogenicity subcohort) and the horizontal red line corresponds to the estimated VE in COVE (follow-up time period from 7 to 100 days post-D57) at a shift of 0, i.e., the observed marker level. BAU, binding antibody units; ID50, 50% inhibitory dilution; ID80, 80% inhibitory dilution; IU, international units; LLOD, lower limit of detection; nAb, neutralizing antibody.

Similar results were obtained for the D29 binding antibody markers (Figure S1A,B), with the D29 neutralizing antibody markers having less-clear trends with marker GM shifts (Figure S1C,D). Figure S2 (D57 antibody markers) and Figure S3 (D29 antibody markers) show estimates of absolute COVID-19 risk of vaccine recipients under the different shift magnitudes, which are simply the numerators of the estimates of stochastic interventional vaccine efficacy.

3.2. Binary Principal Surrogate Analysis Supports Each of the Four Antibody Markers as a Correlate of Protection

For a given immune marker measured at a post-vaccination time point, PS analysis estimates VE for each of two or many subgroups defined by a participant's value of the immune marker if assigned to the vaccine arm (observable in vaccine recipients and a counterfactual in placebo recipients). PS analysis was applied under the assumption that vaccination does not cause an increased risk of COVID-19 prior to the time of immune marker measurement (i.e., the NEH assumption) [20]. First, the approach was applied to estimate VE for each of two subgroups with High or Low immune marker value using the median marker value to define High vs. Low. Results are presented using ignorance intervals (IGIs) and 95% estimated uncertainty intervals (EUIs) [23]. The IGI is a range of VE point estimates, with each estimate calculated under specific fixed values of the sensitivity parameters. A 95% EUI is the union of 95% CIs, where each 95% CI is calculated under specific fixed values of the sensitivity parameters. Whereas 95% CIs only capture uncertainty due to sampling variability, 95% EUIs capture additional uncertainty due to partial non-identifiability of the subgroup VE parameters that occurs because the counterfactual immune marker values if assigned vaccine are not measured for participants who were actually assigned placebo. Results are presented in three ways by specifying three possible ranges for each of three user-specified sensitivity parameters $\beta_2 = \beta_3 = \beta_4$ that are defined in Section 5.4 of Gilbert et al. [20]: [log(1.0), −log(1.0) = 0, 0], [log(0.75), −log(0.75)], or [log(0.50), −log(0.50)], which specify different types and degrees of post-randomization selection bias (see Methods and Supplementary Text for definitions of the sensitivity parameters). For each of the four immune markers at D57, VE point estimates were greater for the High D57 marker subgroup compared to the Low D57 marker subgroup (Table 2, Figure 2). In the special case of setting $\beta_2 = \beta_3 = \beta_4 = 0$, IGIs collapse to point estimates and EUIs collapse to CIs, in which case estimated VE (95% CI) for Low vs. High D57 Spike IgG was 88% (81, 92%) vs. 95% (92, 97%), and results for the other three immune markers were similar. To assess whether VE differed between the Low and High subgroups, we also estimated ratios (1 − VE(Low))/(1 − VE(High)). For example, when setting $\beta_2 = \beta_3 = \beta_4 = 0$, the point estimate (95% CI) for D57 Spike IgG was 2.54 (1.31, 4.93), supporting higher VE for the High marker subgroup. This inference is robust to a moderate amount of allowed uncertainty (95% EUI 1.11, 5.83 when each sensitivity parameter is specified to range over [log(0.75), −log(0.75)]), but not robust to the higher amount of allowed uncertainty (95% EUI 0.72, 9.69 for each sensitivity parameter specified to range over [log(0.50), −log(0.50)]). Similar results were obtained for the four D29 markers (Table S1, Figure S4), with the lower limit of the IGI for the (1 − VE(Low))/(1 − VE(High)) ratio exceeding one under a moderate amount of allowed uncertainty for all four markers.

Table 2. Principal surrogate correlates of vaccine efficacy (VE) results by Gilbert et al. method [20] for High (above median) vs. Low (below median) D57 antibody marker vaccinated subgroups under the No Early Harm (NEH) assumption with sensitivity analysis scenarios.

		VE(0)		**VE(1)**		**(1 − VE(0))/(1 − VE(1))**	
		Low Marker Vaccine Subgroup		**High Marker Vaccine Subgroup**		**Relative Risk Ratio**	
Marker	**Sens. ***	**Ignorance Interval**	**95% Estimated Uncertainty Interval**	**Ignorance Interval**	**95% Estimated Uncertainty Interval**	**Ignorance Interval**	**95% Estimated Uncertainty Interval**
D57 Spike IgG	None	(0.88, 0.88)	(0.81, 0.92)	(0.95, 0.95)	(0.92, 0.97)	(2.54, 2.54)	(1.31, 4.93)
D57 Spike IgG	Med	(0.85, 0.90)	(0.78, 0.93)	(0.95, 0.96)	(0.92, 0.97)	(1.93, 3.35)	(1.11, 5.83)
D57 Spike IgG	High	(0.80, 0.92)	(0.70, 0.95)	(0.94, 0.96)	(0.91, 0.98)	(1.30, 5.08)	(0.72, 9.69)
D57 RBD IgG	None	(0.89, 0.89)	(0.81, 0.93)	(0.95, 0.95)	(0.92, 0.97)	(2.26, 2.26)	(1.17, 4.37)
D57 RBD IgG	Med	(0.86, 0.90)	(0.79, 0.94)	(0.94, 0.95)	(0.92, 0.97)	(1.72, 2.97)	(0.99, 5.17)
D57 RBD IgG	High	(0.81, 0.93)	(0.72, 0.95)	(0.93, 0.96)	(0.90, 0.98)	(1.15, 4.50)	(0.64, 8.48)
D57 nAb-ID50	None	(0.90, 0.90)	(0.84, 0.93)	(0.95, 0.95)	(0.92, 0.97)	(2.25, 2.25)	(1.14, 4.46)
D57 nAb-ID50	Med	(0.88, 0.91)	(0.83, 0.94)	(0.95, 0.96)	(0.92, 0.97)	(1.71, 2.97)	(0.97, 5.26)
D57 nAb-ID50	High	(0.84, 0.93)	(0.78, 0.95)	(0.94, 0.97)	(0.90, 0.98)	(1.15, 4.50)	(0.63, 8.95)
D57 nAb-ID80	None	(0.91, 0.91)	(0.86, 0.95)	(0.94, 0.94)	(0.91, 0.96)	(1.46, 1.46)	(0.76, 2.82)
D57 nAb-ID80	Med	(0.90, 0.93)	(0.85, 0.95)	(0.93, 0.95)	(0.90, 0.97)	(1.11, 1.92)	(0.64, 3.34)
D57 nAb-ID80	High	(0.87, 0.94)	(0.80, 0.96)	(0.92, 0.95)	(0.88, 0.97)	(0.74, 2.90)	(0.41, 5.37)

* Sensitivity parameter settings were: None, β sensitivity parameters β_2, β_3, β_4 set to zero; Med, β sensitivity parameters β_2, β_3, β_4 ranging from log(0.75) to −log(0.75); High, β sensitivity parameters β_2, β_3, β_4 ranging from log(0.5) to −log(0.5).

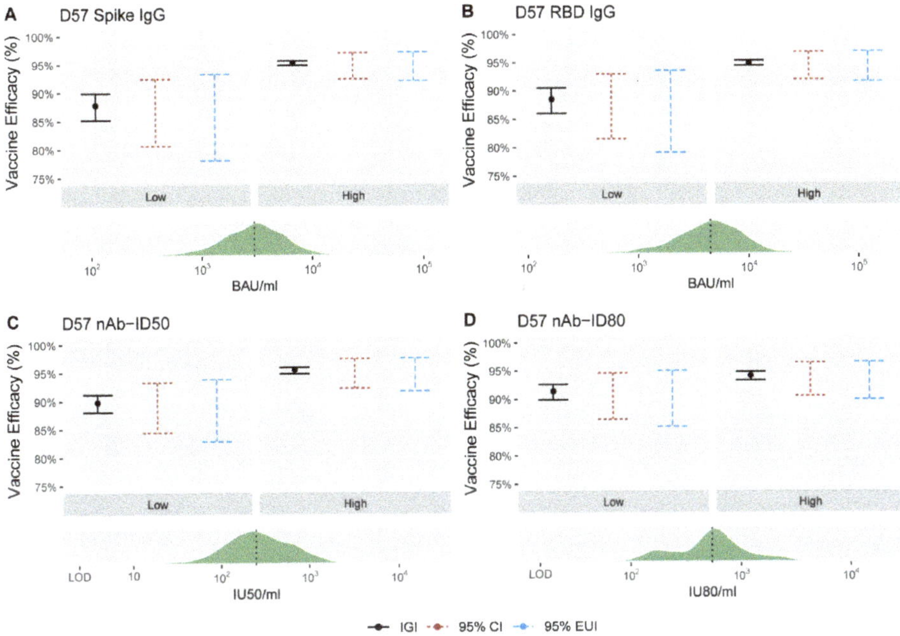

E

			For RR Ratio = [1 - VE(Low)]/[1 - VE(High)]			
		Cut Point	Point Estimate	IGI	95% CI	95% EUI
A)	D57 Spike IgG	2094 BAU/ml	2.54	1.93, 3.35	1.31, 4.93	1.11, 5.83
B)	D57 RBD IgG	3094 BAU/ml	2.26	1.72, 2.97	1.17, 4.37	0.99, 5.17
C)	D57 nAb-ID50	219 IU50/ml	2.25	1.71, 2.97	1.14, 4.46	0.97, 5.26
D)	D57 nAb-ID80	495 IU80/ml	1.46	1.11, 1.92	0.76, 2.82	0.64, 3.34

Figure 2. Binary principal surrogate vaccine efficacy (VE) against COVID-19 by D57 antibody marker greater than vs. less than or equal to the designated cut-point (median value). The black dot in each panel corresponds to the VE estimate for the relevant D57 antibody marker subgroup (Low or High) for (**A**) Spike IgG, (**B**) RBD IgG, (**C**) nAb-ID50, or (**D**) nAb-ID80 when β sensitivity parameters are set to zero. The vertical black line denotes the ignorance interval (IGI) when β sensitivity parameters range from log(0.75) to −log(0.75), the vertical red dashed line denotes the 95% confidence interval (CI) when β sensitivity parameters are set to zero, and the vertical blue dashed line denotes the 95% estimated uncertainty interval (EUI) when β sensitivity parameters range from log(0.75) to −log(0.75). The green histogram on each lower panel denotes the distribution of the D57 antibody marker, with the vertical black dashed line placed at the cut-point separating a Low D57 antibody marker response from a High D57 antibody marker response. This cut-point was the median marker value in baseline negative per-protocol vaccine recipients in the immunogenicity subcohort. (**E**) For each antibody marker, cut-point, relative risk (RR) ratio point estimate, IGI, 95% CI, and 95% EUI. RR ratio = $(1 - VE(0))/(1 - VE(1))$. BAU, binding antibody units; ID50, 50% inhibitory dilution; ID80, 80% inhibitory dilution; IU, international units; nAb, neutralizing antibody.

3.3. Continuous Principal Surrogate Analysis Supports Each of the Four Antibody Markers as a Correlate of Protection

Second, the PS method of Huang, Zhuang, and Gilbert [22] was applied to estimate the VE curve that describes how VE varies by subgroups defined by each possible value of a quantitative immune marker if assigned to the vaccine arm. Estimated VE curves for the D57 and D29 markers with *s* values (hereafter "VE(*s*)") ranging from the 2.5th to 97.5th percentiles are shown in Figures 3 and S5, respectively. Corresponding VE estimates (IGIs and EUIs) are presented in Tables 3 and S2, respectively. For all four antibody markers and both time points, the IGI for VE(s) under each specified value of the sensitivity parameter β varying over the pre-specified range of $[-\log(4), 0]$ (see Methods for a definition of β). For all D57 markers, the lower 95% EUI limit for VE(s) exceeded 80% for all observed s values above its 2.5th percentile. For example, the IGI and EUI for VE(s) at the 2.5th vs. 97.5th percentile of s were [92.6%, 93.4%] (89.2%, 95.2%) vs. [94.3%, 94.6%] (89.7%, 97%) for D57 Spike IgG, and results were similar for the other three immune markers (Table 3).

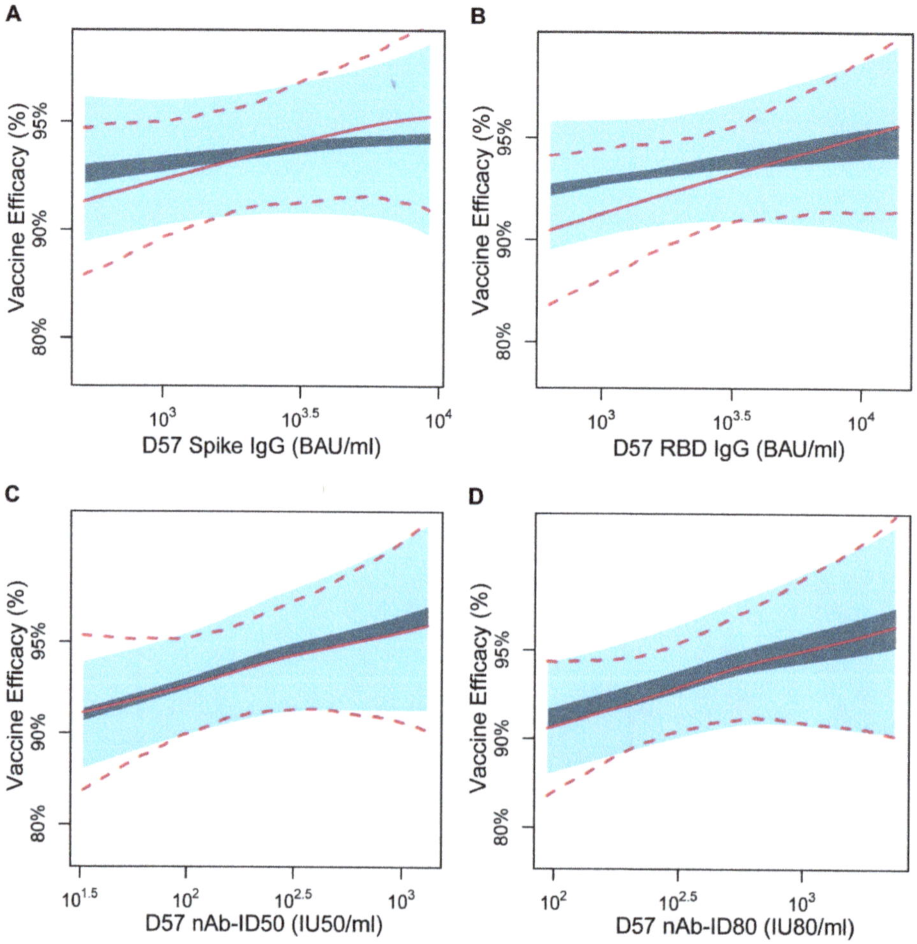

Figure 3. Continuous principal surrogate vaccine efficacy (VE) against COVID-19 by D57 marker level, with ignorance intervals (dark blue) and 95% estimated uncertainty intervals (light blue) under the No Early Harm (NEH) assumption shown for the sensitivity parameter β assumed to fall in the range $[-\log(4), 0]$. Results are shown for (**A**) Spike IgG, (**B**) RBD IgG, (**C**) nAb-ID50, or (**D**) nAb-ID80. In each panel, the solid and dashed lines are the estimated VE curve and 95% perturbation confidence intervals with the Equal Early Clinical Risk (EECR) assumption. The curves are plotted over the marker range of the 2.5th to 97.5th percentile (Spike IgG: 519 to 9263 BAU/mL, RBD IgG: 638 to 13,794 BAU/mL, nAb-ID50: 33 to 1321 IU50/mL, nAb-ID80: 95 to 2385 IU80/mL). BAU, binding antibody units; ID50, 50% inhibitory dilution; ID80, 80% inhibitory dilution; IU, international units; nAb, neutralizing antibody.

Table 3. Principal surrogate correlates of vaccine efficacy results by the Huang, Zhuang, and Gilbert method [22] for D57 antibody marker at various levels under the No Early Harm (NEH) or Equal Early Clinical Risk (EECR) assumption.

			Vaccine Efficacy (S_alpha)						
Marker	**Assumption**		**Alpha = 0.025**	**0.05**	**0.1**	**0.5**	**0.9**	**0.95**	**0.975**
D57 Spike IgG		Concentration (BAU/mL)	519.4	862.1	1224	2926.2	6169.4	7724.8	9262.9
	EECR	Estimate (%)	91.7	92.5	93	94.2	95.0	95.1	95.2
		CI (%)	(86.7, 94.8)	(88.8, 95)	(90.1, 95.1)	(91.5, 96)	(91.9, 96.9)	(91.6, 97.1)	(91.2, 97.4)
	NEH	IGI (%)	[92.6, 93.4]	[93, 93.7]	[93.3, 93.9]	[93.9, 94.3]	[94.2, 94.5]	[94.2, 94.6]	[94.3, 94.6]
		EUI (%)	(89.2, 95.7)	(90.1, 95.7)	(90.5, 95.7)	(91, 96.1)	(90.7, 96.6)	(90.3, 96.8)	(89.7, 97)
D57 RBD IgG		Concentration (BAU/mL)	637.9	1093.5	1670.9	4423.3	9361.8	11,560.8	13,793.5
	EECR	Estimate (%)	90.6	91.8	92.6	94.1	95	95.2	95.4
		CI (%)	(84.4, 94.4)	(87.4, 94.7)	(89.4, 94.8)	(91.4, 95.9)	(91.7, 97)	(91.8, 97.2)	(91.8, 97.5)
	NEH	IGI (%)	[92.6, 93.2]	[93.2, 93.6]	[93.5, 93.9]	[93.9, 94.7]	[94.2, 95.2]	[94.3, 95.4]	[94.3, 95.5]
		EUI (%)	(89.3, 95.5)	(90.4, 95.6)	(90.9, 95.6)	(91.1, 96.4)	(90.7, 97)	(90.5, 97.2)	(90.1, 97.3)
D57 nAb-ID50		Titer (IU50/mL)	33	60.8	88.7	248.1	786.5	1100.8	1320.8
	EECR	Estimate (%)	91.5	92.3	92.9	94.2	95.2	95.5	95.6
		CI (%)	(84.6, 95.3)	(88, 95.1)	(89.5, 95.1)	(91.6, 96)	(91.2, 97.4)	(90.7, 97.8)	(90.3, 98)
	NEH	IGI (%)	[90.9, 91.7]	[92, 92.7]	[92.6, 93.2]	[94.2, 94.7]	[95.2, 95.7]	[95.5, 96]	[95.6, 96.2]
		EUI (%)	(87, 94.2)	(88.7, 94.8)	(89.6, 95.2)	(91.3, 96.4)	(91.6, 97.5)	(91.6, 97.8)	(91.6, 98)
D57 nAb-ID80		Titer (IU80/mL)	94.7	130.6	161.7	544.9	1248.9	1871.8	2385
	EECR	Estimate (%)	90.8	91.5	92	94.3	95.2	95.6	95.9
		CI (%)	(84.4, 94.6)	(86.7, 94.6)	(88, 94.6)	(91.4, 96.2)	(91, 97.4)	(90.7, 97.9)	(90.3, 98.2)
	NEH	IGI (%)	[90.9, 92.1]	[91.5, 92.7]	[91.9, 93]	[94, 94.9]	[94.6, 95.8]	[94.9, 96.2]	[95.1, 96.4]
		EUI (%)	(86.9, 94.5)	(87.8, 94.8)	(88.4, 95.1)	(91, 96.6)	(90.9, 97.5)	(90.6, 97.9)	(90.4, 98.1)

CI, 95% confidence interval; EUI, 95% estimated uncertainty interval; IGI, ignorance interval. alpha = percentile of marker in vaccine recipients.

As a comparison, in Figures 3 and S5, Tables 3 and S2, we also present corresponding analyses under the simpler but less realistic Equal Early Clinical Risk (EECR) assumption, under which there is no sensitivity parameter β. EECR assumes no vaccine effect on risk of COVID-19 prior to the time of immune response measurement and is thus not a reasonable assumption given the observed early vaccine effect in the COVE trial [1,2]. For the two binding antibody markers, the results under EECR show greater moderation of VE across the range of marker values than the results under NEH, whereas for the two neutralizing antibody markers, the results are similar. These results are mostly of academic interest given that the NEH assumption is well justified, whereas the EECR assumption is violated. The overall conclusion is that all PS analyses support that VE increases across subgroups with increasing D29 and D57 binding and neutralizing antibody levels.

4. Conclusions

Results from the stochastic interventional and principal surrogate statistical frameworks for assessing immune markers as correlates of protection (CoP) supported all four antibody markers measured at D29 and D57 as CoPs for two-dose mRNA-1273 vaccine protection against COVID-19 through the COVE blinded phase with a median follow up of 5.3 months post dose two. These findings add to our previous findings supporting CoPs via the controlled VE and mediation frameworks [7,8], thus adding to the body of evidence characterizing these markers as CoPs.

The SVE results suggested that the D57 neutralizing antibody markers were potentially stronger CoPs than the D57 binding antibody markers. For example, there was a greater increase in VE on the relevant multiplicative scale across the range of hypothetical shifts (from a 10-fold decrease to a 10-fold increase) for D57 nAb-ID80 compared to D57 Spike IgG: For D57 nAb-ID80, estimated VE increased from 74.5% (56.5% to 85.1%) to 99.7% (99.6% to 99.7%) (an 85.0-fold increase in the amount of vaccine protection on the multiplicative scale), whereas for D57 Spike IgG, estimated VE increased from 84.2% (95% CI, 79.0% to 88.1%) to 97.6% (97.4% to 97.7%) from the lowest to the highest shift (a 6.6-fold increase in the amount of vaccine protection on the multiplicative scale). This difference between the D57 neutralizing antibody and D57 binding antibody markers was less pronounced for the continuous marker principal surrogate analyses. For example, the IGIs for VE at the 2.5th and 97.5th percentile of marker values were [90.9%, 92.1%] and [95.1%, 96.4%] for D57 nAb-ID80 compared to [92.6%, 93.4%] and [94.3%, 94.6%] for D57 Spike IgG. Moreover, in the binary principal surrogate analysis, there was about a 2-fold multiplicative-scale increase in the amount of vaccine protection going from the Low to High marker subgroup for both antibody markers.

For the D29 markers, the SVE results indicated a similar change in VE across shifts of binding antibody and neutralizing antibody markers, where only shifts upward can be compared (up to 10-fold higher). For example, for D29 Spike IgG, estimated VE increased from 92.9% (95% CI: 91.4%, 93.9%) at no shift to 97.3% (96.0%, 98.2%) at the highest shift (a 2.6-fold multiplicative-scale increase), whereas for D29 nAb-ID50, these values were 92.6% (91.5% to 93.5%) to 96.3% (93.0% to 98.1%) (a 2.0-fold multiplicative-scale increase). By the binary principal surrogate analysis, the change in VE from Low to High marker subgroups was also similar for binding and neutralizing antibody markers: 88% (81% to 92%) vs. 95% (95% CI: 92% to 96%) for D29 Spike IgG compared to 89% (84% to 92%) vs. 95% (92% to 97%) for D29 nAb-ID80 (at sensitivity parameters set to zero).

While we have presented evidence in multiple papers for D29 and D57 antibody markers as CoPs, these markers do not fully represent the entire repertoire of immune responses that can provide protection and did not show perfect mediation of the vaccine's efficacy against COVID-19. This can be partly seen in the PS analyses where vaccine efficacy estimates are well above zero at the lowest end of the biomarker level, which implies the average causal necessity condition for a perfect principal surrogate does not hold. It would be interesting to pursue the identification of improved CoPs by studying other immune responses such as Fc effector and T-cell responses. Technical advances for measuring neutralization more sensitively would also be of interest.

This study shares many of the strengths and limitations of our previous COVID-19 vaccine immune correlates studies, including the constrained scope to a study population naïve to SARS-CoV-2 who received two [5,6,24,25] vaccine or placebo doses, i.e., no boosters. Future work is planned to apply the SVE analysis framework to the COVE trial to assess CoPs for both SARS-CoV-2-naïve and -positive populations and for recipients of a third dose. Additionally, during the follow-up period for correlates analysis, the study population was exposed to predominantly ancestral lineage SARS-CoV-2 (i.e., the D614G B.1/B.1.2 strain) and secondarily to minor genetic drift variants [26]. The antibody markers assessed as correlates in this and our previous studies were measured against the index strain (binding antibodies) or the D614G strain (neutralizing antibodies), thus essentially being matched to the exposing D614G strain viruses. Future work is ongoing to assess these markers and counterpart markers measured against the Omicron BA.1 strain as correlates of Omicron BA.1 COVID-19.

The SVE and PS methods have specific strengths and weaknesses distinguishing them from the previously applied controlled VE and mediation methods. SVE analysis has advantages over both previously applied frameworks in that its hypothetical interventions on an immune marker are more conceivable and can be guided by data. For example, data on how a refined vaccine regimen changes the distribution of an antibody marker can be used to empirically specify a marker shift of interest, and SVE analysis applied to estimate how VE would change had the refined vaccine regimen been evaluated in the VE trial. PS analysis, by not involving any hypothetical intervention on an immune marker, advantageously avoids any issues with the conceivability of the causal estimand. However, the absence of an intervention on the marker implies that the PS framework does not provide results that can be interpreted in terms of an immune marker's causal effect on disease risk (fitting one perspective in causal inference on "no causation without manipulation" [27]); PS analysis is "vanilla subgroup analysis" that obtains separate VE estimates across a range of subgroups. That is, whereas the previously applied SVE CoP frameworks rely on the key assumption that the immune marker is randomized after accounting for other measured participant characteristics, the PS framework does not require this assumption. However, PS analysis replaces this challenge with the equally challenging issue of missing data on the potential immune markers of placebo recipients, which is generally tackled by crossing over placebo recipients to the vaccine arm and measuring their immune markers and/or by measuring pre-vaccination characteristics that predict the post-vaccination immune marker, as well as by specifying pattern mixture models with sensitivity parameters to express the type and degree of post-randomization selection bias.

The PS and SVE methods can be generally applied to any vaccine efficacy trial for which the required immune marker data are available; for instance, the PS framework has been applied to vaccine efficacy trial data for dengue [28], herpes zoster [29], HIV [30], RSV [31], and influenza [32], and the SVE framework has been applied to HIV [10]. Both methods can be readily applied when all participants in the analyzed cohort have no evidence of prior infection with the pathogen of interest, as in the current analysis. Some additional considerations for the PS method are needed if the analyzed cohort includes participants who were previously infected with the pathogen of interest, e.g., the SARS-CoV-2-non-naïve population. In these scenarios, the need to estimate a VE curve is expanded to the need to estimate a VE surface conditional on the pair of potential immune markers under both treatment assignments, to the vaccine and to the placebo [11].

Supplementary Materials: The following supporting information can be downloaded at: https://www.mdpi.com/article/10.3390/v15102029/s1.

Author Contributions: Y.H., N.S.H., B.B., D.B., D.C.M., A.B.M., Y.F., H.E.J., D.F., R.O.D., R.A.K. and P.B.G. conceptualized the study. Y.H., N.S.H., B.B. and P.B.G. developed methodology used in the study. Y.H., N.S.H., B.B., D.B., Y.F., W.D., H.Z., Y.L., C.Y. and P.B.G. curated the data. Y.H., N.S.H., B.B., D.B., Y.F., Y.L., C.Y., A.K., M.C. and P.B.G. conducted the analyses. Y.H., N.S.H., B.B., D.B., D.C.M., A.B.M., Y.F., H.E.J., K.M., L.J., B.F., B.C.L., S.O., C.M., A.E., M.S.-K., Y.L., C.Y., J.M., H.M.E.S., L.R.B., L.A.J., T.B.C., J.C., M.P.A., J.G.K., L.C., K.M.N., R.P., R.O.D., R.A.K. and P.B.G. contributed resources to the project. Y.H., N.S.H., B.B., Y.L., C.Y. and P.B.G. developed software used in the analyses. C.R.H., S.O., M.S.-K., C.H., R.O.D., R.A.K. and P.B.G. performed project administration. Y.H., N.S.H., B.B. and P.B.G. validated the analysis results. Y.H., N.S.H., B.B., L.N.C. and P.B.G. wrote the original draft. The Immune Assays Team performed the binding antibody and neutralizing antibody experiments. The Moderna Inc. team curated the data and contributed resources to the project. The Coronavirus Vaccine Prevention Network (CoVPN)/Coronavirus Efficacy (COVE) Team collected the clinical data in the study and contributed resources to the project. The U.S. government (USG)/CoVPN Biostatistics Teams curated the data, developed the software used in the analyses, conducted the analyses, and validated the results of the analyses. ICMJE guidelines for authorship have been adhered to. All authors have read and agreed to the published version of the manuscript.

Funding: This work was supported by awards R37AI054165 and UM1AI068635 to investigator PBG from the National Institute of Allergy and Infectious Diseases of the National Institutes of Health. This work was supported in whole or part with federal funds from the Department of Health and Human Services; Administration for Strategic Preparedness and Response, Biomedical Advanced Research and Development Authority, under Contract No. 75A50120C00034 (P3001 study). This work was also supported by award R01CA277133 to investigator YH from the National Cancer Institute of the National Institutes of Health. The findings and conclusions herein are those of the authors and do not necessarily represent the views of the Department of Health and Human Services or its components. The content is solely the responsibility of the authors and does not necessarily represent the official views of the National Institutes of Health.

Institutional Review Board Statement: All relevant ethical guidelines were followed, and all necessary IRB and/or ethics committee approvals were obtained. The mRNA-1273-P301 study was conducted in accordance with the International Council for Harmonisation of Technical Requirements for Pharmaceuticals for Human Use, Good Clinical Practice guidelines, and applicable government regulations. The Central Institutional Review Board approved the mRNA-1273-P301 protocol and the consent forms. Central IRB services for the mRNA-1273-P301 study were provided by Advarra, Inc., 6100 Merriweather Dr., Suite 600, Columbia, MD 21044.

Informed Consent Statement: All participants provided written informed consent before enrollment.

Data Availability Statement: Access to patient-level data and supporting clinical documents with qualified external researchers may be available upon request and subject to review.

Conflicts of Interest: W.D., H.Z., J.M. and R.P. are employed by Moderna Inc. and have stock or stock options in Moderna Inc.

References

1. Baden, L.R.; El Sahly, H.M.; Essink, B.; Kotloff, K.; Frey, S.; Novak, R.; Diemert, D.; Spector, S.A.; Rouphael, N.; Creech, C.B.; et al. Efficacy and Safety of the mRNA-1273 SARS-CoV-2 Vaccine. *N. Engl. J. Med.* **2021**, *384*, 403–416. [CrossRef] [PubMed]
2. El Sahly, H.M.; Baden, L.R.; Essink, B.; Doblecki-Lewis, S.; Martin, J.M.; Anderson, E.J.; Campbell, T.B.; Clark, J.; Jackson, L.A.; Fichtenbaum, C.J.; et al. Efficacy of the mRNA-1273 SARS-CoV-2 Vaccine at Completion of Blinded Phase. *N. Engl. J. Med.* **2021**, *385*, 1774–1785. [CrossRef]
3. Koup, R.A.; Donis, R.O.; Gilbert, P.B.; Li, A.W.; Shah, N.A.; Houchens, C.R. A government-led effort to identify correlates of protection for COVID-19 vaccines. *Nat. Med.* **2021**, *27*, 1493–1494. [CrossRef] [PubMed]
4. USG COVID-19 Response Team/Coronavirus Prevention Network (CoVPN) Biostatistics Team. USG COVID-19 Response Team/CoVPN Vaccine Efficacy Trial Immune Correlates Statistical Analysis Plan. Figshare. Last updated 18 April 2022. Available online: https://figshare.com/articles/online_resource/CoVPN_OWS_COVID-19_Vaccine_Efficacy_Trial_Immune_Correlates_SAP/13198595/13 (accessed on 10 June 2022).
5. Gilbert, P.B.; Montefiori, D.C.; McDermott, A.B.; Fong, Y.; Benkeser, D.; Deng, W.; Zhou, H.; Houchens, C.R.; Martins, K.; Jayashankar, L.; et al. Immune correlates analysis of the mRNA-1273 COVID-19 vaccine efficacy clinical trial. *Science* **2022**, *375*, 43–50. [CrossRef] [PubMed]
6. Benkeser, D.; Montefiori, D.C.; McDermott, A.B.; Fong, Y.; Janes, H.E.; Deng, W.; Zhou, H.; Houchens, C.R.; Martins, K.; Jayashankar, L.; et al. Comparing antibody assays as correlates of protection against COVID-19 in the COVE mRNA-1273 vaccine efficacy trial. *Sci. Transl. Med.* **2023**, *15*, eade9078. [CrossRef] [PubMed]
7. Gilbert, P.B.; Fong, Y.; Kenny, A.; Carone, M. A Controlled Effects Approach to Assessing Immune Correlates of Protection. *Biostatistics* **2022**, kxac024. [CrossRef]
8. Benkeser, D.; Díaz, I.; Ran, J. Inference for natural mediation effects under case-cohort sampling with applications in identifying COVID-19 vaccine correlates of protection. *arXiv* **2021**, arXiv:2103.02643. [CrossRef]
9. Frangakis, C.E.; Rubin, D.B. Principal stratification in causal inference. *Biometrics* **2002**, *58*, 21–29. [CrossRef]
10. Hejazi, N.S.; van der Laan, M.J.; Janes, H.E.; Gilbert, P.B.; Benkeser, D.C. Efficient nonparametric inference on the effects of stochastic interventions under two-phase sampling, with applications to vaccine efficacy trials. *Biometrics* **2021**, *77*, 1241–1253. [CrossRef]
11. Gilbert, P.B.; Hudgens, M.G. Evaluating candidate principal surrogate endpoints. *Biometrics* **2008**, *64*, 1146–1154. [CrossRef]
12. Prentice, R.L. A case-cohort design for epidemiologic cohort studies and disease prevention trials. *Biometrika* **1986**, *73*, 1–11. [CrossRef]
13. Huang, Y.; Borisov, O.; Kee, J.J.; Carpp, L.N.; Wrin, T.; Cai, S.; Sarzotti-Kelsoe, M.; McDanal, C.; Eaton, A.; Pajon, R.; et al. Calibration of two validated SARS-CoV-2 pseudovirus neutralization assays for COVID-19 vaccine evaluation. *Sci. Rep.* **2021**, *11*, 23921. [CrossRef] [PubMed]
14. National Institute for Biological Standards and Control (NIBSC). Instructions for Use of First WHO International Standard for Anti-SARS-CoV-2 Immunoglobulin (Version 3.0, Dated 17 December 2020) NIBSC Code: 20/136. Available online: https://www.nibsc.org/science_and_research/idd/cfar/covid-19_reagents.aspx (accessed on 29 July 2021).
15. Hejazi, N.S.; Benkeser, D. txshift: Efficient estimation of the causal effects of stochastic interventions in R. *J. Open Source Softw.* **2020**, *5*, 2447. [CrossRef]
16. Hejazi, N.S.; Benkeser, D. txshift: Efficient Estimation of the Causal Effects of Stochastic Interventions. 2020. Available online: https://zenodo.org/record/4070043#.Yl61nNPMLq4. (accessed on 28 April 2022).
17. Coyle, J.R.; Hejazi, N.S.; Malenica, I.; Sofrygin, O. sl3: Modern Pipelines for Machine Learning and Super Learning. 2021. Available online: https://zenodo.org/record/5802288 (accessed on 28 April 2022).
18. Ihaka, R.; Gentleman, R. *R: A Language for Data Analysis and Graphics*; Taylor & Francis: Abingdon, UK, 1996; pp. 299–314.
19. R Core Team. *R: A Language and Environment for Statistical Computing;* R Foundation for Statistical Computing: Vienna, Austria, 2022.
20. Gilbert, P.B.; Blette, B.S.; Shepherd, B.E.; Hudgens, M.G. Post-randomization Biomarker Effect Modification Analysis in an HIV Vaccine Clinical Trial. *J. Causal Inference* **2020**, *8*, 54–69. [CrossRef]
21. Blette, B.S. Psbinary. Package Implementing Methods for Assessing Effect Modification by a Binary Post-Randomization Variable, as Described in Gilbert et al. *2020, Journal of Causal Inference.* Available online: https://github.com/bblette1/psbinary (accessed on 28 April 2022).
22. Huang, Y.; Zhuang, Y.; Gilbert, P. Sensitivity analysis for evaluating principal surrogate endpoints relaxing the equal early clinical risk assumption. *Ann. Appl. Stat.* **2022**, *16*, 1774–1794. [CrossRef]
23. Vansteelandt, S.; Goetghebeur, E.; Kenward, M.G.; Molenberghs, G. Ignorance and uncertainty regions as inferential tools in a sensitivity analysis. *Stat. Sin.* **2006**, *16*, 953–979.

24. Benkeser, D.; Fong, Y.; Janes, H.E.; Kelly, E.J.; Hirsch, I.; Sproule, S.; Stanley, A.M.; Maaske, J.; Villafana, T.; Houchens, C.R.; et al. Immune correlates analysis of a phase 3 trial of the AZD1222 (ChAdOx1 nCoV-19) vaccine. *NPJ Vaccines* **2023**, *8*, 36. [CrossRef]
25. Fong, Y.; Huang, Y.; Benkeser, D.; Carpp, L.N.; Áñez, G.; Woo, W.; McGarry, A.; Dunkle, L.M.; Cho, I.; Houchens, C.R.; et al. Immune correlates analysis of the PREVENT-19 COVID-19 vaccine efficacy clinical trial. *Nat. Commun.* **2023**, *14*, 331. [CrossRef]
26. Pajon, R.; Paila, Y.D.; Girard, B.; Dixon, G.; Kacena, K.; Baden, L.R.; El Sahly, H.M.; Essink, B.; Mullane, K.M.; Frank, I.; et al. Initial analysis of viral dynamics and circulating viral variants during the mRNA-1273 Phase 3 COVE trial. *Nat. Med.* **2022**, *28*, 823–830. [CrossRef]
27. Holland, P.W. Statistics and Causal Inference. *J. Am. Stat. Assoc.* **1986**, *81*, 945–960. [CrossRef]
28. Moodie, Z.; Juraska, M.; Huang, Y.; Zhuang, Y.; Fong, Y.; Carpp, L.N.; Self, S.G.; Chambonneau, L.; Small, R.; Jackson, N.; et al. Neutralizing Antibody Correlates Analysis of Tetravalent Dengue Vaccine Efficacy Trials in Asia and Latin America. *J. Infect. Dis.* **2018**, *217*, 742–753. [CrossRef] [PubMed]
29. Gilbert, P.B.; Gabriel, E.E.; Miao, X.; Li, X.; Su, S.C.; Parrino, J.; Chan, I.S. Fold rise in antibody titers by measured by glycoprotein-based enzyme-linked immunosorbent assay is an excellent correlate of protection for a herpes zoster vaccine, demonstrated via the vaccine efficacy curve. *J. Infect. Dis.* **2014**, *210*, 1573–1581. [CrossRef] [PubMed]
30. Haynes, B.F.; Gilbert, P.B.; McElrath, M.J.; Zolla-Pazner, S.; Tomaras, G.D.; Alam, S.M.; Evans, D.T.; Montefiori, D.C.; Karnasuta, C.; Sutthent, R.; et al. Immune-correlates analysis of an HIV-1 vaccine efficacy trial. *N. Engl. J. Med.* **2012**, *366*, 1275–1286. [CrossRef] [PubMed]
31. Fong, Y.; Huang, Y.; Borate, B.; van der Laan, L.W.; Zhang, W.; Carpp, L.N.; Cho, I.; Glenn, G.; Fries, L.; Gottardo, R.; et al. Antibody Correlates of Protection from Severe Respiratory Syncytial Virus Disease in a Vaccine Efficacy Trial. *Open Forum Infect. Dis.* **2023**, *10*, ofac693. [CrossRef] [PubMed]
32. Gilbert, P.B.; Fong, Y.; Juraska, M.; Carpp, L.N.; Monto, A.S.; Martin, E.T.; Petrie, J.G. HAI and NAI titer correlates of inactivated and live attenuated influenza vaccine efficacy. *BMC Infect. Dis.* **2019**, *19*, 453. [CrossRef]

viruses

MDPI

Review

Immunogenicity and Efficacy of Vaccination in People Living with Human Immunodeficiency Virus

Eeva Tortellini [1,*], Yann Collins Fosso Ngangue [1], Federica Dominelli [1], Mariasilvia Guardiani [1], Carmen Falvino [1], Fabio Mengoni [1], Anna Carraro [1], Raffaella Marocco [2], Patrizia Pasculli [1], Claudio Maria Mastroianni [1], Maria Rosa Ciardi [1], Miriam Lichtner [2,3] and Maria Antonella Zingaropoli [1]

1 Department of Public Health and Infectious Diseases, Sapienza University of Rome, 00185 Rome, Italy; yanncollins.fosso@uniroma1.it (Y.C.F.N.); federica.dominelli@uniroma1.it (F.D.); mariasilvia.guardiani@uniroma1.it (M.G.); carmen.falvino@uniroma1.it (C.F.); fabio.mengoni@uniroma1.it (F.M.); anna.carraro@uniroma1.it (A.C.); patrizia.pasculli@uniroma1.it (P.P.); claudio.mastroianni@uniroma1.it (C.M.M.); maria.ciardi@uniroma1.it (M.R.C.); mariaantonella.zingaropoli@uniroma1.it (M.A.Z.)

2 Infectious Diseases Unit, SM Goretti Hospital, Sapienza University of Rome, 00185 Latina, Italy; raffaella.marocco@uniroma1.it (R.M.); miriam.lichtner@uniroma1.it (M.L.)

3 Department of Neurosciences, Mental Health, and Sense Organs, NESMOS, Sapienza University of Rome, 00185 Rome, Italy

* Correspondence: eeva.tortellini@uniroma1.it

Abstract: People living with HIV (PLWH) remain at high risk of mortality and morbidity from vaccine-preventable diseases, even though antiretroviral therapy (ART) has restored life expectancy and general well-being. When, which, and how many doses of vaccine should be administered over the lifetime of PLWH are questions that have become clinically relevant. Immune responses to most vaccines are known to be impaired in PLWH. Effective control of viremia with ART and restored CD4+ T-cell count are correlated with an improvement in responsiveness to routine vaccines. However, the presence of immune alterations, comorbidities and co-infections may alter it. In this article, we provide a comprehensive review of the literature on immune responses to different vaccines in the setting of HIV infection, emphasizing the potential effect of HIV-related factors and presence of comorbidities in modulating such responses. A better understanding of these issues will help guide vaccination and prevention strategies for PLWH.

Keywords: HIV; PLWH; ART; vaccination; immune responses; CD4; COVID-19; HPV; influenza

Citation: Tortellini, E.; Fosso Ngangue, Y.C.; Dominelli, F.; Guardiani, M.; Falvino, C.; Mengoni, F.; Carraro, A.; Marocco, R.; Pasculli, P.; Mastroianni, C.M.; et al. Immunogenicity and Efficacy of Vaccination in People Living with Human Immunodeficiency Virus. *Viruses* **2023**, *15*, 1844. https://doi.org/10.3390/v15091844

Academic Editors: Pietro Hiram Guzzi, Marianna Milano and Jayanta Kumar Das

Received: 27 July 2023
Revised: 17 August 2023
Accepted: 21 August 2023
Published: 30 August 2023

1. Introduction

In people living with HIV (PLWH), reduced immune responses to most vaccines are known [1,2]. Antiretroviral therapy (ART) restores life expectancy and general well-being, reducing the risk of severe outcomes after infection in PLWH [3,4].

HIV infection induces a profound disruption of both the innate and adaptive immune systems leading to immunological alterations and persistent immune dysfunction [5]. Primary infection elicits systemic immune activation and inflammation followed by a progressive loss in CD4+ T-cell count and a persistent expansion of circulating CD8+ T cells [6]. Furthermore, exhaustion of T cells often recurs, together with an alteration of the innate immune cell functions [6,7]. Indeed, alterations of B-cell activity such as abnormal activation and lower antibody responses have been described [8].

ART-induced suppression of HIV replication is associated with a significant increase in absolute CD4+ T-cell and B-cell counts, including naïve and memory cells that are essential for humoral and cellular immunity to T-cell-dependent and independent immunogens [9]. However, despite effective virological suppression, chronic activation persists and antigen-specific T- and B-cell responses, including T follicular helper cell (Tfh) functions, are still impaired. Furthermore, PLWH continue to have higher levels of inflammatory mediators, such as interleukin (IL)-6, tumor necrosis factor-alpha (TNF-a), soluble (s) CD163, sCD14 and C-reactive protein (CRP)-accelerated aging, and some other comorbidities may accompany this therapy [10,11]. Some molecules with immunomodulatory properties have been shown to have some beneficial effects on this residual inflammation [12].

However, in PLWH, this impairment of the immune system may affects the quantity, quality and persistence of protective immune responses induced by natural infection or vaccination, reducing responsiveness to vaccines and their effectiveness [13–15]. In addition, vaccine-induced antibodies may decline more rapidly than in the general population [16].

As reported in Table 1, guidelines recommend a proactive approach for immunizing PLWH who are susceptible to vaccine-preventable infections and at risk of exposure, including those who have received previously contraindicated live attenuated vaccines such as those against measles, mumps and rubella [17]. On the contrary, the bacillus Calmette–Guérin (BCG) continues to be contraindicated in PLWH due to its unfavorable benefit/risk profile [18,19].

Table 1. Recommended PLWH Immunization.

Pathogen	Vaccine Platform	Absolute CD4+ T Cell Count	
		<200	>200
HAV	Inactivated	2–3 doses (varies by formulation)	
HBV	Recombinant	2–4 doses (varies by formulation)	
HPV	Recombinant	3 doses through age 26	
Influenza	Inactivated (IIV) Recombinant (RIV) Live, attenuated (LAIV)	1 dose annually 1 dose annually not recommended	
MPXV	Live, attenuated	not recommended	2 doses
SARS-CoV-2	mRNA-based Viral vector Recombinant	2 doses + booster 2 doses 2 doses	
Streptococcus pneumoniae	PCV15 PCV20 PPSV23	1 dose PCV15 followed ≥8 weeks by 1 dose PPSV23 or 1 dose PCV20	
VZV	Live, attenuated (ZVL) Recombinant (RZV)	not recommended 2 doses for 18 and older	

HAV: Hepatitis A virus, HBV: Hepatitis B virus, HPV: Human Papilloma virus, MPXV: Monkeypox virus, VZV: Varicella zoster virus, SARS-CoV-2: Severe acute respiratory syndrome coronavirus 2, PCV: pneumococcal protein-conjugated vaccine, PPSV: pneumococcal polysaccharide vaccine.

In general, the effective viremia control by ART and the improvement in the absolute CD4+ T-cell counts are correlated with an enhancement in responsiveness to routine vaccines, although this issue continues to be of concern. Together with current CD4+ T-cell absolute count, the CD4/CD8 ratio has proved to be an accurate predictor of vaccine success [20] (Figure 1).

Finally, co-infections represent an additional factor that may influence immune responses to vaccination, due to their contribution to a persistent immune activation state and induction of immune senescence [21,22].

Overall, vaccination of PLWH remains challenging and with the present review, we summarized recent works in the literature on different vaccine responses in the setting of HIV infection (Table 1).

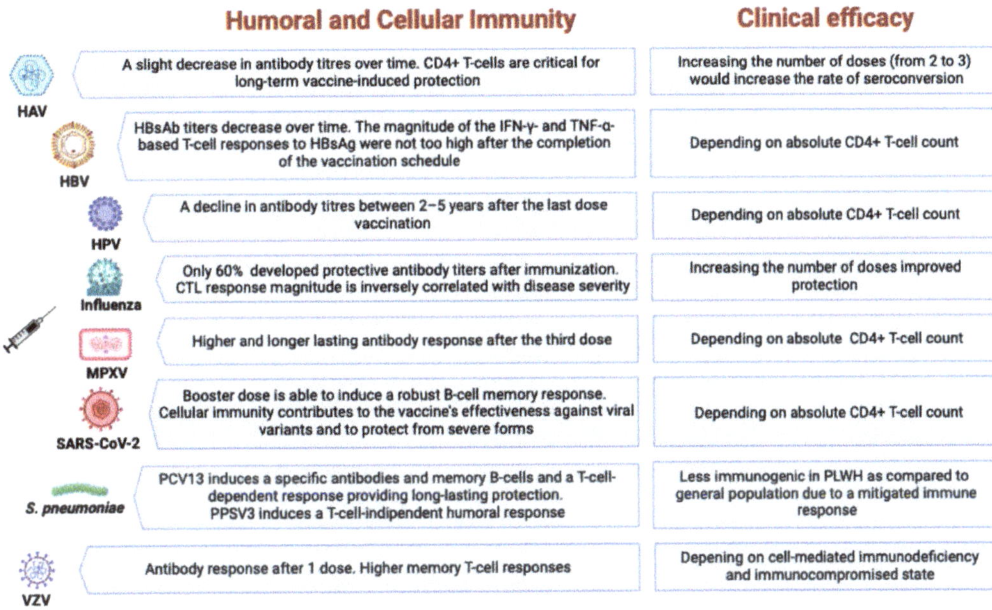

Figure 1. Summary of humoral and cellular immunity in PLWH to recommended vaccinations.

2. Hepatitis A Virus Vaccination

Hepatitis A is a viral infection caused by the Hepatitis A virus (HAV), a single-stranded RNA virus from the *Picornaviridae* family [23]. HAV is commonly transmitted through the fecal–oral route (through ingestion of contaminated food or water), person-to-person contact and men who have sex with men (MSM) [24,25] and may be responsible for forms of acute hepatitis that may progress to fulminant hepatic failure in non-immune adult populations [26,27].

Hepatitis A occurs worldwide and is highly endemic in the most precarious areas of low-income countries [28,29]. It was estimated to have caused, in 2005, apart from approximately 200 million subclinical and oligo-symptomatic HAV infections, 33 million cases of symptomatic illness and 35,000 deaths [30,31].

In Europe, the HAV seroprevalence is low and observational data suggest that PLWH, especially MSM and injecting drug users (IDUs), are at increased risk of contracting HAV [24,32–34]. Additionally, a small study conducted on 15 PLWH with acute hepatitis A showed that the duration of HAV viremia was prolonged compared to the general population with acute hepatitis A, which may increase the likelihood of contracting HAV and transmission to others [35]. Overall, older age, IDUs and MSM have been identified as independent factors associated with HAV seropositivity in PLWH [36–39].

Among the preventive measures to reduce the spread of hepatitis A, vaccination against HAV remains the most effective [40,41]. Currently, two types of HAV vaccines are available: the live attenuated vaccine and the inactivated HAV vaccine [42]. Only the latter is recommended for PLWH [43] including different adult, adolescent and pediatric formulations [43].

In the literature, several studies have evaluated the effectiveness of vaccination against HAV in PLWH. A low rate of seroconversion compared to the general population was observed. Absolute CD4+ T-cell count < 200 cells/μL, viral load, old age, CD4/CD8 ratio, hepatitis C co-infection and gender were identified as factors of poor response after vaccination [44–47].

Several previous studies evaluating the efficacy of HAV vaccination in PLWH have shown lower seroconversion rates (vaccine efficacy) than their seronegative counterparts [48–53]. However, a recent prospective observational study in the setting of an epidemic of acute hepatitis A among MSM in Taiwan observed an overall seroconversion rate among PLWH MSM of 39.7% and 93.4% after receiving one dose and completing a two-dose series of HAV vaccination, respectively, and despite the delayed serological response, HAV vaccination resulted in a 93% reduction in the risk of acute HAV infection in HIV-positive MSM over the course of the epidemic [43]. Higher absolute CD4+ T-cell counts were consistently correlated with higher seroconversion rates [43].

In another study evaluating the efficacy of the vaccine against HAV in a group of 29 children, including 6 children living with HIV and having lost their HAV seropositivity 7 years after being vaccinated with an inactivated vaccine against HAV; after revaccination (two doses), 83% of these PLWH had a seroconversion after the first dose [54].

Other studies evaluating the efficacy of the HAV vaccine according to the increase in the number of vaccine doses have shown that increasing the number of doses from two to three increases the rate of seroconversion in PLWH [52,53,55–57].

Various studies evaluating the persistence of immune memory have demonstrated that in healthy adults following a primary two-dose regimen, anti-HAV antibodies can persist in >90% of vaccines for 40 years or more [58]. In PLWH, on the other hand, a slight decrease was observed over time. In other prospective studies investigating PLWH (all with an inactivated vaccine against HAV), better results of the persistence of immune memory were observed with a duration of seroprotection in PLWH patients equivalent to 7 years in 79% of 29 adolescents, 94 for 6–10 years in 85% of 116 adults, 117 for 3.7 years in 85% of 52 adults [59], and for 5 years in 75.5% of 49 adults [47].

Different studies assessing the persistence of immune memory have observed that in healthy adults following a primary two-dose regimen, anti-HAV antibodies can persist in >90% of vaccines for 40 years or more [58]. In prospective studies of successfully vaccinated PLWH (all with an inactivated HAV vaccine), there was a slight increase in the persistence of seroprotection, which in some of these PLWH (both adults and adolescents) oscillated between 5 and 10 years old [47,54,59,60].

The persistence of immune memory was also confirmed by a recent study comparing three-dose and two-dose HAV vaccination schedules, where a slightly higher seroprotection rate of 94% versus 88% was found after 5 years in 155 and 95 adults, respectively [61].

Despite the evidence of efficacy conferred by the HAV vaccine, PLWH remain susceptible to HAV infection in high-income countries, due to low compliance with recommended HAV vaccination guidelines, at-risk sexual behaviors and injecting drug use.

3. Hepatitis B Virus Vaccination

Hepatitis B is a liver viral infection caused by the Hepatitis B virus (HBV), a DNA virus belonging to the *Hepadnaviridae* family [62].

Because HIV and HBV share similar routes of transmission, co-infection with the two viruses is common [63,64]. HBV infection in PLWH is a global public health problem [65]. Globally, nearly 10% of PLWH are infected with HBV [66,67]. HBV infection in PLWH is generally characterized by an increased rate of cirrhosis (10–20%), a higher risk of hepatocellular carcinoma [63,68–70] and lastly a higher risk of liver-related death [70].

In general, HBV is not cytopathic. It causes damage through the induction of immune mechanisms. Cytotoxic CD8 cells recognize expressed HBV antigens and destroy infected hepatocytes, resulting in increased aminotransferases [71]. Thus, HBV establishes a persistent infection with a stable reservoir of genetic material in the form of circularized DNA in the cell nucleus [65]. The different phases of HBV infection are characterized by the presence of certain viral and immunological markers allowing the orientation of the therapeutic decision and the evaluation of the response to treatment. Surface antigen (HBsAg) is the first marker detected in serum [72]. Its presence indicates HBV infection [73]. The disappearance of HBsAg is followed by the appearance of anti-HBs antibodies. Anti-HBs is considered as a neutralizing antibody and is recognized as a marker of disease protection and cure [73]. In most patients, anti-HBs persists for life, confirming long-term immunity.

Anti-HBs is only serological marker in individuals who have an immune response after vaccination against hepatitis B [74].

HBV vaccination is recommended in PLWH as the most important method of prevention [75]. Despite these recommendations in 2015, only 2/3 of PLWH receive at least one dose of HBV vaccine [75]. The efficacy rate of this vaccine in terms of immune response is generally defined by seroconversion with anti-HBs antibodies > 10 μL/mL in these subjects [76].

The first studies on the efficacy of the vaccine against HBV applying the "classical" schedule (20 μg of HBs antigen at months 0–1–6) showed relatively low seroconversion rates in PLWH, with only 20–70% overprotected against 90–95% in the general population [77,78]. Nevertheless, in these studies, a low rate of response to vaccination could be correlated with various risk predictors of poor response, including viral load and absolute CD4+ T-cell count, CD4/CD8 ratio, co-infection with HCV, poor general health and occult hepatitis B [79–83]. In addition, the female sex [84,85], younger age [86–88], alcohol consumption [86] and smoking [88] are all factors of negative response to vaccination.

Recent studies have shown improvements in the effectiveness of the HBV vaccine in PLWH, particularly with seroconversion rates [89–91]. In Uganda, to assess the efficacy of HBV vaccines in PLWH, a cohort of 132 participants received both the base dose of the vaccine and the 1-month dose, and 127 received the 6-month dose. The 132 participants who entered the study were predominantly female and 52% had received ART for ≥3 months and 94% had undetectable HIV RNA. The median (IQR) CD4+ T-cell count was 426 cells/μL (261–583). A high humoral response rate in PLWH was seen. Nevertheless, in this study, a variation was observed in the immune response of these PLWH to the vaccine, with 86% of participants that were high-level responders with anti-HBs titer levels ≥ 100 IU/L, while 6% were low-level responders (anti-HBs levels 10 to 99 IU/L) and 10 (8%) were non-responders (titer levels < 10 IU/L) [89].

This high response rate differs from some studies that have shown suboptimal seroconversion rates in response to the standard series of HBV vaccines in PLWH. These results are like a study in China that also found high response rates to the HBV vaccine in PLWH. Nevertheless, compared to the general population in China [90], the response to HB vaccination was diminished in PLWH.

In a study in Thailand, high rates of response to the HBV vaccine were also observed in PLWH with fully suppressed HIV viral load and absolute CD4+ T-cell count ≥ 200 cells/μL [88].

Given the importance of immune status in vaccine response, it is possible that participants' degree of immune reconstitution differed between studies, despite apparently similar current absolute CD4+ T-cell counts documented in these studies. On the other hand, in the literature, we find results of the efficacy of the vaccine against HBV, correlated with the increase in vaccination schedules against HBV. Launay et al. [92], in one study, found that PLWH vaccinated with a four-double-dose schedule had higher anti-HBV titers and stronger immune responses than those vaccinated with the standard three-dose schedule (82% versus 65%, $p < 0.05$). Chaiklang et al. compared the immunogenicity and safety of three standard doses and four double doses versus four standard doses in one RCT [88]. A large randomized trial evaluating three-dose 20 μg and four-dose 40 μg HB vaccination regimens in PLWH reported seroconversion rates of 65 and 82%, respectively [92]. Prospective studies and randomized trials have reported similar response rates of 50–62% [93,94], while other studies have reported similar or better rates, ranging from 84 to 92% [88,89,95]. However, these studies with higher seroconversion rates exclusively enrolled patients with absolute CD4+ T-cell counts of 200 cells/μL or higher or included patients regardless of whether they had received an HBV vaccine in the past.

In another study evaluating the persistence of vaccination, O'Bryan et al. [96] followed 186 HIV patients for 5 years and found that the persistence of the HBV vaccine response was longer when these patients had an undetectable or low viral load [97].

These data demonstrate an improvement in the efficacy conferred by the HBV vaccine, but achieving a long-lasting and protective level of immunity remains a challenge in patients with detectable HIV RNA or low CD4+ T-cell counts at the time of vaccination.

4. Human Papillomavirus Vaccination

Human papillomavirus (HPV) infection represents the most prevalent sexually transmitted infection in the world [98]. HPV is a small, non-enveloped DNA virus infecting skin or mucosal cells that belongs to the *Papillomaviridae* family. The genome encodes for six early proteins responsible for virus replication and two late proteins, L1 and L2, which are the viral structural proteins [99]. At least 13 of more than 100 known HPV genotypes can cause cancer of the cervix and are associated with other anogenital cancers and cancers of the head and neck [99]. HPV types are divided into high risk, associated with the development of anogenital cancer, and low risk, rarely associated with the development of cancers [100]. The two most common "high-risk" genotypes (HPV 16 and 18) cause approximately 70% of all cervical cancers [99].

In immunocompetent individuals, most HPV infections spontaneously resolve; however, the persistent infection with oncogenic HPV genotypes is associated with cancers of the cervix, vulva, vagina, anus, penis and the oropharynx [100]. In immunosuppressed individuals, including PLWH, HPV infection often becomes chronic [101]. In particular, PLWH have a higher incidence of HPV infection, abnormal pap smears and persistent HPV infection due to the less efficient viral clearance, leading to a high risk of HPV-related cancers [101].

Because of the more common and persistent HPV-related complications in PLWH, HPV vaccination programs are encouraged in this population. HPV vaccination represents the main preventive tool for HPV-related cancers as well as other HPV-related diseases [102]. HPV vaccines are based on the recombinant protein virus-like particle (L1 VLP) with a proprietary adjuvant. Currently, there are three licensed prophylactic L1 VLP-based vaccines that provide protection against two (bivalent, commercialized in 2008), four (quadrivalent) and nine (nonavalent) HPV genotypes [103]. Given the higher vulnerability of PLWH, particularly in men who have sex with men (MSM) who are also PLWH, to acquire multisite infections mostly characterized by various genotype combinations and the ability of the nonavalent vaccine to prevent 80% of HPV infections, vaccination programs with this nine-genotype protection should be implemented, especially among MSM [104,105].

HPV vaccines have proved their safety, efficacy and effectiveness in immunocompetent young persons, leading to a standard vaccination regime for young girls and boys (aged 9–14 years) reduced from the originally licensed three-dose regimen to two doses [106,107].

Serum neutralizing antibodies are thought to be the major protective branch of adaptative immunity afforded by L1 VLP-based vaccines, although CD4+ T cells are involved in the induction and re(activation) of antigen-specific memory B cells, leading to high antibody levels and, therefore, are critical for long-term vaccine-induced protection [108].

The immune response elicited by the quadrivalent HPV vaccine seems to persist in vaccinated individuals up to 5 years post-vaccination [109].

Concerning PLWH, licensed HPV vaccines have proven to be generally safe and well tolerated [110]. Lower rates of antibody levels elicited by similar vaccine constructs have been observed, raising concern about the efficacy of HPV vaccines in PLWH [111]. However, several studies have reported that the immune response induced in PLWH is similar to that found in general population, with high rates of seroconversion and a cellular immunogenicity comparable to that of general population [98,110,112–114]. Furthermore, antibody levels following vaccination appear to be stable over time [115].

Significant positive correlations between T-cell responses and current absolute CD4+ T-cell count together with negative correlations between such responses and HIV viremia have been observed [116,117]. Furthermore, higher seroconversion rates among PLWH with current absolute CD4+ T-cell counts >200 cells/μL compared with ≤200 cells/μL have also been reported [116,117]. However, a possible decline in B-cell memory responses between 2 and 5 years after the last vaccination dose has been described in PLWH [116]. This is consistent with observations focused on the characterization of the immunogenicity of other vaccines, but few immunogenicity studies of HPV vaccines among PLWH include participants with low absolute CD4+ T-cell counts, supporting the need to further elucidate their immune capacity. In addition, to the best of our knowledge, few studies are focused on the immunogenicity of the nonavalent HPV vaccine, and information is lacking about the quality of such response.

5. Influenza Vaccination

Influenza is an infectious respiratory disease with annual estimations of approximately 1 billion infections, 3–5 million cases of severe illness and 300,000–500,000 deaths, according to WHO [118]. Influenza viruses belong to the *Orthomyxoviridae* family and account for three different types: influenza A, B and C. All three types share common characteristics, such as the segmented genome, made of negative-sense single-stranded RNA, and the presence of an envelope (derived from host cell membrane) with glycoproteins, essential for viral entry in target cells [119]. Viral particles are enveloped, and surface glycoproteins hemagglutinin (HA) and neuraminidase (NA) represent the major antigenic determinants [120]. Influenza viruses are characterized by antigenic variation, based on two different mechanisms: antigenic drift (present in all influenza types) and antigenic shift (characteristic of influenza A only) [120].

Influenza B and C have a narrower host range (humans only and humans and swine, respectively) than influenza A, which can infect humans, swine, equine, other mammals such as ferrets, felids, mink, dogs, civets, marine mammals and avians [119]. Influenza A and B viruses are most relevant clinically, since they cause severe respiratory infections in humans [121]. Sequencing has confirmed that these viruses share a common genetic ancestry. They have genetically diverged, and an exchange of viral RNA segments between viruses has been reported occurring within each genus or type, but not across types [120]. Influenza A viruses are further characterized by the subtype of their surface glycoproteins, the HA and the NA. There are 18 different HA subtypes and 11 different NA subtypes [122]. Unlike influenza A, influenza B is not further divided into subtypes [123].

During infection, the epithelial cells are the primary targets for influenza viruses. These cells line the respiratory tract and initiate an antiviral immune response upon influenza virus detection. The first line of defense is represented by the innate immune system and constituted by physical barriers and the innate immune cellular responses [124]. A critical role is performed by the adaptive immunity in the clearance of viral pathogens during the later stages of infection. Furthermore, respiratory mucosal immunity is induced in the related mucosal tissues during influenza infection and involved in antiviral defense [125].

Generally, infections occur in children, although most of the severe cases involve very young or elderly individuals and individuals affected by chronic pulmonary or cardiac conditions, diabetes mellitus or immunocompromising conditions [126]. Considering the periodical recurrence of influenza infection and the severe complications occurring in the elderly and in patients with concomitant chronic diseases, the influenza vaccine represents an essential tool for preventing infection and limiting the burden of the disease. The choice of relevant antigens remains of paramount importance in developing influenza vaccines which are formulated every year to match the circulating strains.

Currently, there are three kinds of vaccines, inactivated, live attenuated and recombinant HA vaccines, licensed in different countries [127]. The WHO recommends seasonal influenza vaccination to children 6 months to 5 years of age, elderly individuals (>65), all persons with chronic medical conditions and pregnant women. All available influenza virus vaccines are injected intramuscularly, except for the live attenuated influenza virus vaccines, which are administered intranasally [128].

Debate is still ongoing on the efficacy and effectiveness of the licensed vaccines, although most studies find a positive effect of vaccination on vaccinated individuals [129]. The effectiveness of influenza vaccines has been found to be related predominantly to the age and immune competence of the vaccinated individual and the antigenic relatedness of vaccine strains to circulating strains [130].

Currently licensed influenza vaccines focus on the production of antibodies against the viral HA, which binds host receptors to mediate viral entry, neutralizing the virus and preventing the infection [131]. However, the decline in vaccine-specific antibodies and the antigenic drift of influenza viruses over time leads to the necessity of annual revaccination. Targeting T-cell responses seems to be a promising technique to ameliorate influenza vaccines, although it does not prevent infection, but it can reduce the severity of the infection [131].

Indeed, the role of T-cell immunity was demonstrated during the 2001 H1N1 pandemic, where the magnitude of the pre-existing cytotoxic T-lymphocyte (CTL) response inversely correlated with disease severity in individuals without detectable neutralizing antibody [132].

PLWH experience prolonged duration of influenza infection and increased severity of illness, together with higher rates of hospitalizations compared to the general population [133]. As a result, annual vaccination against seasonal influenza is recommended by many national immunization guidelines [134].

Again, the success of influenza vaccination is related to current absolute CD4+ T-cell count [135]. Indeed, weaker response rates were observed in PLWH with lower current absolute CD4+ T-cell count, probably due to impaired function of the peripheral blood Tfh and B-cell functions [136,137].

However, the data are not always in accordance, as demonstrated by the study conducted by Tebas et al. [138], aimed at evaluating safety and immunogenicity of the H1N1 2009 vaccine in PLWH. The authors showed that only 60% of the participants developed protective antibody titers after immunization [138]. On the contrary, a clinical trial (P1088) launched by the International Maternal Pediatric and Adolescent Clinical Trials (IMPAACT) Network evaluated safety and efficacy of a monovalent pandemic H1N1 (pH1N1) vaccine in perinatally HIV-1-infected children and adolescents, showing that two doses of double-strength pH1N1 vaccine are safe and immunogenic and may provide improved protection against influenza in this population [139].

Concerns remain about the efficacy in elderly PLWH, as observed in the general population. Alternative vaccines, dosing, adjuvants or schedule strategies may be needed to achieve effective immunization of this vulnerable population.

6. Monkeypox Virus Vaccination

Monkeypox virus infection (MPXV), also commonly known as "monkey pox" or simian orthopoxvirus, is an infectious disease caused by an *Orthopoxvirus* (family *Poxviridae*). There are two genetically distinct MPXV clades that exhibit different lethality rates. Clade II comprising the first cases of infections was reported in West Africa and clade I in Central Africa. Clade IIb was responsible for the global epidemic outbreak in May 2022. Monkeypox virus infection presents a clinical picture that can vary according to the clades: clades I and IIa resemble smallpox and clade IIb is characterized by atypical presentations. During the 2022 epidemic which raged in countries where the disease was not endemic, the symptoms were very polymorphic (cutaneous and mucous membrane involvement, painful lymphadenopathy, angina, anitis or proctitis, etc.) which could lead to more complicated forms (ocular involvement, encephalitis or encephalopathies, multiorgan involvement in otherwise immunocompromised patients). Transmission to humans occurs from an animal reservoir or from human to human via direct or indirect physical contact (contaminated objects).

Despite the announcement of the end of the MPXV epidemic by the WHO in May 2023 [140], questions around MPXV remain topical in the scientific community, especially in immunocompromised subjects such as PLWH who are considered as a population at risk [141]. Nearly 957 cases of monkeypox virus (MPXV) infection in Italy and 25,887 cases of infection in Europe have been reported so far [142].

Recent studies have described the fatal nature of the MPXV infection in a subpopulation of these PLWH characterized by absolute CD4+ T-cell counts of >200 cells/μL, presenting a clinical picture marked by massive necrotizing skin and cutaneous, genital and non-genital mucosal lesions, which can sometimes be accompanied by pulmonary involvement with multifocal nodular opacities or respiratory failure and severe cutaneous and blood bacterial sequelae which, in 15% of cases, led to death [141,143,144].

Most cases of MPXV infections reported in Europe and North America since May 2022 were mainly transmitted among men who have sex with men (MSM) with evidence of an increased prevalence of HIV and other sexually transmitted infections (STIs). Given the morbidity and lethality in PLWH, a strong evolution of the therapeutic arsenal against MPXV with many vaccines has been made available to stem the epidemic [145].

The third-generation vaccine contains the live modified attenuated virus of vaccinia Ankara. MVA-BN is currently the only approved vaccine in areas where sufficient vaccine stocks are available. Jynneos/MVA-BN is used for pre-exposure prophylaxis to MPXV in HIV-infected individuals; it is also indicated for the prevention of MPXV in individuals 18 years of age and older who are at high risk of infection [146,147].

To date, there are very few data on an immune response of the MVA-BN vaccine against MPXV in PLWH. In a study that focused on evaluating the safety and immunogenicity of MVA-BN in immunocompromised subjects [148], a phase II trial was conducted between 2006 and 2009 in the United States and Puerto Rico; a total of 579 volunteers were recruited into the study: 439 vaccine-naïve subjects (88 immunocompetent subjects, 351 PLWH) and 140 vaccine-experienced subjects (9 immunocompetent subjects, 131 PLWH) received at least one vaccination. The results of this study demonstrated that the MVA-BN vaccine presents a better safety and tolerance profile in PLWH with absolute CD4+ T-cell counts < 200 cells/μL than in immunocompetent subjects, regardless of their previous smallpox vaccination status. In other subsequent studies, the safety profile of the MVA-BN vaccine in immunocompromised subjects, particularly those infected with HIV, was comparable or even better in terms of local reactions than in subjects not infected with HIV [149,150].

The results of the studies showed that the safety profile of the MVA-BN vaccine in immunocompromised subjects, particularly PLWH, considered at risk for conventional smallpox vaccination, was comparable or even better in terms of local reactions than in the general population [148,150]. Antibody responses were also comparable between immunocompetent subjects and PLWH.

7. SARS-CoV-2 Vaccination

With more than 757 million confirmed cases, the coronavirus disease 2019 (COVID-19), caused by severe acute respiratory syndrome coronavirus 2 (SARS-CoV-2), is the third coronavirus disease in the past 20 years [151,152].

SARS-CoV-2 is an enveloped, positive-sense, single-stranded RNA virus that belongs to the *Coronaviridae* family [153]. Its genome encodes for four major structural proteins, namely the spike surface glycoprotein (S), which is responsible for the binding to the host receptor *angiotensin-converting enzyme 2* (ACE2), the small envelope protein (E), the matrix protein (M) and the nucleocapsid protein (N), and other non-structural proteins [154]. Viral transmission can occur by direct, indirect or close contact by infected people through secretions (saliva or respiratory droplets) [155]. SARS-CoV-2 infects bronchial epithelial cells, pneumocytes and upper respiratory tract cells in humans, developing into severe, life-threatening respiratory diseases and lung injuries [156].

Many countries have launched vaccination campaigns to prevent SARS-CoV-2 infection, and several vaccines have been approved by the World Health Organization (WHO), although many obstacles to global vaccination remain [157]. Among the vaccines based on different technologies that have been developed during the health emergency, messenger RNA (mRNA)-based vaccines have been widely used to contain the pandemic [158].

In the general population, mRNA-based vaccines have proven to elicit a robust and protective humoral and cellular response against the SARS-CoV-2 S protein, reducing mortality and morbidity related to SARS-CoV-2 infection [159–161]. In addition, the specific T-cell response induced by the vaccine, with the ability to recognize different regions of the S protein, contributes to the vaccine's effectiveness against viral variants and protects individuals from severe forms of COVID-19 [162,163].

It has been estimated that PLWH represent 1% of total hospitalized cases and, differently from HIV infection which, in the absence of ART, is invariably fatal, COVID-19 disease is highly variable, ranging from mild to severe and critical forms of illness [164,165].

During the health emergency, vaccination of PLWH became of vital significance and strongly recommended by health authorities because of the potentially worse outcomes after SARS-CoV-2 infection, although reports about the increased risk of severe COVID-19 in this population are in some cases contradictory [166]. However, PLWH may experience a higher burden of various comorbidities, many of which have emerged as risk factors for severe COVID-19. Key risk factors for severe COVID-19 include both non-HIV comorbidities known to be associated with severe disease like older age, diabetes, obesity and cardiovascular disease as well as HIV-specific risk factors such as low absolute CD4+ T-cell count, viremia and *Mycobacterium tuberculosis* co-infection [167]. Furthermore, the suboptimal responses to other vaccines have raised concerns about the efficacy of vaccines against SARS-CoV-2 in this potentially more vulnerable population.

Some HIV viral blips following mRNA vaccinations have been reported; however, licensed vaccines have proven to be safe and efficacious in PLWH with stable absolute CD4+ T-cell counts and well-controlled viremia [168].

In particular, published data on the immunogenicity of mRNA vaccines show values of anti-S antibodies, neutralizing antibody activity and cellular immune responses in PLWH on ART and with current absolute CD4+ T-cell counts above 200 cells/μL comparable to those observed in the general population after a primary vaccination cycle [168–170]. Such a response was found to be significantly inferior in PLWH with current absolute CD4+ T-cell count < 200 cells/μL compared to those with >500 cell/mm^3 and the general population, suggesting that the immunogenicity at the time of vaccination is related to the current absolute CD4+ T-cell count [171,172].

In September 2021, the administration of an additional booster of anti-SARS-CoV-2 mRNA vaccine was approved in Italy to be given after >28 days after completion of the primary vaccination cycle in PLWH depending on current absolute CD4+ T-cell count and/or detectable HIV viremia [173]. The third dose improved the responsiveness particularly in PLWH on ART with current absolute CD4+ T-cell counts < 200 cells/μL, improving both the rate and the magnitude of the response and supporting the additional dose strategy in this category of patients with severe immune impairment [172,174].

Some strategies aimed at increasing the tolerability of the mRNA vaccine, such as the use of pidotimod, which was able to reduce vaccination-related adverse events, could be useful to encourage people to received vaccination [175].

In PLWH with a current absolute CD4+ T-cell count > 200 cells/μL, T-cell response elicited by the third dose was like that induced by the primary vaccination cycle, suggesting that the first two doses were able to achieve full T-cell immunization. Furthermore, the increased humoral response is consistent with the hypothesis that the third dose is able to induce a robust B-cell memory response, previously elicited by the primary vaccination series [176].

However, questions remain about mRNA vaccine's immunogenicity in PLWH with ongoing immunosuppression and viremia who represent a particularly vulnerable group that is poorly represented in vaccine trials. Furthermore, recent studies have shown a lower polyfunctional capacity in this population, as already described in the setting of other co-infections, raising issues about the real capability of their immune response [174,177].

8. Streptococcus Pneumoniae Vaccination

Streptococcus pneumoniae (*S. pneumoniae*), a Gram-positive bacterium, is the most significant cause of bacterial disease in humans. A variety of clinical syndromes are related to its infection, including pneumonia, meningitis, bacteremia, acute otitis media and sinusitis [178]. Despite the availability of a broad arsenal of antibiotics and a vaccine, worldwide, approximately 14.5 million cases of serious pneumococcal diseases per year have been reported, leading to approximately 826,000 deaths [179].

In PLWH, invasive pneumococcal disease (IPD) and pneumococcal pneumonia continue to pose a challenge with high recurrence rates [180], significant public health impact, morbidity and a high mortality rate of up to 25% [181].

The introduction of pneumococcal vaccines has significantly reduced morbidity, although PLWH still remain at a 30-times-higher risk of IPD as compared to the general population [182]. Specifically, an absolute CD4+ T-cell count < 200 cells/µL and high levels of HIV RNA have been strongly associated with the risk of IPD [182].

Four pneumococcal vaccines are currently available: PCV13, PPSV23, PCV15 and PCV20 [183]. The PCV13 contains protein-conjugated polysaccharides of 13 serotypes of pneumococci [184]. The PCV15 and PCV20 contain all the PCV13 serotypes, with two additional serotypes in PCV15 and seven additional serotypes in PCV20 [183]. The PPSV23 contains 23-valent pneumococcal polysaccharides [185]. The PCV15 is administered as a single dose with one PPSV23 follow-up dose given at least 8 weeks later; no additional doses are recommended after that. The PCV20 requires one dose only; there are no additional doses needed [183].

There are limited data on the efficacy of pneumococcal vaccination in PLWH. Retrospective studies indicate that PPSV23 alone has modest clinical benefit, if any, in reducing rates of pneumococcal infections [186,187]. The immune response induced by PPSV23 is a T-cell-independent humoral response, while PCV13 induces a T-cell-dependent response producing pneumococcal serotype-specific antibodies and memory B cells which provide long-lasting protection [188]. Sequential vaccination with PCV 13 followed by PPSV23 provides a prime boost effect on inducing and maintaining protective immunity [189].

These vaccines are less immunogenic in PLWH compared to the general population due to a mitigated immune response. The combination of PPSV23 and PCV 13 has been shown to be more immunogenic than either of the vaccines alone and is recommended internationally for prevention of IPD in PLWH [190]. Moreover, other strategies to improve the immunogenicity of pneumococcal vaccines in PLWH were performed. Indeed, in a double-blind, placebo-controlled study, the addition of adjuvant CPG 7909, a toll-like receptor agonist, significantly enhanced the proportion of high responders to the vaccine [191].

According to the Italian Vaccination Plan (2017–2019), the tetanus, diphtheria and pertussis (Tdap) vaccine is co-administered with PCV and HBV vaccines. For PLWH of ≥11 years old who have never received any vaccine, three doses of Tdap are administered, with an interval of 0, 1 month, 6–12 months. In individuals with advanced-stage HIV, the response is suboptimal for both tetanus and diphtheria, while in subjects with CD4 > 300 cells/µL, the response against tetanus is optimal, comparable to subjects without HIV while, for diphtheria, it can remain markedly lower [192].

9. Varicella Zoster Virus Vaccination

The Varicella zoster virus (VZV), a double-stranded DNA ubiquitous human alphaherpesvirus [193], causes varicella, establishes lifelong latency in ganglionic neurons and reactivates later in life to cause herpes zoster, commonly associated with chronic pain [193–196].

Varicella and herpes zoster are more common and more severe in the elderly, the female sex [197] and in people who are immunocompromised, such as PLWH and people taking immunosuppressive drugs and chemotherapy [198]. The incidence of herpes zoster is more than 15 times higher in PLWH compared to age-matched immunocompetent subjects. Herpes zoster can occur in PLWH at any absolute CD4+ T-cell count, but disease frequency is highest when absolute CD4+ T-cell counts are below 200 cells/µL [199–201].

Despite the mandatory vaccination of children against chickenpox in the early 1995s in the United States [202] and from 2003 in Europe [203] (this led to immunization (83% to 95%) of the general population [204]), the risk of VZV reactivation remains particularly high in seropositive adults [205,206]. The incidence of herpes zoster is approximately 4–7 cases/1000 person-years and without vaccination, the lifetime herpes zoster risk is 20–30% [207,208].

Vaccination offers an option that could overcome the challenges associated with conventional antiviral prophylaxis while potentially providing longer-lasting protection against shingles [209]. The live zoster vaccine (ZVL) is a live attenuated vaccine approved for people aged ≥60 years [210]. The effectiveness of the ZVL vaccine may decrease with age and it is generally contraindicated in immunocompromised subjects due to its potential infection risks [211,212].

Recently, the recombinant zoster vaccine (RZV), an adjuvanted subunit vaccine recommended for use in adults ≥ 50 years of age since 2017 by the Advisory Committee on Immunization Practices (ACIP) [213], was also approved by the ACIP for the prevention of herpes zoster in adults aged 19 and older who have or will have an increased risk of shingles due to immunodeficiency or immunosuppression caused by diseases or treatment [210].

The efficacy of RZV in immunocompromised subjects is lower than in immunocompetent subjects, reflecting cell-mediated immunodeficiency and a weaker immune response due to an underlying immunocompromised state [214]. Clinical trial data comparing memory T-cell responses to both vaccines mentioned (ZVL and RZV) found higher responses in RZV recipients, and only RZV recipients had five-year persistence of higher responses [215,216]. The efficacy of RZV is high, even in people aged ≥70 years [217]. Pooled analyses also showed that the vaccine was 91.3% effective against shingles in participants over the age of 70 [218,219]. The clinical efficacy of the RZV vaccine has also been demonstrated in various phase II and III, placebo-controlled, observer-blinded studies conducted in immunocompromised adults aged 18 years and older with two doses administered 1–2 months apart [220,221].

Regarding the safety of the RZV vaccine, results from an observational study showed no difference between immunocompetent and immunocompromised groups, indicating that immunosuppression may not be a determinant of adverse vaccine effects [221].

These preliminary data confirm the efficacy conferred by the RZV vaccine against herpes zoster. However, these data should be interpreted with caution and require in-depth studies.

10. Conclusions

Despite ART-induced virologic suppression, PLWH remain at increased risk of mortality and morbidity from vaccine-preventable diseases, in part because of persistent immunopathology, resulting in a compromised response to vaccination, and vaccine-induced antibodies may fade more rapidly in PLWH than in the general population [1,16].

Moreover, besides the primary response, long-term persistence of protection has been poorly documented and recommendations on the timing of booster injections are based on data collected in the general population, although patterns of antibody decay may differ. In this regard, it is necessary to estimate how seroprotection declines over time among patients who initially responded to immunization.

Many efforts have been made during the SARS-CoV-2 pandemic to evaluate vaccine efficacy in PLWH. The findings obtained further confirm the critical role of CD4+ T cells as a key factor of effective humoral responses and predictor of vaccine success. In addition, evidence about the complementary role of T cell-specific responses in mediating protection has emerged, particularly in individuals with low seroconversion rates, reducing mortality and morbidity related to SARS-CoV-2 infection. However, what constitutes protective immunity is still discussed, making it difficult to define protective efficacy of vaccines. In determining vaccine scheduling and efficacy, CD4+ T-cell count, CD4/CD8 ratio and viremia should be considered, with the awareness that it will not capture the full immune profile of this population.

In fact, it is becoming increasingly clear that PLWH represent a diverse population in terms of immune phenotype, with the consequence that different subgroups require different vaccination strategies to improve their immunological responses.

Furthermore, the setting of co-infection poses additional concerns, particularly regarding T-cell immunity, since with the intersecting of SARS-CoV-2, HIV and TB epidemics, SARS-CoV-2-specific CD4+ T cells have shown a lower polyfunctional capacity.

In our opinion, a better understanding of these issues will help guide vaccination and prevention strategies for PLWH.

We should also consider that male adults living in Europe and in the United States are the most represented participants in the studies, which poorly reflects the global prevalence of PLWH, and that, with the pandemic, a reduction in the access to ART and in vaccine coverage may leave PLWH potentially more vulnerable.

To date, studies assessing long-term immunogenicity, planned with scientific rigor, are needed. An improvement in the field of vaccine development could bring changes in the lives of PLWH. In conclusion, the main preventive tool for many infectious diseases remains vaccination, together with counseling and screening programs. However, greater attention needs to be paid to PLWH with uncontrolled viral infection and/or low CD4+ T-cell counts and to the effects of aging and comorbidities.

Author Contributions: Conceptualization, M.L. and M.A.Z.; resources, R.M., A.C., P.P., C.F., M.G., F.D., F.M., M.R.C. and C.M.M.; writing—original draft preparation, E.T., Y.C.F.N. and M.A.Z.; writing—review and editing, supervision, M.L. and M.A.Z.; project administration, M.L. All authors have read and agreed to the published version of the manuscript.

Funding: This research received no external funding.

Institutional Review Board Statement: Not applicable.

Informed Consent Statement: Not applicable.

Data Availability Statement: Not applicable.

Acknowledgments: We would like to acknowledge the Department of Public Health and Infectious Diseases of Sapienza University of Rome.

Conflicts of Interest: Miriam Lichtner received department grants from Gilead outside the submitted work and support from advisory boards for attending conferences from MSD, Abbvie, Gilead, GSK, Angelini and ViiV. All other authors have no conflicts of interest to declare.

References

1. Geretti, A.M.; Doyle, T. Immunization for HIV-Positive Individuals. *Curr. Opin. Infect. Dis.* **2010**, *23*, 32–38. [CrossRef]
2. Abzug, M.J. Vaccination in the Immunocompromised Child: A Probe of Immune Reconstitution. *Pediatr. Infect. Dis. J.* **2009**, *28*, 233–236. [CrossRef] [PubMed]
3. Samji, H.; Cescon, A.; Hogg, R.S.; Modur, S.P.; Althoff, K.N.; Buchacz, K.; Burchell, A.N.; Cohen, M.; Gebo, K.A.; Gill, M.J.; et al. Closing the Gap: Increases in Life Expectancy among Treated HIV-Positive Individuals in the United States and Canada. *PLoS ONE* **2013**, *8*, e81355. [CrossRef] [PubMed]
4. Antiretroviral Therapy Cohort Collaboration. Life Expectancy of Individuals on Combination Antiretroviral Therapy in High-Income Countries: A Collaborative Analysis of 14 Cohort Studies. *Lancet* **2008**, *372*, 293–299. [CrossRef] [PubMed]
5. Mullender, C.; da Costa, K.A.S.; Alrubayyi, A.; Pett, S.L.; Peppa, D. SARS-CoV-2 Immunity and Vaccine Strategies in People with HIV. *Oxf. Open Immunol.* **2022**, *3*, iqac005. [CrossRef] [PubMed]
6. Fenwick, C.; Joo, V.; Jacquier, P.; Noto, A.; Banga, R.; Perreau, M.; Pantaleo, G. T-Cell Exhaustion in HIV Infection. *Immunol. Rev.* **2019**, *292*, 149–163. [CrossRef] [PubMed]
7. Sabado, R.L.; O'Brien, M.; Subedi, A.; Qin, L.; Hu, N.; Taylor, E.; Dibben, O.; Stacey, A.; Fellay, J.; Shianna, K.V.; et al. Evidence of Dysregulation of Dendritic Cells in Primary HIV Infection. *Blood* **2010**, *116*, 3839–3852. [CrossRef]
8. Bussmann, B.M.; Reiche, S.; Bieniek, B.; Krznaric, I.; Ackermann, F.; Jassoy, C. Loss of HIV-Specific Memory B-Cells as a Potential Mechanism for the Dysfunction of the Humoral Immune Response against HIV. *Virology* **2010**, *397*, 7–13. [CrossRef]

9. Moir, S.; Malaspina, A.; Ho, J.; Wang, W.; Dipoto, A.C.; O'Shea, M.A.; Roby, G.; Mican, J.M.; Kottilil, S.; Chun, T.-W.; et al. Normalization of B Cell Counts and Subpopulations after Antiretroviral Therapy in Chronic HIV Disease. *J. Infect. Dis.* **2008**, *197*, 572–579. [CrossRef]
10. Guaraldi, G.; Orlando, G.; Zona, S.; Menozzi, M.; Carli, F.; Garlassi, E.; Berti, A.; Rossi, E.; Roverato, A.; Palella, F. Premature Age-Related Comorbidities among HIV-Infected Persons Compared with the General Population. *Clin. Infect. Dis. Off. Publ. Infect. Dis. Soc. Am.* **2011**, *53*, 1120–1126. [CrossRef]
11. Meir-Shafrir, K.; Pollack, S. Accelerated Aging in HIV Patients. *Rambam Maimonides Med. J.* **2012**, *3*, e0025. [CrossRef] [PubMed]
12. Ucciferri, C.; Falasca, K.; Reale, M.; Tamburro, M.; Auricchio, A.; Vignale, F.; Vecchiet, J. Pidotimod and Immunological Activation in Individuals Infected with HIV. *Curr. HIV Res.* **2021**, *19*, 260–268. [CrossRef] [PubMed]
13. Pallikkuth, S.; De Armas, L.R.; Pahwa, R.; Rinaldi, S.; George, V.K.; Sanchez, C.M.; Pan, L.; Dickinson, G.; Rodriguez, A.; Fischl, M.; et al. Impact of Aging and HIV Infection on Serologic Response to Seasonal Influenza Vaccination. *AIDS* **2018**, *32*, 1085–1094. [CrossRef]
14. Bonetti, T.C.S.; Succi, R.C.M.; Weckx, L.Y.; Tavares-Lopes, L.; de Moraes-Pinto, M.I. Tetanus and Diphtheria Antibodies and Response to a Booster Dose in Brazilian HIV-1-Infected Women. *Vaccine* **2004**, *22*, 3707–3712. [CrossRef]
15. Avelino-Silva, V.I.; Miyaji, K.T.; Hunt, P.W.; Huang, Y.; Simoes, M.; Lima, S.B.; Freire, M.S.; Caiaffa-Filho, H.H.; Hong, M.A.; Costa, D.A.; et al. CD4/CD8 Ratio and KT Ratio Predict Yellow Fever Vaccine Immunogenicity in HIV-Infected Patients. *PLoS Negl. Trop. Dis.* **2016**, *10*, e0005219. [CrossRef] [PubMed]
16. Kernéis, S.; Launay, O.; Turbelin, C.; Batteux, F.; Hanslik, T.; Boëlle, P.-Y. Long-Term Immune Responses to Vaccination in HIV-Infected Patients: A Systematic Review and Meta-Analysis. *Clin. Infect. Dis. Off. Publ. Infect. Dis. Soc. Am.* **2014**, *58*, 1130–1139. [CrossRef]
17. Geretti, A.M.; BHIVA Immunization Writing Committee; Brook, G.; Cameron, C.; Chadwick, D.; Heyderman, R.S.; MacMahon, E.; Pozniak, A.; Ramsay, M.; Schuhwerk, M. British HIV Association Guidelines for Immunization of HIV-Infected Adults 2008. *HIV Med.* **2008**, *9*, 795–848. [CrossRef] [PubMed]
18. Mansoor, N.; Scriba, T.J.; de Kock, M.; Tameris, M.; Abel, B.; Keyser, A.; Little, F.; Soares, A.; Gelderbloem, S.; Mlenjeni, S.; et al. HIV-1 Infection in Infants Severely Impairs the Immune Response Induced by Bacille Calmette-Guérin Vaccine. *J. Infect. Dis.* **2009**, *199*, 982–990. [CrossRef]
19. Hesseling, A.C.; Cotton, M.F.; Fordham von Reyn, C.; Graham, S.M.; Gie, R.P.; Hussey, G.D. Consensus Statement on the Revised World Health Organization Recommendations for BCG Vaccination in HIV-Infected Infants. *Int. J. Tuberc. Lung Dis. Off. J. Int. Union Tuberc. Lung Dis.* **2008**, *12*, 1376–1379.
20. Hadrup, S.R.; Strindhall, J.; Køllgaard, T.; Seremet, T.; Johansson, B.; Pawelec, G.; thor Straten, P.; Wikby, A. Longitudinal Studies of Clonally Expanded CD8 T Cells Reveal a Repertoire Shrinkage Predicting Mortality and an Increased Number of Dysfunctional Cytomegalovirus-Specific T Cells in the Very Elderly. *J. Immunol.* **2006**, *176*, 2645–2653. [CrossRef]
21. Perello, R.; Vergara, A.; Monclus, E.; Jimenez, S.; Montero, M.; Saubi, N.; Moreno, A.; Eto, Y.; Inciarte, A.; Mallolas, J.; et al. Cytomegalovirus Infection in HIV-Infected Patients in the Era of Combination Antiretroviral Therapy. *BMC Infect. Dis.* **2019**, *19*, 1030. [CrossRef]
22. Lichtner, M.; Cicconi, P.; Vita, S.; Cozzi-Lepri, A.; Galli, M.; Lo Caputo, S.; Saracino, A.; De Luca, A.; Moioli, M.; Maggiolo, F.; et al. Cytomegalovirus Coinfection Is Associated with an Increased Risk of Severe Non–AIDS-Defining Events in a Large Cohort of HIV-Infected Patients. *J. Infect. Dis.* **2015**, *211*, 178–186. [CrossRef] [PubMed]
23. Feinstone, S.M.; Kapikian, A.Z.; Purceli, R.H. Hepatitis A: Detection by Immune Electron Microscopy of a Viruslike Antigen Associated with Acute Illness. *Science* **1973**, *182*, 1026–1028. [CrossRef] [PubMed]
24. McFarlane, E.S.; Embil, J.A.; Manuel, F.R.; Thiébaux, H.J. Antibodies to Hepatitis A Antigen in Relation to the Number of Lifetime Sexual Partners in Patients Attending an STD Clinic. *Br. J. Vener. Dis.* **1981**, *57*, 58–61. [CrossRef]
25. Corey, L.; Holmes, K.K. Sexual Transmission of Hepatitis A in Homosexual Men: Incidence and Mechanism. *N. Engl. J. Med.* **1980**, *302*, 435–438. [CrossRef] [PubMed]
26. Chen, C.-M.; Chen, S.C.-C.; Yang, H.-Y.; Yang, S.-T.; Wang, C.-M. Hospitalization and Mortality Due to Hepatitis A in Taiwan: A 15-Year Nationwide Cohort Study. *J. Viral Hepat.* **2016**, *23*, 940–945. [CrossRef] [PubMed]
27. Boucher, A.; Meybeck, A.; Alidjinou, K.; Huleux, T.; Viget, N.; Baclet, V.; Valette, M.; Alcaraz, I.; Sauser, E.; Bocket, L.; et al. Clinical and Virological Features of Acute Hepatitis A during an Ongoing Outbreak among Men Who Have Sex with Men in the North of France. *Sex. Transm. Infect.* **2019**, *95*, 75–77. [CrossRef]
28. Lin, K.-Y.; Sun, H.-Y.; Chen, Y.-H.; Lo, Y.-C.; Hsieh, S.-M.; Sheng, W.-H.; Chuang, Y.-C.; Pan, S.-C.; Cheng, A.; Hung, C.-C.; et al. Effect of a Hepatitis A Vaccination Campaign during a Hepatitis A Outbreak in Taiwan, 2015–2017: A Modeling Study. *Clin. Infect. Dis.* **2020**, *70*, 1742–1749. [CrossRef]
29. Aggarwal, R.; Goel, A. Hepatitis A: Epidemiology in Resource-Poor Countries. *Curr. Opin. Infect. Dis.* **2015**, *28*, 488–496. [CrossRef]
30. Rein, D.B.; Stevens, G.; Flaxman, A.; Wittenborn, J.S.; Timothy, N.; Wiktor, S.Z.; Wiersma, S.T. P703 the Global Burden of Hepatitis a Virus in 1990 and 2005. *J. Hepatol.* **2014**, *1*, S303. [CrossRef]
31. WHO Position Paper on Hepatitis A Vaccines—June 2012. *Releve Epidemiol. Hebd.* **2012**, *87*, 261–276. Available online: https://www.who.int/publications-detail-redirect/who-wer8728-29-261-276 (accessed on 6 June 2012).

32. Ochnio, J.J.; Patrick, D.; Ho, M.; Talling, D.N.; Dobson, S.R. Past Infection with Hepatitis A Virus among Vancouver Street Youth, Injection Drug Users and Men Who Have Sex with Men: Implications for Vaccination Programs. *CMAJ Can. Med. Assoc. J. J. Assoc. Medicale Can.* **2001**, *165*, 293–297.
33. Crofts, N.; Cooper, G.; Stewart, T.; Kiely, P.; Coghlan, P.; Hearne, P.; Hocking, J. Exposure to Hepatitis A Virus among Blood Donors, Injecting Drug Users and Prison Entrants in Victoria. *J. Viral Hepat.* **1997**, *4*, 333–338. [CrossRef]
34. Corona, R.; Stroffolini, T.; Giglio, A.; Cotichini, R.; Tosti, M.E.; Prignano, G.; Di Carlo, A.; Maini, A.; Mele, A. Lack of Evidence for Increased Risk of Hepatitis A Infection in Homosexual Men. *Epidemiol. Infect.* **1999**, *123*, 89–93. [CrossRef] [PubMed]
35. Ida, S.; Tachikawa, N.; Nakajima, A.; Daikoku, M.; Yano, M.; Kikuchi, Y.; Yasuoka, A.; Kimura, S.; Oka, S. Influence of Human Immunodeficiency Virus Type 1 Infection on Acute Hepatitis A Virus Infection. *Clin. Infect. Dis. Off. Publ. Infect. Dis. Soc. Am.* **2002**, *34*, 379–385. [CrossRef] [PubMed]
36. Sun, H.-Y.; Kung, H.-C.; Ho, Y.-C.; Chien, Y.-F.; Chen, M.-Y.; Sheng, W.-H.; Hsieh, S.-M.; Wu, C.-H.; Liu, W.-C.; Hung, C.-C.; et al. Seroprevalence of Hepatitis A Virus Infection in Persons with HIV Infection in Taiwan: Implications for Hepatitis A Vaccination. *Int. J. Infect. Dis. IJID Off. Publ. Int. Soc. Infect. Dis.* **2009**, *13*, e199–e205. [CrossRef]
37. Lee, H.-C.; Ko, N.-Y.; Lee, N.-Y.; Chang, C.-M.; Ko, W.-C. Seroprevalence of Viral Hepatitis and Sexually Transmitted Disease among Adults with Recently Diagnosed HIV Infection in Southern Taiwan, 2000-2005: Upsurge in Hepatitis C Virus Infections among Injection Drug Users. *J. Formos. Med. Assoc.* **2008**, *107*, 404–411. [CrossRef]
38. Kourkounti, S.; Paparizos, V.; Leuow, K.; Kyriakis, K.; Antoniou, C. Prevalence and Titre of Antibodies against Hepatitis A Virus in HIV-Infected Men Having Sex with Men in Greece. *Infez. Med.* **2014**, *22*, 206–212.
39. Linkins, R.W.; Chonwattana, W.; Holtz, T.H.; Wasinrapee, P.; Chaikummao, S.; Varangrat, A.; Tongtoyai, J.; Mock, P.A.; Curlin, M.E.; Sirivongrangson, P.; et al. Hepatitis A and Hepatitis B Infection Prevalence and Associated Risk Factors in Men Who Have Sex with Men, Bangkok, 2006–2008. *J. Med. Virol.* **2013**, *85*, 1499–1505. [CrossRef]
40. NIH. Immunizations for Preventable Diseases in Adults and Adolescents with HIV. Available online: https://clinicalinfo.hiv.gov/en/guidelines/hiv-clinical-guidelines-adult-and-adolescent-opportunistic-infections/immunizations (accessed on 10 June 2023).
41. Doshani, M.; Weng, M.; Moore, K.L.; Romero, J.R.; Nelson, N.P. Recommendations of the Advisory Committee on Immunization Practices for Use of Hepatitis A Vaccine for Persons Experiencing Homelessness. *MMWR Morb. Mortal. Wkly. Rep.* **2019**, *68*, 153–156. [CrossRef]
42. Infection Par Le Virus de l'hépatite A: Traitement et Prévention—UpToDate. Available online: https://www.uptodate.com/contents/hepatitis-a-virus-infection-treatment-and-prevention (accessed on 11 June 2023).
43. Lin, K.-Y.; Chen, G.-J.; Lee, Y.-L.; Huang, Y.-C.; Cheng, A.; Sun, H.-Y.; Chang, S.-Y.; Liu, C.-E.; Hung, C.-C. Hepatitis A Virus Infection and Hepatitis A Vaccination in Human Immunodeficiency Virus-Positive Patients: A Review. *World J. Gastroenterol.* **2017**, *23*, 3589–3606. [CrossRef]
44. Zimmermann, P.; Curtis, N. Factors That Influence the Immune Response to Vaccination. *Clin. Microbiol. Rev.* **2019**, *32*, e00084-18. [CrossRef] [PubMed]
45. Neukam, K.; Fernández, M.D.; Quero, J.H.; Rivero-Juárez, A.; Llaves-Flores, S.; Oñate, F.J.; Gutiérrez-Valencia, A.; Espinosa, N.; Viciana, P.; López-Cortés, L.-F. Brief report: Response to hepatitis a virus vaccine in HIV-infected patients within a retrospective, multicentric cohort: Facing hepatitis a outbreaks in the clinical practice. *JAIDS J. Acquir. Immune Defic. Syndr.* **2019**, *81*, e1–e5. [CrossRef] [PubMed]
46. Fritzsche, C.; Bergmann, L.; Loebermann, M.; Glass, A.; Reisinger, E.C. Immune Response to Hepatitis A Vaccine in Patients with HIV. *Vaccine* **2019**, *37*, 2278–2283. [CrossRef]
47. Jabłonowska, E.; Kuydowicz, J. Durability of Response to Vaccination against Viral Hepatitis A in HIV-Infected Patients: A 5-Year Observation. *Int. J. STD AIDS* **2014**, *25*, 745–750. [CrossRef] [PubMed]
48. Santagostino, E.; Gringeri, A.; Rocino, A.; Zanetti, A.; de Biasi, R.; Mannucci, P.M. Patterns of Immunogenicity of an Inactivated Hepatitis A Vaccine in Anti-HIV Positive and Negative Hemophilic Patients. *Thromb. Haemost.* **1994**, *72*, 508–510. [CrossRef]
49. Hess, G.; Clemens, R.; Bienzle, U.; Schönfeld, C.; Schunck, B.; Bock, H.L. Immunogenicity and Safety of an Inactivated Hepatitis A Vaccine in Anti-HIV Positive and Negative Homosexual Men. *J. Med. Virol.* **1995**, *46*, 40–42. [CrossRef]
50. Weissman, S.; Feucht, C.; Moore, B.A. Response to Hepatitis A Vaccine in HIV-Positive Patients. *J. Viral Hepat.* **2006**, *13*, 81–86. [CrossRef]
51. Neilsen, G.A.; Bodsworth, N.J.; Watts, N. Response to Hepatitis A Vaccination in Human Immunodeficiency Virus-Infected and -Uninfected Homosexual Men. *J. Infect. Dis.* **1997**, *176*, 1064–1067. [CrossRef]
52. Kemper, C.A.; Haubrich, R.; Frank, I.; Dubin, G.; Buscarino, C.; McCutchan, J.A.; Deresinski, S.C.; California Collaborative Treatment Group. Safety and Immunogenicity of Hepatitis A Vaccine in Human Immunodeficiency Virus-Infected Patients: A Double-Blind, Randomized, Placebo-Controlled Trial. *J. Infect. Dis.* **2003**, *187*, 1327–1331. [CrossRef]
53. Wallace, M.R.; Brandt, C.J.; Earhart, K.C.; Kuter, B.J.; Grosso, A.D.; Lakkis, H.; Tasker, S.A. Safety and Immunogenicity of an Inactivated Hepatitis A Vaccine among HIV-Infected Subjects. *Clin. Infect. Dis. Off. Publ. Infect. Dis. Soc. Am.* **2004**, *39*, 1207–1213. [CrossRef]

54. de Fátima Thomé Barbosa Gouvêa, A.; de Moraes Pinto, M.I.; Miyamoto, M.; Machado, D.M.; Pessoa, S.D.; do Carmo, F.B.; de Vasconcelos Beltrão, S.C.; de Menezes Succi, R.C. Persistence of Hepatitis A Virus Antibodies after Primary Immunization and Response to Revaccination in Children and Adolescents with Perinatal HIV Exposure. *Rev. Paul. Pediatr.* **2015**, *33*, 142–149. [CrossRef]
55. Tseng, Y.-T.; Chang, S.-Y.; Liu, W.-C.; Sun, H.-Y.; Wu, C.-H.; Wu, P.-Y.; Lu, C.-L.; Hung, C.-C.; Chang, S.-C. Comparative Effectiveness of Two Doses versus Three Doses of Hepatitis A Vaccine in Human Immunodeficiency Virus-Infected and -Uninfected Men Who Have Sex with Men. *Hepatology* **2013**, *57*, 1734–1741. [CrossRef] [PubMed]
56. Overton, E.T.; Nurutdinova, D.; Sungkanuparph, S.; Seyfried, W.; Groger, R.K.; Powderly, W.G. Predictors of Immunity after Hepatitis A Vaccination in HIV-Infected Persons. *J. Viral Hepat.* **2007**, *14*, 189–193. [CrossRef] [PubMed]
57. Launay, O.; Grabar, S.; Gordien, E.; Desaint, C.; Jegou, D.; Abad, S.; Girard, P.-M.; Bélarbi, L.; Guérin, C.; Dimet, J.; et al. Immunological Efficacy of a Three-Dose Schedule of Hepatitis A Vaccine in HIV-Infected Adults: HEPAVAC Study. *J. Acquir. Immune Defic. Syndr.* **2008**, *49*, 272–275. [CrossRef] [PubMed]
58. Theeten, H.; Van Herck, K.; Van Der Meeren, O.; Crasta, P.; Van Damme, P.; Hens, N. Long-Term Antibody Persistence after Vaccination with a 2-Dose Havrix (Inactivated Hepatitis A Vaccine): 20 Years of Observed Data, and Long-Term Model-Based Predictions. *Vaccine* **2015**, *33*, 5723–5727. [CrossRef]
59. Kernéis, S.; Desaint, C.; Brichler, S.; Rey, D.; Belarbi, L.; Gordien, E.; Pacanowski, J.; Lortholary, O.; Abgrall, S.; Boëlle, P.-Y.; et al. Long-Term Persistence of Humoral Immunity after Hepatitis A Vaccination in HIV-Infected Adults. *J. Acquir. Immune Defic. Syndr.* **2011**, *57*, e63–e66. [CrossRef]
60. Crum-Cianflone, N.F.; Wilkins, K.; Lee, A.W.; Grosso, A.; Landrum, M.L.; Weintrob, A.; Ganesan, A.; Maguire, J.; Klopfer, S.; Brandt, C.; et al. Long-Term Durability of Immune Responses after Hepatitis A Vaccination among HIV-Infected Adults. *J. Infect. Dis.* **2011**, *203*, 1815–1823. [CrossRef]
61. Cheng, A.; Chang, S.-Y.; Sun, H.-Y.; Tsai, M.-S.; Liu, W.-C.; Su, Y.-C.; Wu, P.-Y.; Hung, C.-C.; Chang, S.-C. Long-Term Durability of Responses to 2 or 3 Doses of Hepatitis A Vaccination in Human Immunodeficiency Virus-Positive Adults on Antiretroviral Therapy. *J. Infect. Dis.* **2017**, *215*, 606–613. [CrossRef]
62. Liang, T.J. Hepatitis B: The Virus and Disease. *Hepatology* **2009**, *49*, S13–S21. [CrossRef]
63. Kellerman, S.E.; Hanson, D.L.; McNaghten, A.D.; Fleming, P.L. Prevalence of Chronic Hepatitis B and Incidence of Acute Hepatitis B Infection in Human Immunodeficiency Virus-Infected Subjects. *J. Infect. Dis.* **2003**, *188*, 571–577. [CrossRef] [PubMed]
64. Bruguera, M.; Cremades, M.; Salinas, R.; Costa, J.; Grau, M.; Sans, J. Impaired Response to Recombinant Hepatitis B Vaccine in HIV-Infected Persons. *J. Clin. Gastroenterol.* **1992**, *14*, 27–30. [CrossRef] [PubMed]
65. Soriano, V.; Puoti, M.; Bonacini, M.; Brook, G.; Cargnel, A.; Rockstroh, J.; Thio, C.; Benhamou, Y. Care of Patients with Chronic Hepatitis B and HIV Co-Infection: Recommendations from an HIV-HBV International Panel. *AIDS* **2005**, *19*, 221–240. [CrossRef] [PubMed]
66. Thio, C.L. Hepatitis B in the Human Immunodeficiency Virus-Infected Patient: Epidemiology, Natural History, and Treatment. *Semin. Liver Dis.* **2003**, *23*, 125–136. [CrossRef] [PubMed]
67. Homann, C.; Krogsgaard, K.; Pedersen, C.; Andersson, P.; Nielsen, J.O. High Incidence of Hepatitis B Infection and Evolution of Chronic Hepatitis B Infection in Patients with Advanced HIV Infection. *J. Acquir. Immune Defic. Syndr.* **1991**, *4*, 416–420. [PubMed]
68. Sterling, R.K.; Wahed, A.S.; King, W.C.; Kleiner, D.E.; Khalili, M.; Sulkowski, M.; Chung, R.T.; Jain, M.K.; Lisker-Melman, M.; Wong, D.K.; et al. Spectrum of Liver Disease in Hepatitis B Virus (HBV) Patients Co-Infected with Human Immunodeficiency Virus (HIV): Results of the HBV-HIV Cohort Study. *Am. J. Gastroenterol.* **2019**, *114*, 746–757. [CrossRef]
69. Kottilil, S.; Jackson, J.O.; Polis, M.A. Hepatitis B & Hepatitis C in HIV-Infection. *Indian J. Med. Res.* **2005**, *121*, 424–450.
70. Thio, C.L.; Seaberg, E.C.; Skolasky, R.; Phair, J.; Visscher, B.; Muñoz, A.; Thomas, D.L. Multicenter AIDS Cohort Study HIV-1, Hepatitis B Virus, and Risk of Liver-Related Mortality in the Multicenter Cohort Study (MACS). *Lancet* **2002**, *360*, 1921–1926. [CrossRef]
71. Zarski, J.P.; Thelu, M.A.; Rachail, M.; Seigneurin, J.M. Molecular biology of the hepatitis B virus. II: Importance of the detection of DNA of the hepatitis B virus in serum and liver. *Gastroenterol. Clin. Biol.* **1991**, *15*, 497–508.
72. Krugman, S.; Overby, L.R.; Mushahwar, I.K.; Ling, C.M.; Frösner, G.G.; Deinhardt, F. Viral Hepatitis, Type B. Studies on Natural History and Prevention Re-Examined. *N. Engl. J. Med.* **1979**, *300*, 101–106. [CrossRef]
73. Bowden, S. Serological and Molecular Diagnosis. *Semin. Liver Dis.* **2006**, *26*, 97–103. [CrossRef]
74. Tsang, T.K.; Blei, A.T.; O'Reilly, D.J.; Decker, R. Clinical Significance of Concurrent Hepatitis B Surface Antigen and Antibody Positivity. *Dig. Dis. Sci.* **1986**, *31*, 620–624. [CrossRef] [PubMed]
75. Mast, E.E.; Weinbaum, C.M.; Fiore, A.E.; Alter, M.J.; Bell, B.P.; Finelli, L.; Rodewald, L.E.; Douglas, J.M.; Janssen, R.S.; Ward, J.W.; et al. A Comprehensive Immunization Strategy to Eliminate Transmission of Hepatitis B Virus Infection in the United States: Recommendations of the Advisory Committee on Immunization Practices (ACIP) Part II: Immunization of Adults. *MMWR Recomm. Rep. Morb. Mortal. Wkly. Rep. Recomm. Rep.* **2006**, *55*, 1–33.
76. Martins, S.; do Livramento, A.; Andrigueti, M.; Kretzer, I.F.; Machado, M.J.; Spada, C.; Treitinger, A. Vaccination Coverage and Immunity against Hepatitis B among HIV-Infected Patients in South Brazil. *Braz. J. Infect. Dis. Off. Publ. Braz. Soc. Infect. Dis.* **2015**, *19*, 181–186. [CrossRef] [PubMed]

77. Rey, D.; Krantz, V.; Partisani, M.; Schmitt, M.P.; Meyer, P.; Libbrecht, E.; Wendling, M.J.; Vetter, D.; Nicolle, M.; Kempf-Durepaire, G.; et al. Increasing the Number of Hepatitis B Vaccine Injections Augments Anti-HBs Response Rate in HIV-Infected Patients. Effects on HIV-1 Viral Load. *Vaccine* **2000**, *18*, 1161–1165. [CrossRef] [PubMed]
78. Wilson, C.M.; Ellenberg, J.H.; Sawyer, M.K.; Belzer, M.; Crowley-Nowick, P.A.; Puga, A.; Futterman, D.C.; Peralta, L.; Adolescent Medicine HIV/AIDS Research Network. Serologic Response to Hepatitis B Vaccine in HIV Infected and High-Risk HIV Uninfected Adolescents in the REACH Cohort. Reaching for Excellence in Adolescent Care and Health. *J. Adolesc. Health Off. Publ. Soc. Adolesc. Med.* **2001**, *29*, 123–129. [CrossRef]
79. Laurence, J.C. Hepatitis A and B Immunizations of Individuals Infected with Human Immunodeficiency Virus. *Am. J. Med.* **2005**, *118* (Suppl. 10A), 75S–83S. [CrossRef] [PubMed]
80. Wong, E.K.; Bodsworth, N.J.; Slade, M.A.; Mulhall, B.P.; Donovan, B. Response to Hepatitis B Vaccination in a Primary Care Setting: Influence of HIV Infection, CD4+ Lymphocyte Count and Vaccination Schedule. *Int. J. STD AIDS* **1996**, *7*, 490–494. [CrossRef]
81. Collier, A.C.; Corey, L.; Murphy, V.L.; Handsfield, H.H. Antibody to Human Immunodeficiency Virus (HIV) and Suboptimal Response to Hepatitis B Vaccination. *Ann. Intern. Med.* **1988**, *109*, 101–105. [CrossRef]
82. Gandhi, R.T.; Wurcel, A.; Lee, H.; McGovern, B.; Shopis, J.; Geary, M.; Sivamurthy, R.; Sax, P.E.; Ukomadu, C. Response to Hepatitis B Vaccine in HIV-1-Positive Subjects Who Test Positive for Isolated Antibody to Hepatitis B Core Antigen: Implications for Hepatitis B Vaccine Strategies. *J. Infect. Dis.* **2005**, *191*, 1435–1441. [CrossRef] [PubMed]
83. Tedaldi, E.M.; Baker, R.K.; Moorman, A.C.; Wood, K.C.; Fuhrer, J.; McCabe, R.E.; Holmberg, S.D.; HIV Outpatient Study (HOPS) Investigators. Hepatitis A and B Vaccination Practices for Ambulatory Patients Infected with HIV. *Clin. Infect. Dis. Off. Publ. Infect. Dis. Soc. Am.* **2004**, *38*, 1478–1484. [CrossRef] [PubMed]
84. Bauer, S.; Kirschning, C.J.; Häcker, H.; Redecke, V.; Hausmann, S.; Akira, S.; Wagner, H.; Lipford, G.B. Human TLR9 Confers Responsiveness to Bacterial DNA via Species-Specific CpG Motif Recognition. *Proc. Natl. Acad. Sci. USA* **2001**, *98*, 9237–9242. [CrossRef] [PubMed]
85. Kim, J.H.; Psevdos, G.; Groce, V.; Sharp, V. Persistence of Protective Hepatitis B Surface Antibody Titers after Successful Double-Dose Hepatitis B Virus Rescue Vaccination in HIV-Infected Patients. *Gut Liver* **2012**, *6*, 86–91. [CrossRef] [PubMed]
86. Shafran, S.D.; Mashinter, L.D.; Lindemulder, A.; Taylor, G.D.; Chiu, I. Poor Efficacy of Intradermal Administration of Recombinant Hepatitis B Virus Immunization in HIV-Infected Individuals Who Fail to Respond to Intramuscular Administration of Hepatitis B Virus Vaccine. *HIV Med.* **2007**, *8*, 295–299. [CrossRef]
87. de Vries-Sluijs, T.E.M.S.; Hansen, B.E.; van Doornum, G.J.J.; Springeling, T.; Evertsz, N.M.; de Man, R.A.; van der Ende, M.E. A Prospective Open Study of the Efficacy of High-Dose Recombinant Hepatitis B Rechallenge Vaccination in HIV-Infected Patients. *J. Infect. Dis.* **2008**, *197*, 292–294. [CrossRef]
88. Chaiklang, K.; Wipasa, J.; Chaiwarith, R.; Praparattanapan, J.; Supparatpinyo, K. Comparison of Immunogenicity and Safety of Four Doses and Four Double Doses vs. Standard Doses of Hepatitis B Vaccination in HIV-Infected Adults: A Randomized, Controlled Trial. *PLoS ONE* **2013**, *8*, e80409. [CrossRef] [PubMed]
89. Seremba, E.; Ocama, P.; Ssekitoleko, R.; Mayanja-Kizza, H.; Adams, S.; Orem, J.; Katabira, E.; Reynolds, S.; Nabatanzi, R.; Casper, C.; et al. Immune Response to the Hepatitis B Vaccine Among HIV-Infected Adults in Uganda. *Vaccine* **2021**, *39*, 1265–1271. [CrossRef]
90. Nie, L.; Hua, W.; Liu, X.; Pang, X.; Guo, C.; Zhang, W.; Tian, Y.; Qiu, Q. Associated Factors and Immune Response to the Hepatitis B Vaccine with a Standard Schedule: A Prospective Study of People with HIV in China. *Vaccines* **2023**, *11*, 921. [CrossRef]
91. Xu, L.; Zhang, L.; Kang, S.; Li, X.; Lu, L.; Liu, X.; Song, X.; Li, Y.; Li, X.; Lyu, W.; et al. Immune Responses to HBV Vaccine in People Living with HIV (PLWHs) Who Achieved Successful Treatment: A Prospective Cohort Study. *Vaccines* **2023**, *11*, 400. [CrossRef]
92. Launay, O.; van der Vliet, D.; Rosenberg, A.R.; Michel, M.-L.; Piroth, L.; Rey, D.; Colin de Verdière, N.; Slama, L.; Martin, K.; Lortholary, O.; et al. Safety and Immunogenicity of 4 Intramuscular Double Doses and 4 Intradermal Low Doses vs Standard Hepatitis B Vaccine Regimen in Adults with HIV-1: A Randomized Controlled Trial. *JAMA* **2011**, *305*, 1432–1440. [CrossRef]
93. Fuster, F.; Vargas, J.I.; Jensen, D.; Sarmiento, V.; Acuña, P.; Peirano, F.; Fuster, F.; Arab, J.P.; Martínez, F.; Fuster, F.; et al. CD4/CD8 Ratio as a Predictor of the Response to HBV Vaccination in HIV-Positive Patients: A Prospective Cohort Study. *Vaccine* **2016**, *34*, 1889–1895. [CrossRef] [PubMed]
94. de Vries-Sluijs, T.E.M.S.; Hansen, B.E.; van Doornum, G.J.J.; Kauffmann, R.H.; Leyten, E.M.S.; Mudrikova, T.; Brinkman, K.; den Hollander, J.G.; Kroon, F.P.; Janssen, H.L.A.; et al. A Randomized Controlled Study of Accelerated Versus Standard Hepatitis B Vaccination in HIV-Positive Patients. *J. Infect. Dis.* **2011**, *203*, 984–991. [CrossRef] [PubMed]
95. Feng, Y.; Yao, T.; Chang, Y.; Gao, L.; Shao, Z.; Dong, S.; Wu, Y.; Shi, X.; Shi, J.; Feng, D.; et al. Immunogenicity and Persistence of High-Dose Recombinant Hepatitis B Vaccine in Adults Infected with Human Immunodeficiency Virus in China: A Randomized, Double-Blind, Parallel Controlled Trial. *Vaccine* **2021**, *39*, 3582–3589. [CrossRef]
96. O'Bryan, T.A.; Rini, E.A.; Okulicz, J.F.; Messner, O.; Ganesan, A.; Lalani, T.; Bavaro, M.F.; O'Connell, R.J.; Agan, B.K.; Landrum, M.L. HIV Viraemia during Hepatitis B Vaccination Shortens the Duration of Protective Antibody Levels. *HIV Med.* **2015**, *16*, 161–167. [CrossRef] [PubMed]
97. Fonseca, M.O.; Pang, L.W.; de Paula Cavalheiro, N.; Barone, A.A.; Lopes, M.H. Randomized Trial of Recombinant Hepatitis B Vaccine in HIV-Infected Adult Patients Comparing a Standard Dose to a Double Dose. *Vaccine* **2005**, *23*, 2902–2908. [CrossRef]

98. Zurek Munk-Madsen, M.; Toft, L.; Kube, T.; Richter, R.; Ostergaard, L.; Søgaard, O.S.; Tolstrup, M.; Kaufmann, A.M. Cellular Immunogenicity of Human Papillomavirus Vaccines Cervarix and Gardasil in Adults with HIV Infection. *Hum. Vaccines Immunother.* **2017**, *14*, 909–916. [CrossRef]
99. Doorbar, J.; Egawa, N.; Griffin, H.; Kranjec, C.; Murakami, I. Human Papillomavirus Molecular Biology and Disease Association. *Rev. Med. Virol.* **2015**, *25* (Suppl. S1), 2–23. [CrossRef]
100. Melbye, M.; Frisch, M. The Role of Human Papillomaviruses in Anogenital Cancers. *Semin. Cancer Biol.* **1998**, *8*, 307–313. [CrossRef]
101. Phanuphak, N.; Teeraananchai, S.; Hansudewechakul, R.; Gatechompol, S.; Chokephaibulkit, K.; Dang, H.L.D.; Tran, D.N.H.; Achalapong, J.; Teeratakulpisarn, N.; Chalermchockcharoenkit, A.; et al. Incidence and Persistence of High-Risk Anogenital Human Papillomavirus Infection Among Female Youth with and without Perinatally Acquired Human Immunodefiency Virus Infection: A 3-Year Observational Cohort Study. *Clin. Infect. Dis. Off. Publ. Infect. Dis. Soc. Am.* **2020**, *71*, e270–e280. [CrossRef]
102. Garland, S.M.; Hernandez-Avila, M.; Wheeler, C.M.; Perez, G.; Harper, D.M.; Leodolter, S.; Tang, G.W.K.; Ferris, D.G.; Steben, M.; Bryan, J.; et al. Quadrivalent Vaccine against Human Papillomavirus to Prevent Anogenital Diseases. *N. Engl. J. Med.* **2007**, *356*, 1928–1943. [CrossRef]
103. Mo, Y.; Ma, J.; Zhang, H.; Shen, J.; Chen, J.; Hong, J.; Xu, Y.; Qian, C. Prophylactic and Therapeutic HPV Vaccines: Current Scenario and Perspectives. *Front. Cell. Infect. Microbiol.* **2022**, *12*, 909223. [CrossRef] [PubMed]
104. Ucciferri, C.; Tamburro, M.; Falasca, K.; Sammarco, M.L.; Ripabelli, G.; Vecchiet, J. Prevalence of Anal, Oral, Penile and Urethral Human Papillomavirus in HIV Infected and HIV Uninfected Men Who Have Sex with Men. *J. Med. Virol.* **2018**, *90*, 358–366. [CrossRef] [PubMed]
105. Tartaglia, E.; Falasca, K.; Vecchiet, J.; Sabusco, G.P.; Picciano, G.; Di Marco, R.; Ucciferri, C. Prevalence of HPV Infection among HIV-positive and HIV-negative Women in Central/Eastern Italy: Strategies of Prevention. *Oncol. Lett.* **2017**, *14*, 7629–7635. [CrossRef]
106. Lacey, C.J.N. HPV Vaccination in HIV Infection. *Papillomavirus Res.* **2019**, *8*, 100174. [CrossRef] [PubMed]
107. Olsson, S.-E.; Villa, L.L.; Costa, R.L.R.; Petta, C.A.; Andrade, R.P.; Malm, C.; Iversen, O.-E.; Høye, J.; Steinwall, M.; Riis-Johannessen, G.; et al. Induction of Immune Memory Following Administration of a Prophylactic Quadrivalent Human Papillomavirus (HPV) Types 6/11/16/18 L1 Virus-like Particle (VLP) Vaccine. *Vaccine* **2007**, *25*, 4931–4939. [CrossRef]
108. Einstein, M.H.; Baron, M.; Levin, M.J.; Chatterjee, A.; Fox, B.; Scholar, S.; Rosen, J.; Chakhtoura, N.; Meric, D.; Dessy, F.J.; et al. Comparative Immunogenicity and Safety of Human Papillomavirus (HPV)-16/18 Vaccine and HPV-6/11/16/18 Vaccine: Follow-up from Months 12–24 in a Phase III Randomized Study of Healthy Women Aged 18–45 Years. *Hum. Vaccin.* **2011**, *7*, 1343–1358. [CrossRef]
109. Villa, L.L.; Ault, K.A.; Giuliano, A.R.; Costa, R.L.R.; Petta, C.A.; Andrade, R.P.; Brown, D.R.; Ferenczy, A.; Harper, D.M.; Koutsky, L.A.; et al. Immunologic Responses Following Administration of a Vaccine Targeting Human Papillomavirus Types 6, 11, 16, and 18. *Vaccine* **2006**, *24*, 5571–5583. [CrossRef] [PubMed]
110. Giacomet, V.; Penagini, F.; Trabattoni, D.; Viganò, A.; Rainone, V.; Bernazzani, G.; Bonardi, C.M.; Clerici, M.; Bedogni, G.; Zuccotti, G.V. Safety and Immunogenicity of a Quadrivalent Human Papillomavirus Vaccine in HIV-Infected and HIV-Negative Adolescents and Young Adults. *Vaccine* **2014**, *32*, 5657–5661. [CrossRef] [PubMed]
111. Overton, E.T.; Sungkanuparph, S.; Powderly, W.G.; Seyfried, W.; Groger, R.K.; Aberg, J.A. Undetectable Plasma HIV RNA Load Predicts Success after Hepatitis B Vaccination in HIV-Infected Persons. *Clin. Infect. Dis. Off. Publ. Infect. Dis. Soc. Am.* **2005**, *41*, 1045–1048. [CrossRef] [PubMed]
112. Pinto, L.A.; Wilkin, T.J.; Kemp, T.J.; Abrahamsen, M.; Isaacs-Soriano, K.; Pan, Y.; Webster-Cyriaque, J.; Palefsky, J.M.; Giuliano, A.R. Oral and Systemic HPV Antibody Kinetics Post-Vaccination Among HIV-Positive and HIV-Negative Men. *Vaccine* **2019**, *37*, 2502–2510. [CrossRef]
113. Wilkin, T.; Lee, J.Y.; Lensing, S.Y.; Stier, E.A.; Goldstone, S.E.; Berry, J.M.; Jay, N.; Aboulafia, D.; Cohn, D.L.; Einstein, M.H.; et al. Safety and Immunogenicity of the Quadrivalent Human Papillomavirus Vaccine in HIV-1–Infected Men. *J. Infect. Dis.* **2010**, *202*, 1246–1253. [CrossRef] [PubMed]
114. Denny, L.; Hendricks, B.; Gordon, C.; Thomas, F.; Hezareh, M.; Dobbelaere, K.; Durand, C.; Hervé, C.; Descamps, D. Safety and Immunogenicity of the HPV-16/18 AS04-Adjuvanted Vaccine in HIV-Positive Women in South Africa: A Partially-Blind Randomised Placebo-Controlled Study. *Vaccine* **2013**, *31*, 5745–5753. [CrossRef] [PubMed]
115. Chow, E.P.F.; Fairley, C.K.; Zou, H.; Wigan, R.; Garland, S.M.; Cornall, A.M.; Atchison, S.; Tabrizi, S.N.; Chen, M.Y. Human Papillomavirus Antibody Levels Following Vaccination or Natural Infection Among Young Men Who Have Sex with Men. *Clin. Infect. Dis.* **2022**, *75*, 323–329. [CrossRef] [PubMed]
116. Weinberg, A.; Huang, S.; Moscicki, A.-B.; Saah, A.; Levin, M.J. Persistence of Memory B-Cell and T-Cell Responses to the Quadrivalent HPV Vaccine in HIV-Infected Children. *AIDS* **2018**, *32*, 851–860. [CrossRef]
117. Kojic, E.M.; Kang, M.; Cespedes, M.S.; Umbleja, T.; Godfrey, C.; Allen, R.T.; Firnhaber, C.; Grinsztejn, B.; Palefsky, J.M.; Webster-Cyriaque, J.Y.; et al. Immunogenicity and Safety of the Quadrivalent Human Papillomavirus Vaccine in HIV-1-Infected Women. *Clin. Infect. Dis. Off. Publ. Infect. Dis. Soc. Am.* **2014**, *59*, 127–135. [CrossRef] [PubMed]
118. Krammer, F.; Smith, G.J.D.; Fouchier, R.A.M.; Peiris, M.; Kedzierska, K.; Doherty, P.C.; Palese, P.; Shaw, M.L.; Treanor, J.; Webster, R.G.; et al. Influenza. *Nat. Rev. Dis. Primer* **2018**, *4*, 3. [CrossRef] [PubMed]
119. Sandrock, C.; Kelly, T. Clinical Review: Update of Avian Influenza A Infections in Humans. *Crit. Care* **2007**, *11*, 209. [CrossRef]

120. Bouvier, N.M.; Palese, P. The Biology of Influenza Viruses. *Vaccine* **2008**, *26* (Suppl. 4), D49–D53. [CrossRef]
121. Hampson, A.W.; Mackenzie, J.S. The Influenza Viruses. *Med. J. Aust.* **2006**, *185*, S39–S43. [CrossRef] [PubMed]
122. Kosik, I.; Yewdell, J.W. Influenza Hemagglutinin and Neuraminidase: Yin−Yang Proteins Coevolving to Thwart Immunity. *Viruses* **2019**, *11*, 346. [CrossRef]
123. Biere, B.; Bauer, B.; Schweiger, B. Differentiation of Influenza B Virus Lineages Yamagata and Victoria by Real-Time PCR. *J. Clin. Microbiol.* **2010**, *48*, 1425–1427. [CrossRef] [PubMed]
124. Wei, H.; Wang, S.; Chen, Q.; Chen, Y.; Chi, X.; Zhang, L.; Huang, S.; Gao, G.F.; Chen, J.-L. Suppression of Interferon Lambda Signaling by SOCS-1 Results in Their Excessive Production during Influenza Virus Infection. *PLoS Pathog.* **2014**, *10*, e1003845. [CrossRef]
125. Mifsud, E.J.; Kuba, M.; Barr, I.G. Innate Immune Responses to Influenza Virus Infections in the Upper Respiratory Tract. *Viruses* **2021**, *13*, 2090. [CrossRef]
126. Jain, S.; Kamimoto, L.; Bramley, A.M.; Schmitz, A.M.; Benoit, S.R.; Louie, J.; Sugerman, D.E.; Druckenmiller, J.K.; Ritger, K.A.; Chugh, R.; et al. Hospitalized Patients with 2009 H1N1 Influenza in the United States, April–June 2009. *N. Engl. J. Med.* **2009**, *361*, 1935–1944. [CrossRef] [PubMed]
127. Yamayoshi, S.; Kawaoka, Y. Current and Future Influenza Vaccines. *Nat. Med.* **2019**, *25*, 212–220. [CrossRef] [PubMed]
128. Influenza (Seasonal). Available online: https://www.who.int/health-topics/influenza-seasonal?gclid=CjwKCAjwwb6lBhBJEiwAbuVUSm7fCcN6ZB7s7RGpY4-ANtolRWn13UBuJy8xRZgEyrQyj-xsF6eZqRoCQoEQAvD_BwE#tab=tab_1 (accessed on 13 July 2023).
129. DiazGranados, C.A.; Denis, M.; Plotkin, S. Seasonal Influenza Vaccine Efficacy and Its Determinants in Children and Non-Elderly Adults: A Systematic Review with Meta-Analyses of Controlled Trials. *Vaccine* **2012**, *31*, 49–57. [CrossRef]
130. Fiore, A.E.; Bridges, C.B.; Cox, N.J. Seasonal Influenza Vaccines. In *Vaccines for Pandemic Influenza*; Compans, R.W., Orenstein, W.A., Eds.; Current Topics in Microbiology and Immunology; Springer: Berlin/Heidelberg, Germany, 2009; pp. 43–82, ISBN 978-3-540-92165-3.
131. Houser, K.; Subbarao, K. Influenza Vaccines: Challenges and Solutions. *Cell Host Microbe* **2015**, *17*, 295–300. [CrossRef]
132. Sridhar, S.; Begom, S.; Bermingham, A.; Hoschler, K.; Adamson, W.; Carman, W.; Bean, T.; Barclay, W.; Deeks, J.J.; Lalvani, A. Cellular Immune Correlates of Protection against Symptomatic Pandemic Influenza. *Nat. Med.* **2013**, *19*, 1305–1312. [CrossRef]
133. Tempia, S.; Walaza, S.; Moyes, J.; Cohen, A.L.; von Mollendorf, C.; McMorrow, M.L.; Mhlanga, S.; Treurnicht, F.K.; Venter, M.; Pretorius, M.; et al. The Effects of the Attributable Fraction and the Duration of Symptoms on Burden Estimates of Influenza-associated Respiratory Illnesses in a High HIV Prevalence Setting, South Africa, 2013–2015. *Influenza Other Respir. Viruses* **2018**, *12*, 360–373. [CrossRef]
134. Ministero della Salute. Piano Nazionale Prevenzione Vaccinale. Available online: https://www.salute.gov.it/portale/vaccinazioni/dettaglioContenutiVaccinazioni.jsp?lingua=italiano&id=4828&area=vaccinazioni&menu=vuoto (accessed on 13 July 2023).
135. Kroon, F.P.; van Dissel, J.T.; de Jong, J.C.; Zwinderman, K.; van Furth, R. Antibody Response after Influenza Vaccination in HIV-Infected Individuals: A Consecutive 3-Year Study. *Vaccine* **2000**, *18*, 3040–3049. [CrossRef]
136. Pallikkuth, S.; Parmigiani, A.; Silva, S.Y.; George, V.K.; Fischl, M.; Pahwa, R.; Pahwa, S. Impaired Peripheral Blood T-Follicular Helper Cell Function in HIV-Infected Nonresponders to the 2009 H1N1/09 Vaccine. *Blood* **2012**, *120*, 985–993. [CrossRef]
137. Pallikkuth, S.; Kanthikeel, S.P.; Silva, S.Y.; Fischl, M.; Pahwa, R.; Pahwa, S. Innate Immune Defects Correlate with Failure of Antibody Responses to H1N1/09 Vaccine in HIV-Infected Patients. *J. Allergy Clin. Immunol.* **2011**, *128*, 1279–1285. [CrossRef]
138. Tebas, P.; Frank, I.; Lewis, M.; Quinn, J.; Zifchak, L.; Thomas, A.; Kenney, T.; Kappes, R.; Wagner, W.; Maffei, K.; et al. Poor Immunogenicity of the H1N1 2009 Vaccine in Well Controlled HIV-Infected Individuals. *AIDS* **2010**, *24*, 2187–2192. [CrossRef] [PubMed]
139. Flynn, P.M.; Nachman, S.; Muresan, P.; Fenton, T.; Spector, S.A.; Cunningham, C.K.; Pass, R.; Yogev, R.; Burchett, S.; Heckman, B.; et al. Safety and Immunogenicity of 2009 Pandemic H1N1 Influenza Vaccination in Perinatally HIV-1–Infected Children, Adolescents, and Young Adults. *J. Infect. Dis.* **2012**, *206*, 421–430. [CrossRef] [PubMed]
140. Fifth Meeting of the International Health Regulations (2005) (IHR) Emergency Committee on the Multi-Country Outbreak of Mpox (Monkeypox). Available online: https://www.who.int/news/item/11-05-2023-fifth-meeting-of-the-international-health-regulations-(2005)-(ihr)-emergency-committee-on-the-multi-country-outbreak-of-monkeypox-(mpox) (accessed on 16 May 2023).
141. Mitjà, O.; Alemany, A.; Marks, M.; Mora, J.I.L.; Rodríguez-Aldama, J.C.; Silva, M.S.T.; Herrera, E.A.C.; Crabtree-Ramirez, B.; Blanco, J.L.; Girometti, N.; et al. Mpox in People with Advanced HIV Infection: A Global Case Series. *Lancet* **2023**, *401*, 939–949. [CrossRef]
142. Joint ECDC-WHO Regional Office for Europe Mpox Surveillance Bulletin. Available online: https://monkeypoxreport.ecdc.europa.eu/ (accessed on 16 May 2023).
143. Ortiz-Saavedra, B.; Montes-Madariaga, E.S.; Cabanillas-Ramirez, C.; Alva, N.; Ricardo-Martínez, A.; León-Figueroa, D.A.; Barboza, J.J.; Mohanty, A.; Padhi, B.K.; Sah, R. Epidemiologic Situation of HIV and Monkeypox Coinfection: A Systematic Review. *Vaccines* **2023**, *11*, 246. [CrossRef]
144. Núñez, I.; Valdés-Ferrer, S.I. Fulminant Mpox as an AIDS-Defining Condition: Useful or Stigmatising? *Lancet* **2023**, *401*, 881–884. [CrossRef] [PubMed]
145. Martín-Delgado, M.C.; Martín Sánchez, F.J.; Martínez-Sellés, M.; Molero García, J.M.; Moreno Guillén, S.; Rodríguez-Artalejo, F.J.; Ruiz-Galiana, J.; Cantón, R.; De Lucas Ramos, P.; García-Botella, A.; et al. Monkeypox in Humans: A New Outbreak. *Rev. Esp. Quimioter. Publ. Of. Soc. Espanola Quimioter.* **2022**, *35*, 509–518. [CrossRef]

146. Public Health—European Commission. Available online: https://ec.europa.eu/health/documents/community-register/html/h855.htm (accessed on 12 July 2023).
147. Rizk, J.G.; Lippi, G.; Henry, B.M.; Forthal, D.N.; Rizk, Y. Prevention and Treatment of Monkeypox. *Drugs* **2022**, *82*, 957–963. [CrossRef]
148. Overton, E.T.; Stapleton, J.; Frank, I.; Hassler, S.; Goepfert, P.A.; Barker, D.; Wagner, E.; von Krempelhuber, A.; Virgin, G.; Weigl, J.; et al. Safety and Immunogenicity of Modified Vaccinia Ankara-Bavarian Nordic Smallpox Vaccine in Vaccinia-Naive and Experienced Human Immunodeficiency Virus-Infected Individuals: An Open-Label, Controlled Clinical Phase II Trial. *Open Forum Infect. Dis.* **2015**, *2*, ofv040. [CrossRef]
149. Frey, S.E.; Winokur, P.L.; Salata, R.A.; El-Kamary, S.S.; Turley, C.B.; Walter, E.B.; Hay, C.M.; Newman, F.K.; Hill, H.R.; Zhang, Y.; et al. Safety and Immunogenicity of IMVAMUNE®Smallpox Vaccine Using Different Strategies for a Post Event Scenario. *Vaccine* **2013**, *31*, 3025–3033. [CrossRef]
150. Greenberg, R.N.; Overton, E.T.; Haas, D.W.; Frank, I.; Goldman, M.; von Krempelhuber, A.; Virgin, G.; Bädeker, N.; Vollmar, J.; Chaplin, P. Safety, Immunogenicity, and Surrogate Markers of Clinical Efficacy for Modified Vaccinia Ankara as a Smallpox Vaccine in HIV-Infected Subjects. *J. Infect. Dis.* **2013**, *207*, 749–758. [CrossRef]
151. Cui, J.; Li, F.; Shi, Z.-L. Origin and Evolution of Pathogenic Coronaviruses. *Nat. Rev. Microbiol.* **2019**, *17*, 181–192. [CrossRef] [PubMed]
152. WHO Coronavirus (COVID-19) Dashboard. Available online: https://covid19.who.int (accessed on 10 May 2023).
153. Lu, R.; Zhao, X.; Li, J.; Niu, P.; Yang, B.; Wu, H.; Wang, W.; Song, H.; Huang, B.; Zhu, N.; et al. Genomic Characterisation and Epidemiology of 2019 Novel Coronavirus: Implications for Virus Origins and Receptor Binding. *Lancet* **2020**, *395*, 565–574. [CrossRef] [PubMed]
154. Wu, A.; Peng, Y.; Huang, B.; Ding, X.; Wang, X.; Niu, P.; Meng, J.; Zhu, Z.; Zhang, Z.; Wang, J.; et al. Genome Composition and Divergence of the Novel Coronavirus (2019-NCoV) Originating in China. *Cell Host Microbe* **2020**, *27*, 325–328. [CrossRef]
155. Zhou, L.; Ayeh, S.K.; Chidambaram, V.; Karakousis, P.C. Modes of Transmission of SARS-CoV-2 and Evidence for Preventive Behavioral Interventions. *BMC Infect. Dis.* **2021**, *21*, 496. [CrossRef] [PubMed]
156. V'kovski, P.; Kratzel, A.; Steiner, S.; Stalder, H.; Thiel, V. Coronavirus Biology and Replication: Implications for SARS-CoV-2. *Nat. Rev. Microbiol.* **2021**, *19*, 155–170. [CrossRef] [PubMed]
157. COVID-19 Vaccines. Available online: https://www.who.int/emergencies/diseases/novel-coronavirus-2019/covid-19-vaccines (accessed on 10 May 2023).
158. Zhang, H.; Zhang, L.; Lin, A.; Xu, C.; Li, Z.; Liu, K.; Liu, B.; Ma, X.; Zhao, F.; Jiang, H.; et al. Algorithm for Optimized MRNA Design Improves Stability and Immunogenicity. *Nature* **2023**, 1–3. [CrossRef]
159. Polack, F.P.; Thomas, S.J.; Kitchin, N.; Absalon, J.; Gurtman, A.; Lockhart, S.; Perez, J.L.; Pérez Marc, G.; Moreira, E.D.; Zerbini, C.; et al. Safety and Efficacy of the BNT162b2 MRNA Covid-19 Vaccine. *N. Engl. J. Med.* **2020**, *383*, 2603–2615. [CrossRef]
160. Sahin, U.; Muik, A.; Derhovanessian, E.; Vogler, I.; Kranz, L.M.; Vormehr, M.; Baum, A.; Pascal, K.; Quandt, J.; Maurus, D.; et al. COVID-19 Vaccine BNT162b1 Elicits Human Antibody and TH1 T Cell Responses. *Nature* **2020**, *586*, 594–599. [CrossRef]
161. Kalimuddin, S.; Tham, C.Y.L.; Qui, M.; de Alwis, R.; Sim, J.X.Y.; Lim, J.M.E.; Tan, H.-C.; Syenina, A.; Zhang, S.L.; Le Bert, N.; et al. Early T Cell and Binding Antibody Responses Are Associated with COVID-19 RNA Vaccine Efficacy Onset. *Med* **2021**, *2*, 682–688.e4. [CrossRef]
162. Tarke, A.; Sidney, J.; Methot, N.; Yu, E.D.; Zhang, Y.; Dan, J.M.; Goodwin, B.; Rubiro, P.; Sutherland, A.; Wang, E.; et al. Impact of SARS-CoV-2 Variants on the Total CD4+ and CD8+ T Cell Reactivity in Infected or Vaccinated Individuals. *Cell Rep. Med.* **2021**, *2*, 100355. [CrossRef]
163. Woldemeskel, B.A.; Garliss, C.C.; Blankson, J.N. SARS-CoV-2 MRNA Vaccines Induce Broad CD4+ T Cell Responses That Recognize SARS-CoV-2 Variants and HCoV-NL63. *J. Clin. Investig.* **2021**, *131*, e149335. [CrossRef]
164. Ambrosioni, J.; Blanco, J.L.; Reyes-Urueña, J.M.; Davies, M.-A.; Sued, O.; Marcos, M.A.; Martínez, E.; Bertagnolio, S.; Alcamí, J.; Miro, J.M.; et al. Overview of SARS-CoV-2 Infection in Adults Living with HIV. *Lancet HIV* **2021**, *8*, e294–e305. [CrossRef] [PubMed]
165. Clinical Spectrum. Available online: https://www.covid19treatmentguidelines.nih.gov/overview/clinical-spectrum/ (accessed on 10 May 2023).
166. Eisinger, R.W.; Lerner, A.M.; Fauci, A.S. Human Immunodeficiency Virus/AIDS in the Era of Coronavirus Disease 2019: A Juxtaposition of 2 Pandemics. *J. Infect. Dis.* **2021**, *224*, 1455–1461. [CrossRef]
167. Spinelli, M.A.; Jones, B.L.H.; Gandhi, M. COVID-19 Outcomes and Risk Factors Among People Living with HIV. *Curr. HIV/AIDS Rep.* **2022**, *19*, 425–432. [CrossRef] [PubMed]
168. Levy, I.; Wieder-Finesod, A.; Litchevsky, V.; Biber, A.; Indenbaum, V.; Olmer, L.; Huppert, A.; Mor, O.; Goldstein, M.; Levin, E.G.; et al. Immunogenicity and Safety of the BNT162b2 MRNA COVID-19 Vaccine in People Living with HIV-1. *Clin. Microbiol. Infect. Off. Publ. Eur. Soc. Clin. Microbiol. Infect. Dis.* **2021**, *27*, 1851–1855. [CrossRef]
169. Woldemeskel, B.A.; Karaba, A.H.; Garliss, C.C.; Beck, E.J.; Wang, K.H.; Laeyendecker, O.; Cox, A.L.; Blankson, J.N. The BNT162b2 MRNA Vaccine Elicits Robust Humoral and Cellular Immune Responses in People Living with Human Immunodeficiency Virus (HIV). *Clin. Infect. Dis.* **2021**, *74*, ciab648. [CrossRef]

170. Ruddy, J.A.; Boyarsky, B.J.; Bailey, J.R.; Karaba, A.H.; Garonzik-Wang, J.M.; Segev, D.L.; Durand, C.M.; Werbel, W.A. Safety and Antibody Response to Two-Dose SARS-CoV-2 Messenger RNA Vaccination in Persons with HIV. *AIDS* **2021**, *35*, 2399–2401. [CrossRef]
171. Antinori, A.; Cicalini, S.; Meschi, S.; Bordoni, V.; Lorenzini, P.; Vergori, A.; Lanini, S.; De Pascale, L.; Matusali, G.; Mariotti, D.; et al. Humoral and Cellular Immune Response Elicited by MRNA Vaccination against SARS-CoV-2 in People Living with HIV (PLWH) Receiving Antiretroviral Therapy (ART) According with Current CD4 T-Lymphocyte Count. *Clin. Infect. Dis. Off. Publ. Infect. Dis. Soc. Am.* **2022**, *75*, ciac238. [CrossRef]
172. Vergori, A.; Cozzi Lepri, A.; Cicalini, S.; Matusali, G.; Bordoni, V.; Lanini, S.; Meschi, S.; Iannazzo, R.; Mazzotta, V.; Colavita, F.; et al. Immunogenicity to COVID-19 MRNA Vaccine Third Dose in People Living with HIV. *Nat. Commun.* **2022**, *13*, 4922. [CrossRef]
173. Governo Italiano—Report Vaccini Anti COVID-19. Available online: https://www.governo.it/it/cscovid19/report-vaccini/ (accessed on 8 March 2022).
174. Tortellini, E.; Zingaropoli, M.A.; Mancarella, G.; Marocco, R.; Carraro, A.; Jamhour, M.; Barbato, C.; Guardiani, M.; Dominelli, F.; Pasculli, P.; et al. Quality of T-Cell Response to SARS-CoV-2 MRNA Vaccine in ART-Treated PLWH. *Int. J. Mol. Sci.* **2022**, *23*, 14988. [CrossRef] [PubMed]
175. Ucciferri, C.; Vecchiet, J.; Auricchio, A.; Falasca, K. Improving BNT162b2 MRNA Vaccine Tolerability without Efficacy Loss by Pidotimod Supplementation. *Mediterr. J. Hematol. Infect. Dis.* **2022**, *14*, e2022023. [CrossRef] [PubMed]
176. Turner, J.S.; Kim, W.; Kalaidina, E.; Goss, C.W.; Rauseo, A.M.; Schmitz, A.J.; Hansen, L.; Haile, A.; Klebert, M.K.; Pusic, I.; et al. SARS-CoV-2 Infection Induces Long-Lived Bone Marrow Plasma Cells in Humans. *Nature* **2021**, *595*, 421–425. [CrossRef] [PubMed]
177. Riou, C.; du Bruyn, E.; Stek, C.; Daroowala, R.; Goliath, R.T.; Abrahams, F.; Said-Hartley, Q.; Allwood, B.W.; Hsiao, N.-Y.; Wilkinson, K.A.; et al. Relationship of SARS-CoV-2-Specific CD4 Response to COVID-19 Severity and Impact of HIV-1 and Tuberculosis Coinfection. *J. Clin. Investig.* **2021**, *131*, 149125. [CrossRef]
178. CDC. Clinical Features of Pneumococcal Disease. Available online: https://www.cdc.gov/pneumococcal/clinicians/clinical-features.html (accessed on 10 July 2023).
179. O'Brien, K.L.; Wolfson, L.J.; Watt, J.P.; Henkle, E.; Deloria-Knoll, M.; McCall, N.; Lee, E.; Mulholland, K.; Levine, O.S.; Cherian, T.; et al. Burden of Disease Caused by Streptococcus Pneumoniae in Children Younger than 5 Years: Global Estimates. *Lancet* **2009**, *374*, 893–902. [CrossRef]
180. Grau, I.; Ardanuy, C.; Liñares, J.; Podzamczer, D.; Schulze, M.H.; Pallares, R. Trends in Mortality and Antibiotic Resistance among HIV-Infected Patients with Invasive Pneumococcal Disease. *HIV Med.* **2009**, *10*, 488–495. [CrossRef]
181. van Aalst, M.; Lötsch, F.; Spijker, R.; van der Meer, J.T.M.; Langendam, M.W.; Goorhuis, A.; Grobusch, M.P.; de Bree, G.J. Incidence of Invasive Pneumococcal Disease in Immunocompromised Patients: A Systematic Review and Meta-Analysis. *Travel Med. Infect. Dis.* **2018**, *24*, 89–100. [CrossRef]
182. French, N.; Gordon, S.B.; Mwalukomo, T.; White, S.A.; Mwafulirwa, G.; Longwe, H.; Mwaiponya, M.; Zijlstra, E.E.; Molyneux, M.E.; Gilks, C.F. A Trial of a 7-Valent Pneumococcal Conjugate Vaccine in HIV-Infected Adults. *N. Engl. J. Med.* **2010**, *362*, 812–822. [CrossRef]
183. Kobayashi, M.; Farrar, J.L.; Gierke, R.; Britton, A.; Childs, L.; Leidner, A.J.; Campos-Outcalt, D.; Morgan, R.L.; Long, S.S.; Talbot, H.K.; et al. Use of 15-Valent Pneumococcal Conjugate Vaccine and 20-Valent Pneumococcal Conjugate Vaccine among U.S. Adults: Updated Recommendations of the Advisory Committee on Immunization Practices—United States, 2022. *MMWR Morb. Mortal. Wkly. Rep.* **2022**, *71*, 109–117. [CrossRef]
184. Plosker, G.L. 13-Valent Pneumococcal Conjugate Vaccine: A Review of Its Use in Adults. *Drugs* **2015**, *75*, 1535–1546. [CrossRef] [PubMed]
185. CDC. Pneumococcal Polysaccharide Vaccine Information Statement. Available online: https://www.cdc.gov/vaccines/hcp/vis/vis-statements/ppv.html (accessed on 22 July 2023).
186. Rodriguez-Barradas, M.C.; Alexandraki, I.; Nazir, T.; Foltzer, M.; Musher, D.M.; Brown, S.; Thornby, J. Response of Human Immunodeficiency Virus-Infected Patients Receiving Highly Active Antiretroviral Therapy to Vaccination with 23-Valent Pneumococcal Polysaccharide Vaccine. *Clin. Infect. Dis. Off. Publ. Infect. Dis. Soc. Am.* **2003**, *37*, 438–447. [CrossRef] [PubMed]
187. Dworkin, M.S.; Ward, J.W.; Hanson, D.L.; Jones, J.L.; Kaplan, J.E.; Adult and Adolescent Spectrum of HIV Disease Project. Pneumococcal Disease among Human Immunodeficiency Virus-Infected Persons: Incidence, Risk Factors, and Impact of Vaccination. *Clin. Infect. Dis. Off. Publ. Infect. Dis. Soc. Am.* **2001**, *32*, 794–800. [CrossRef] [PubMed]
188. Clutterbuck, E.A.; Lazarus, R.; Yu, L.-M.; Bowman, J.; Bateman, E.A.L.; Diggle, L.; Angus, B.; Peto, T.E.; Beverley, P.C.; Mant, D.; et al. Pneumococcal Conjugate and Plain Polysaccharide Vaccines Have Divergent Effects on Antigen-Specific B Cells. *J. Infect. Dis.* **2012**, *205*, 1408–1416. [CrossRef]
189. Farmaki, P.F.; Chini, M.C.; Mangafas, N.M.; Tzanoudaki, M.T.; Piperi, C.P.; Lazanas, M.Z.; Spoulou, V.S. Immunogenicity and Immunological Memory Induced by the 13-Valent Pneumococcal Conjugate Followed by the 23-Valent Polysaccharide Vaccine in HIV-Infected Adults. *J. Infect. Dis.* **2018**, *218*, 26–34. [CrossRef]
190. Lopez, A.; Mariette, X.; Bachelez, H.; Belot, A.; Bonnotte, B.; Hachulla, E.; Lahfa, M.; Lortholary, O.; Loulergue, P.; Paul, S.; et al. Vaccination Recommendations for the Adult Immunosuppressed Patient: A Systematic Review and Comprehensive Field Synopsis. *J. Autoimmun.* **2017**, *80*, 10–27. [CrossRef]

191. Søgaard, O.S.; Lohse, N.; Harboe, Z.B.; Offersen, R.; Bukh, A.R.; Davis, H.L.; Schønheyder, H.C.; Østergaard, L. Improving the Immunogenicity of Pneumococcal Conjugate Vaccine in HIV-Infected Adults with a Toll-like Receptor 9 Agonist Adjuvant: A Randomized, Controlled Trial. *Clin. Infect. Dis. Off. Publ. Infect. Dis. Soc. Am.* **2010**, *51*, 42–50. [CrossRef]
192. Kroon, F.P.; van Dissel, J.T.; Labadie, J.; van Loon, A.M.; van Furth, R. Antibody Response to Diphtheria, Tetanus, and Poliomyelitis Vaccines in Relation to the Number of CD4+ T Lymphocytes in Adults Infected with Human Immunodeficiency Virus. *Clin. Infect. Dis. Off. Publ. Infect. Dis. Soc. Am.* **1995**, *21*, 1197–1203. [CrossRef]
193. Gershon, A.A.; Breuer, J.; Cohen, J.I.; Cohrs, R.J.; Gershon, M.D.; Gilden, D.; Grose, C.; Hambleton, S.; Kennedy, P.G.E.; Oxman, M.N.; et al. Varicella Zoster Virus Infection. *Nat. Rev. Dis. Primer* **2015**, *1*, 15016. [CrossRef]
194. Clinical Practice: Herpes Zoster—PubMed. Available online: https://pubmed.ncbi.nlm.nih.gov/23863052/ (accessed on 4 June 2023).
195. CDC. Clinical Overview of Herpes Zoster (Shingles). Available online: https://www.cdc.gov/shingles/hcp/clinical-overview.html (accessed on 11 June 2023).
196. Chawki, S.; Vilcu, A.-M.; Etienne, C.; Finet, F.; Blanchon, T.; Souty, C.; Hanslik, T. Incidence of Complications of Herpes Zoster in Individuals on Immunosuppressive Therapy: A Register-Based Population Study. *J. Infect.* **2022**, *84*, 531–536. [CrossRef]
197. Weitzman, D.; Shavit, O.; Stein, M.; Cohen, R.; Chodick, G.; Shalev, V. A Population Based Study of the Epidemiology of Herpes Zoster and Its Complications. *J. Infect.* **2013**, *67*, 463–469. [CrossRef] [PubMed]
198. Jansen, K.; Haastert, B.; Michalik, C.; Guignard, A.; Esser, S.; Dupke, S.; Plettenberg, A.; Skaletz-Rorowski, A.; Brockmeyer, N.H. Incidence and Risk Factors of Herpes Zoster among Hiv-Positive Patients in the German Competence Network for HIV/AIDS (KompNet): A Cohort Study Analysis. *BMC Infect. Dis.* **2013**, *13*, 372. [CrossRef]
199. Gershon, A.A.; Mervish, N.; LaRussa, P.; Steinberg, S.; Lo, S.H.; Hodes, D.; Fikrig, S.; Bonagura, V.; Bakshi, S. Varicella-Zoster Virus Infection in Children with Underlying Human Immunodeficiency Virus Infection. *J. Infect. Dis.* **1997**, *176*, 1496–1500. [CrossRef] [PubMed]
200. Donahue, J.G.; Choo, P.W.; Manson, J.E.; Platt, R. The Incidence of Herpes Zoster. *Arch. Intern. Med.* **1995**, *155*, 1605–1609. [CrossRef]
201. Buchbinder, S.P.; Katz, M.H.; Hessol, N.A.; Liu, J.Y.; O'Malley, P.M.; Underwood, R.; Holmberg, S.D. Herpes Zoster and Human Immunodeficiency Virus Infection. *J. Infect. Dis.* **1992**, *166*, 1153–1156. [CrossRef]
202. Successes and Challenges in Varicella Vaccine—PMC. Available online: https://www.ncbi.nlm.nih.gov/pmc/articles/PMC3991154/ (accessed on 5 June 2023).
203. Spoulou, V.; Alain, S.; Gabutti, G.; Giaquinto, C.; Liese, J.; Martinon-Torres, F.; Vesikari, T. Implementing Universal Varicella Vaccination in Europe: The Path Forward. *Pediatr. Infect. Dis. J.* **2019**, *38*, 181–188. [CrossRef]
204. Lee, Y.H.; Choe, Y.J.; Lee, J.; Kim, E.; Lee, J.Y.; Hong, K.; Yoon, Y.; Kim, Y.-K. Global Varicella Vaccination Programs. *Clin. Exp. Pediatr.* **2022**, *65*, 555–562. [CrossRef]
205. Zou, J.; Krentz, H.B.; Lang, R.; Beckthold, B.; Fonseca, K.; Gill, M.J. Seropositivity, Risks, and Morbidity from Varicella-Zoster Virus Infections in an Adult PWH Cohort From 2000–2020. *Open Forum Infect. Dis.* **2022**, *9*, ofac395. [CrossRef] [PubMed]
206. Muñoz-Quiles, C.; López-Lacort, M.; Díez-Domingo, J.; Orrico-Sánchez, A. Herpes Zoster Risk and Burden of Disease in Immunocompromised Populations: A Population-Based Study Using Health System Integrated Databases, 2009–2014. *BMC Infect. Dis.* **2020**, *20*, 905. [CrossRef] [PubMed]
207. BMJ Open. Burden of Herpes Zoster in 16 Selected Immunocompromised Populations in England: A Cohort Study in the Clinical Practice Research Datalink 2000–2012. Available online: https://bmjopen.bmj.com/content/8/6/e020528.abstract (accessed on 5 June 2023).
208. Brisson, M.; Edmunds, W.J.; Law, B.; Gay, N.J.; Walld, R.; Brownell, M.; Roos, L.L.; Serres, G.D. Epidemiology of Varicella Zoster Virus Infection in Canada and the United Kingdom. *Epidemiol. Infect.* **2001**, *127*, 305–314. [CrossRef]
209. Sullivan, K.M.; Farraye, F.A.; Winthrop, K.L.; Willer, D.O.; Vink, P.; Tavares-Da-Silva, F. Safety and Efficacy of Recombinant and Live Herpes Zoster Vaccines for Prevention in At-Risk Adults with Chronic Diseases and Immunocompromising Conditions. *Vaccine* **2023**, *41*, 36–48. [CrossRef] [PubMed]
210. Anderson, T.C.; Masters, N.B.; Guo, A.; Shepersky, L.; Leidner, A.J.; Lee, G.M.; Kotton, C.N.; Dooling, K.L. Use of Recombinant Zoster Vaccine in Immunocompromised Adults Aged ≥19 Years: Recommendations of the Advisory Committee on Immunization Practices—United States, 2022. *MMWR Morb. Mortal. Wkly. Rep.* **2022**, *71*, 80–84. [CrossRef]
211. Schmader, K.E.; Levin, M.J.; Gnann, J.W.; McNeil, S.A.; Vesikari, T.; Betts, R.F.; Keay, S.; Stek, J.E.; Bundick, N.D.; Su, S.-C.; et al. Efficacy, Safety, and Tolerability of Herpes Zoster Vaccine in Persons Aged 50-59 Years. *Clin. Infect. Dis. Off. Publ. Infect. Dis. Soc. Am.* **2012**, *54*, 922–928. [CrossRef]
212. Oxman, M.N.; Levin, M.J.; Johnson, G.R.; Schmader, K.E.; Straus, S.E.; Gelb, L.D.; Arbeit, R.D.; Simberkoff, M.S.; Gershon, A.A.; Davis, L.E.; et al. A Vaccine to Prevent Herpes Zoster and Postherpetic Neuralgia in Older Adults. *N. Engl. J. Med.* **2005**, *352*, 2271–2284. [CrossRef]
213. Recommendations of the Advisory Committee on Immunization Practices for Use of Herpes Zoster Vaccines—PubMed. Available online: https://pubmed.ncbi.nlm.nih.gov/29370152/ (accessed on 5 June 2023).
214. Effect of Recombinant Zoster Vaccine on Incidence of Herpes Zoster after Autologous Stem Cell Transplantation: A Randomized Clinical Trial—PubMed. Available online: https://pubmed.ncbi.nlm.nih.gov/31287523/ (accessed on 5 June 2023).
215. Levin, M.J.; Kroehl, M.E.; Johnson, M.J.; Hammes, A.; Reinhold, D.; Lang, N.; Weinberg, A. Th1 Memory Differentiates Recombinant from Live Herpes Zoster Vaccines. *J. Clin. Investig.* **2018**, *128*, 4429–4440. [CrossRef]

216. Johnson, M.J.; Liu, C.; Ghosh, D.; Lang, N.; Levin, M.J.; Weinberg, A. Cell-Mediated Immune Responses After Administration of the Live or the Recombinant Zoster Vaccine: 5-Year Persistence. *J. Infect. Dis.* **2022**, *225*, 1477–1481. [CrossRef]
217. Le, P.; Rothberg, M. Herpes Zoster Infection. *BMJ* **2019**, *364*, k5095. [CrossRef]
218. Lal, H.; Cunningham, A.L.; Godeaux, O.; Chlibek, R.; Diez-Domingo, J.; Hwang, S.-J.; Levin, M.J.; McElhaney, J.E.; Poder, A.; Puig-Barberà, J.; et al. Efficacy of an Adjuvanted Herpes Zoster Subunit Vaccine in Older Adults. *N. Engl. J. Med.* **2015**, *372*, 2087–2096. [CrossRef] [PubMed]
219. Cunningham, A.L.; Lal, H.; Kovac, M.; Chlibek, R.; Hwang, S.-J.; Díez-Domingo, J.; Godeaux, O.; Levin, M.J.; McElhaney, J.E.; Puig-Barberà, J.; et al. Efficacy of the Herpes Zoster Subunit Vaccine in Adults 70 Years of Age or Older. *N. Engl. J. Med.* **2016**, *375*, 1019–1032. [CrossRef] [PubMed]
220. Benson, C.A.; Andersen, J.W.; Macatangay, B.J.C.; Mailliard, R.B.; Rinaldo, C.R.; Read, S.; Bozzolo, D.R.; Purdue, L.; Jennings, C.; Keefer, M.C.; et al. Safety and Immunogenicity of Zoster Vaccine Live in Human Immunodeficiency Virus-Infected Adults with CD4+ Cell Counts > 200 Cells/ML Virologically Suppressed on Antiretroviral Therapy. *Clin. Infect. Dis. Off. Publ. Infect. Dis. Soc. Am.* **2018**, *67*, 1712–1719. [CrossRef]
221. Mok, C.C.; Chan, K.H.; Ho, L.Y.; Fung, Y.F.; Fung, W.F.; Woo, P.C.Y. Safety and Immune Response of a Live-Attenuated Herpes Zoster Vaccine in Patients with Systemic Lupus Erythematosus: A Randomised Placebo-Controlled Trial. *Ann. Rheum. Dis.* **2019**, *78*, 1663–1668. [CrossRef] [PubMed]

viruses MDPI

Article

Mucosal Application of a Low-Energy Electron Inactivated Respiratory Syncytial Virus Vaccine Shows Protective Efficacy in an Animal Model

Valentina Eberlein [1,2], Mareike Ahrends [2,3], Lea Bayer [1], Julia Finkensieper [1,2], Joana Kira Besecke [2,4], Yaser Mansuroglu [2,5], Bastian Standfest [2,6], Franziska Lange [1,2], Simone Schopf [2,4], Martin Thoma [2,6], Jennifer Dressman [2,5], Christina Hesse [2,3], Sebastian Ulbert [1,2] and Thomas Grunwald [1,2,*]

1 Fraunhofer Institute for Cell Therapy and Immunology, 04103 Leipzig, Germany; valentina.eberlein@izi.fraunhofer.de (V.E.)
2 Fraunhofer Cluster of Excellence Immune-Mediated Diseases CIMD, 60596 Frankfurt am Main, Germany; yaser.mansuroglu@itmp.fraunhofer.de (Y.M.)
3 Fraunhofer Institute for Toxicology and Experimental Medicine, 30625 Hannover, Germany
4 Fraunhofer Institute for Organic Electronics, Electron Beam and Plasma Technology FEP, 01277 Dresden, Germany
5 Fraunhofer Institute for Translational Medicine and Pharmacology, 60596 Frankfurt, Germany
6 Fraunhofer Institute for Manufacturing Engineering and Automation, 70569 Stuttgart, Germany
* Correspondence: thomas.grunwald@izi.fraunhofer.de; Tel.: +49-341-35536-5423

Abstract: Respiratory syncytial virus (RSV) is a leading cause of acute lower respiratory tract infections in the elderly and in children, associated with pediatric hospitalizations. Recently, first vaccines have been approved for people over 60 years of age applied by intramuscular injection. However, a vaccination route via mucosal application holds great potential in the protection against respiratory pathogens like RSV. Mucosal vaccines induce local immune responses, resulting in a fast and efficient elimination of respiratory viruses after natural infection. Therefore, a low-energy electron irradiated RSV (LEEI-RSV) formulated with phosphatidylcholine-liposomes (PC-LEEI-RSV) was tested ex vivo in precision cut lung slices (PCLSs) for adverse effects. The immunogenicity and protective efficacy in vivo were analyzed in an RSV challenge model after intranasal vaccination using a homologous prime-boost immunization regimen. No side effects of PC-LEEI-RSV in PCLS and an efficient antibody induction in vivo could be observed. In contrast to unformulated LEEI-RSV, the mucosal vaccination of mice with PC formulated LEEI-RSV showed a statistically significant reduction in viral load after challenge. These results are a proof-of-principle for the use of LEEI-inactivated viruses formulated with liposomes to be administered intranasally to induce a mucosal immunity that could also be adapted for other respiratory viruses.

Keywords: respiratory syncytial virus; RSV; mucosal vaccine; inactivated vaccine; low-energy electron irradiation; LEEI; PC formulation; PCLS

Citation: Eberlein, V.; Ahrends, M.; Bayer, L.; Finkensieper, J.; Besecke, J.K.; Mansuroglu, Y.; Standfest, B.; Lange, F.; Schopf, S.; Thoma, M.; et al. Mucosal Application of a Low-Energy Electron Inactivated Respiratory Syncytial Virus Vaccine Shows Protective Efficacy in an Animal Model. *Viruses* **2023**, *15*, 1846. https://doi.org/10.3390/v15091846

Academic Editors: Pietro Hiram Guzzi, Marianna Milano and Jayanta Kumar Das

Received: 9 August 2023
Revised: 28 August 2023
Accepted: 29 August 2023
Published: 30 August 2023

1. Introduction

Human respiratory syncytial virus (RSV) is a highly infectious and seasonally occurring member of the Pneumoviridae family that can lead to upper and lower respiratory tract infections (LRTI) [1–3]. Patients such as infants, especially in the first six months of life, pre-term born, the elderly over 60 years of age, or patients suffering from additional lung pathologies are at high risk of severe lung disease after RSV infection [1,2,4]. In 2019, the global burden of RSV was approximately 33 million associated LRTI and 101,400 RSV-attributed deaths in children under six years of age [4]. In the elderly, RSV has a similar or even greater burden than influenza, as evidenced in prolonged hospital stays, more intensive care unit admissions, and higher mortality [5–8].

The vaccine development against RSV has been faced with several drawbacks. In the 1960s, a formalin-inactivated RSV (FI-RSV) vaccine trial in children enhanced the severity of the disease after natural reinfection, which was associated with bronchopneumonia, pulmonary eosinophilia, and extensive monocyte infiltrations, with two fatal cases [9,10].

Since 1986, passive immunization with the monoclonal antibody palivizumab [1,11–13], and more recently, the improved nirsevimab, are available [14,15]. Several vaccine approaches have currently entered clinical trials and the first vaccines were approved by the Food and Drug Administration (FDA) in 2023 [8,16,17]. Two recombinant subunit vaccines containing stabilized RSV prefusion-F protein have been FDA-approved for adults 60 years and above. The immunizations showed an efficacy against RSV-related LRTI of 82.6% (AReSVi-006) and 85.7% (RSVPreF), respectively [17–22]. The latter prefusion F protein vaccine (RSVPreF) approved for older adults (NCT05035212) also showed protection in infants after maternal vaccination (NCT04424316) [16,21–25]. In addition, an mRNA-vaccine expressing the stabilized RSV pre-F is currently in phase III clinical trial [26,27]. Further vaccine approaches include vector-based vaccines such as MVA-BN-RSV, a vector expressing the F, G, M2, and N protein of RSV in phase III [8,27], and live-attenuated vaccines, virus-like particles, and nanoparticles [8,27].

Despite these great breakthroughs, all setups are based on intramuscular applications whereby a mucosal vaccination route could be beneficial against respiratory viruses. The mucosal application is atraumatic and may enhance the vaccination acceptance in the community over needle-based applications [28,29]. Most importantly, mucosal vaccinations against respiratory viruses trigger the mucosal immune system, promoting local immune responses, resulting in a fast and efficient elimination of viruses directly in the respiratory tract [28,30]. The induction of the mucosal immunity can be advantageous over systemic vaccination approaches, especially in RSV-naïve children, as the latter could lead to the overwhelming immune pathology known after natural RSV infection [31,32]. Only a few vaccine candidates have the potential to be safe and effective after mucosal application. In particular, vector vaccines and live-attenuated vaccines are known candidates, as promising results in preclinical trials have been shown and may have favorable outcomes for patients [30,33,34]. Even though live-attenuated and vector vaccines have great potential, reactogenicity, reversion to a virulent pathogen, or the possibility of retrograde transport into the brain, are undesirable risks [35–37]. Inactivated vaccines can circumvent these biosafety risks, since pathogens are no longer capable of the replication in the vaccinees.

Virus inactivation approaches, especially those using physical methods, are safe and have low production costs [38–40]. We have shown previously that low-energy electron irradiation (LEEI) is a safe, non-toxic, and non-probe harming inactivation method [41–46]. The advantage over other irradiation methods is that the emission of secondary photon radiation is minimal, reducing the need for extensive shielding and making LEEI-technologies applicable in standard laboratories [43,47]. Since the penetration depth of low-energy electrons is highly limited [48,49], we have developed automated processes that generate thin liquid films, enabling the efficient LEEI of pathogens in suspension up to multi-liter scales [41–43,46]. LEEI has advantages over other radiation types such as ultraviolet (UV) light. It was previously shown that UV light is harmful to viral proteins including RSV-F, the most relevant protein for the induction of protective immune responses against RSV [50,51]. This is in line with the observation that RSV-F is also damaged by treating RSV with formalin [52,53]. It has been shown that this misfolding of surface RSV-F proteins was one major reason for the unfortunate outcome of the vaccine trial in the 1960s, resulting in poor protective [10,53,54] and unbalanced Th2 immune responses [55,56].

In contrast to these studies with misfolded RSV-F protein in formalin-inactivated RSV vaccine preparations, we have previously shown that in LEEI-inactivated RSV (LEEI-RSV), at least 70% of the RSV-protein is present in the Pre-F-conformation [44]. In the associated preclinical trial, intramuscular application of LEEI-RSV with Alhdydrogel showed high protection upon RSV challenge, which was associated with no significant immune pathology in the lungs of infected mice [44].

In the present study, we investigated the mucosal application of LEEI-RSV for the first time. As the mucosa forms a physical and enzymatic barrier, inactivated vaccines applied mucosally should be formulated for resorption and immune stimulatory reasons [28]. LEEI-RSV was formulated with liposomes as they are known to be safe, low, or non-toxic and easy to produce [29,57]. As a liposome, phosphatidylcholine (PC) was chosen to provide a broad mucosal and systemic immune response after instillation [58–61].

RSV is one main reason for hospitalizations in infants and the elderly. Even though first vaccines directed against RSV-F have been approved, a mucosal whole virus vaccine could be beneficial with a broader immune response, especially in combinational use with the approved vaccines. Therefore, we developed an inactivated vaccine using LEEI and tested it in a mucosal vaccination model. The formulated PC-LEEI-RSV showed no safety concerns in the ex vivo murine precision cut lung slice (PCLS) model. The intranasal application of PC-LEEI-RSV was well-tolerated and reduced the viral load in mice. Our results indicate that the formulation of LEEI inactivated RSV mediates protective efficacy after mucosal homologous prime and boost vaccination.

2. Materials and Methods

1. Cell Culture and Virus Production

Type 2 human epithelial cells (HEp-2; ATCC, Manassas, VA, USA) were used for all RSV production and in vitro assays. Cells were maintained in Dulbecco's modified Eagle's medium (DMEM) with GlutaMAX (Thermo Fisher Scientific, Dreieich, Germany), containing 10% heat inactivated fetal calf serum (FCS) and antibiotics (100 IU/mL penicillin with 100 µg/mL Streptomycin, Thermo Fisher Scientific, Dreieich, Germany) at 37 °C with 5% CO_2.

RSV laboratory strain A long was obtained from ATCC (VR-26). M. Peeples and P. Collins (NIH, Bethesda, MD, USA) kindly provided the recombinant RSV expressing GFP (rgRSV). Virus propagation and titer determination was performed as previously described [34,62]. In short, the different virus strains were propagated on 90% confluent monolayers of HEp-2 cells. Cells were infected with a multiplicity of infection (MOI) of 0.1 in FCS-free medium for 3 h at 37 °C with 5% CO_2. Unbound virus was removed by replacing the medium with fresh 1% FCS containing medium and incubated for 72 h at 37 °C with 5% CO_2. The infected cell supernatant was clarified by centrifugation for 5 min at 2000× *g* and 4 °C, followed by filtration through a 0.45 µm-filter and ultracentrifugation at 21,000× *g* through a 20% (*w*/*v*) sucrose cushion in PBS for 3 h at 4 °C in a SureSpin 630 swing-out rotor (Thermo Fisher Scientific, Dreieich, Germany). The pelleted virus was resuspended in 10% (*w*/*v*) sucrose in PBS and titrated using a focus forming assay or a tissue culture infectious dose 50 (TCID50) assay. For the focus forming assay, confluent monolayers of HEp-2 cells were infected with serial dilutions of the virus and incubated for 48 h at 37 °C with 5% CO_2. Viral rgRSV foci were visualized by fluorescence and RSV-long was analyzed by immunocytochemical (ICC) staining with an anti-RSV antibody (AB1128; Sigma Aldrich, Taufkirchen, Germany) [63]. For the TCID50-assay, viral stocks were diluted in 10-fold increments and incubated on confluent HEp-2 cell monolayers in a 96-well microwell plate for a period of five to six days. The cells were monitored for cytopathic effects (CPE) and the titer was calculated using the Reed–Muench method [64].

2. Virus Inactivation

a. Low-energy electron irradiation (LEEI)

RSV samples were irradiated in a custom-built irradiation device situated in a BSL2 laboratory at the Fraunhofer Institute for Cell Therapy and Immunology [41]. This can be equipped with different modules constructed as research prototypes to enable the automated LEEI of liquid samples. RSV was inactivated in a system using disposable bags. Briefly, bags were filled with 10 mL each of an RSV-solution diluted in PBS with 10% (*w*/*v*) sucrose and sealed. Based on previous experiences, the samples were treated with LEEI of 300 keV, 1.2 mA [41,44]. Controls underwent the same process without applying LEEI.

Afterward, the bags were reopened and the liquid collected for further testing. Inactivation of RSV was confirmed in a cell culture assay as described previously [41]. In short, HEp-2 cells in a 6-well cell culture plate were inoculated with 100 µL per well of irradiated samples and controls. The development of cytopathic effects (CPE) was monitored for five to six days before the supernatant was passaged to fresh cells. After an additional five to six days of incubation, the samples were considered inactivated when no CPE was visible.

b. Dosimetric analysis

The absorbed dose applied was estimated by using a liquid dosimeter based on 2,3,5-triphenyl-tetrazolium chloride (TTC, Carl Roth, Karlsruhe, Germany), which undergoes colorization due to a dose-dependent reaction to red formazan, as described in the study of Schopf et al. [65]. Previously, a calibration function to convert absorbance values into dose was found with a combined uncertainty of 11.8% in the dose range of 6.5–38 kGy. That calibration was performed at the radiation plant REAMODE of Fraunhofer FEP using a standard reference film dosimeter Risø B3 (DTU Health Tech, Lyngby, Denmark).

The TTC was filled in the bag and irradiated with a constant acceleration voltage of 300 kV. The beam current was varied from 0 to 2.0 mA in intervals of 0.5 mA, and a linear regression was conducted to calculate the resulting dose in the unit gray (Gy).

3. ELISA RSV Conservation after LEEI

To examine the conservation of the antigenicity of RSV, enzyme-linked immunosorbent assays (ELISAs) were performed as previously described [41,44]. Briefly, 5 µL irradiated RSV samples and controls were coated on black NUNC 96-well MicroWell™ PolySorp® plates (Thermo Fisher Scientific, Dreieich, Germany) in carbonate coating buffer (35 mM $NaHCO_3$ (Carl Roth, Karlsruhe, Germany), 15 mM Na_2CO_3 (Carl Roth, Karlsruhe, Germany), pH 9.6) in a total volume of 100 µL/well overnight at 4 °C. For a quantitative standard curve, RSV was coated in fivefold dilutions in concentrations ranging from 5×10^2 to 1×10^5 FFU and processed as the LEEI-RSV. To convert the FFU values to TCID50, the correlation factor 0.7 was used [66]. The plate was washed three times with PBS containing 0.05% Tween 20 (PBS-T, Carl Roth, Karlsruhe, Germany) and blocked with 5% (*w*/*v*) skimmed milk powder (Carl Roth, Karlsruhe, Germany) in PBS (Bio&Sell, Feucht, Germany) for 2 h at room temperature. A monoclonal antibody recognizing RSV-F (18F12, [62]), diluted 1:200 in 2% (*w*/*v*) skim milk in PBS, was added and incubated for 2 h at room temperature, followed by the polyclonal Peroxidase AffiniPure Sheep Anti-Mouse IgG (H + L) antibody (Dianova, Hamburg, Germany) diluted 1:500. After 1 h of incubation, the readout was performed using a Centro XS3 luminometer (Berthold Technologies, Bad Wildbad, Germany). Enhanced chemiluminescent substrate (ECL, Pierce, Waltham, MA, USA) was diluted 1:10 in PBS and 100 µL of the substrate was injected into each well after 1.5 s of delay. Relative light units (RLU) were counted for 1 s.

4. Lipid Production and Virus Packaging

Phosphatidycholine (PC, LIPOID GmbH, Ludwigshafen, Germany) liposome formulations were prepared by a thin-film hydration method, followed by size reduction via manual extrusion. Forty mg of lipids was dissolved in 1.5 mL methanol (VWR, Darmstadt, Germany) and evaporated in a rotavapor (Büchi R-114, Büchi Labortechnik AG, Flawil, Schweiz) by reducing the pressure to 500 mPa, followed by further reductions to 150 mPa and finally to 50 mPa. The pressure was held at each of these values for 30 min before proceeding to the next pressure reduction. The whole process was conducted at 37 °C. The resulting lipid film was rehydrated by adding 1 mL of PBS pH 7.4 buffer (Fisher Scientific GmbH, Schwerte, Germany) and vortexing for 20 min. The liposome mixture was then extruded manually through a 200-nm polycarbonate membrane (Avestin, Mannheim, Germany) using a two-chamber manual extruder (Avestin, Mannheim, Germany). This procedure resulted in unilamellar liposomes. In order to prevent the loss of activity, the vaccine was incorporated into the vesicles after manufacturing the liposome formulation. The inactivated virus material was mixed in a ratio of 1:5 (liposomes:LEEI-RSV) with the liposome formulation and vortexed for 5 min. This material was kept at 4 °C prior to testing.

5. Precision Cut Lung Slices (PCLSs)

a. Preparation and treatment of murine, precision cut lung slices (PCLSs)

Female BALB/c mice (Charles River Laboratories, Sulzfeld, Germany) were sacrificed at three months of age. Lungs were resected and filled with a warm 2% agarose (Sigma Aldrich, Taufkirchen, Germany) solution in DMEM/F12 (Gibco, Waltham, MA, USA). Solidified lung lobes were cut into slices with a 300 µm thickness on a vibratome (Krumdieck Tissue Slicer, Alabama Research and Development, Muniford, AL, USA) in EBSS (Th Geyer, Renningen, Germany). The generated murine precision cut lung slices were treated with different concentrations of the vaccine candidates and cultured in DMEM/F12 supplemented with penicillin and streptomycin (10,000 U/mL, Gibco, Waltham, MA, USA) at 37 °C with 5% CO_2 for 24 h.

b. Viability testing

To assess the viability of the murine precision cut lung slices (mPCLSs), the amount of released lactate dehydrogenase (LDH) was analyzed using the Cytotoxicity Detection Kit (Roche, Basel, Switzerland). The metabolic activity was measured using the Cell Proliferation Reagent WST-1 (Roche, Basel, Switzerland). Furthermore, cells were marked using the LIVE/DEAD Viability/Cytotoxicity Kit (Invitrogen, Waltham, MA, USA) to visualize the viable and dead cells by confocal laser scanning microscopy (LSM 800, Zeiss, Jena, Germany). All kits and reagents were applied at the manufacturer's recommendations.

c. Cytokine secretion

To assume the immune response of the vaccine-treated mPCLS, the concentration of TNF-α was measured in the supernatants by ELISA (RnD, DuoSet, Minneapolis, MN, USA). Furthermore, released IFN-α, -β, -γ, IL-1β, IL-6, IL-10, IP-10, KC, MCP-1, MIP-1α, and RANTES were quantified by the U-Plex assay (Mesoscale Discovery, Rockville, MD, USA). All assays were performed following the manufacturer's recommendations at appropriate sample dilutions.

6. Immunization and RSV Challenge in Mice

Female BALB/c mice were purchased at Charles River Laboratories (Sulzfeld, Germany) or breed at the Center for Experimental Medicine at the Fraunhofer Institute of Cell Therapy and Immunology and maintained under a specific pathogen-free environment in isolated ventilated cages. Seven- to eight-week-old mice were included in the experiment. Animal experiments were carried out according to the EU Directive 2010/63/EU for animal experiments and were approved by local authorities. Groups of five mice each were vaccinated in a homologous prime-boost manner at a four-week regime. Vaccination was intramuscular (i.m.) with LEEI-RSV adjuvanted with Alhdydrogel (LEEI-RSV i.m.) with 50 µL per hind leg (3×10^6 TCID50/mL) or intranasal (i.n.) with PC-formulated LEEI-RSV (2×10^5 TCID50/mL) or non-formulated LEEI-RSV with 40 µL per immunization. Blood for serum samples was collected one week before the first vaccination and three weeks after prime and boost vaccination. Mice were challenged four weeks after boost immunization with 10^6 FFU RSV per animal intranasally as previously published [44]. Five days after infection, mice were euthanized and lungs were isolated for analysis of the viral load. In detail, lungs were homogenized in gentlMACS™ M Tubes (Miltenyi Biotec., Bergisch Gladbach, Germany) with 2 mL ice-cold PBS using a gentlMACS Dissociator (Miltenyi Biotec., Bergisch Gladbach, Germany). After centrifugation at 2000× *g* for 5 min at 4 °C, the cleared supernatant was stored at −80 °C before viral RNA isolation [63].

7. RSV RNA Copy Analysis with qRT-PCR

Viral RNA was isolated from 140 µL of the cell-free lung homogenate supernatant using the QIAamp-Viral-RNA-Mini-Kit (Qiagen, Hilden, Germany) according to the manufacturer's instructions. To determine the RSV-copy numbers, 45 ng of RNA was reverse transcribed and analyzed with the QIAGEN QuantiTECT RT-qPCR Kit using the RSV sense primer (5′-AGATCAACTTCTGTCATCCAGCAA-3′), RSV antisense primer (5′-GCACATCATAATTAGGAGTATCAAT-3′), and SYBR Green for detection in LightCycler® 480 (Roche, Basel, Switzerland). Synthetic RSV-RNA of T7-transcripts served as the standard for the quantification of viral genome copy numbers [34].

8. Analyzing RSV-Specific Neutralizing Antibodies

RSV neutralizing antibody titers in the sera were determined by co-incubation with rgRSV. After sixfold dilution of the mouse sera in Hanks' balanced salt solution (HBSS, Thermo Fisher Scientific, Dreieich, Germany), the complement was inactivated by incubation at 56 °C for 30 min. Serial twofold dilutions in DMEM with GlutaMAX (Thermo Fisher Scientific, Dreieich, Germany) containing 10% FBS and penicillin/streptomycin (Thermo Fisher Scientific, Dreieich, Germany) or a negative control without serum was incubated with around 100 FFU rgRSV per well for 1 h at 37 °C at 5% CO_2. After 1 h, the serum containing virus was transferred on pre-seeded Hep2-cells and incubated for 48 h at 37 °C with 5% CO_2. The analysis was performed by counting the green fluorescent viral foci using a FlouroSpot reader (AID Diagnistika, Straßberg, Germany). The neutralizing-antibody titer was defined as the highest serum dilution inhibiting rgRSV infection by more than 50% in comparison to the negative control (PRNT50 = 50% plaque neutralization titers). The detection limit of the neutralizing antibody was set at the lowest serum dilution (1:6) [34,44].

9. Analysis of RSV-Binding Antibodies in Mouse Sera

To examine the amount of RSV-binding IgG-antibodies in the sera of vaccinated animals, ELISA analyses were performed as previously described [41,44]. Briefly, purified and heat-inactivated (56 °C, 30 min) RSV was coated at a concentration of 5×10^5 FFU/well on black NUNC 96-well MicroWell™ PolySorp® plates (Thermo Fisher Scientific, Dreieich, Germany) in carbonate coating buffer (35 mM $NaHCO_3$, 15 mM Na_2CO_3, pH 9.6) in a total volume of 100 µL/well at 4 °C for 24 h. The plate was washed three times with PBS containing 0.05% Tween 20 (PBS-T) and blocked with 5% (*w*/*v*) milk powder in PBS for 1 h at room temperature. The sera were diluted 1:2000 in 2% (*w*/*v*) skim milk in PBS and added in 100 µL to each well. After incubation for 2 h at room temperature, secondary antibodies for the total IgG, horse radish peroxidase (HRP) conjugated anti-mouse IgG antibody (Peroxidase AffiniPure Sheep Anti-Mouse IgG (H + L), polyclonal, Dianova, Hamburg, Germany) was added at a 1:1000 dilution and incubated for 1 h at room temperature. The readout was performed using enhanced chemiluminescent substrate (ECL, Pierce, Waltham, MA, USA) diluted 1:10 in PBS and 100 µL of the substrate was injected into each well. After 1.5 s of delay, the relative light units (RLU) were counted for 1 s at the Centro XS3 luminometer (Berthold Technologies, Bad Wildbad, Germany).

10. Statistical Analysis

Statistical analysis was performed using GraphPad Prism Version 6.07. Data were analyzed using the Mann–Whitney U-test. Level of statistical significance is indicated as follows: *: $p \leq 0.05$, **: $p \leq 0.01$, ***: $p \leq 0.001$.

3. Results

3.1. LEEI Inactivation and Formulation of RSV

For pathogen inactivation, LEEI is a potent method, which we described earlier in detail [41]. For the vaccination of mice, we inactivated RSV with LEEI, therefore confirming the dosimetry and success of the inactivation process. LEEI-RSV was then formulated with PC for further testing.

3.1.1. LEEI Inactivation in the Bag Module Leads to Sufficient Surface Conservation

To determine the current to use for an irradiation with 25 kGy, which we had already defined as a safe inactivation dose of RSV [44], a dosimetry of the custom-built prototype with the use of the bag module was performed. Therefore, the radiochromic liquid dosimeter TTC was used for an estimation of the applied dose [41]. Through an interpolation of the applied beam current (Figure 1a), a dose of 25 kGy was defined at 1.2 mA

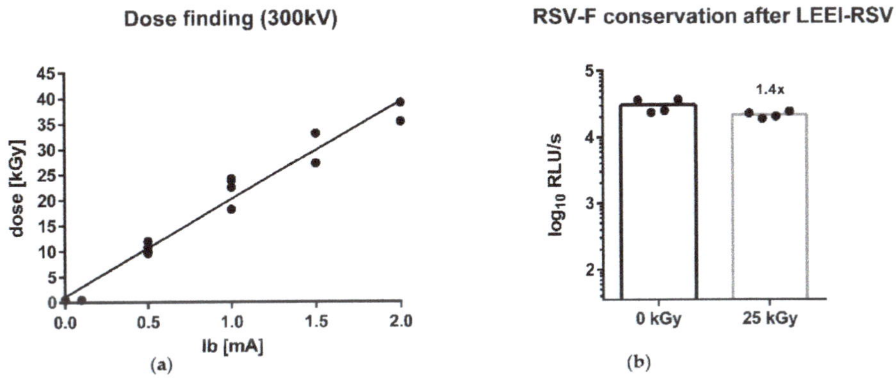

Figure 1. Dose finding and conservation of the RSV-F surface protein after irradiation. To determine the dose distribution, dosimetry of the low-energy electron irradiation (LEEI) was performed in the bag module with TTC. Two to three independent runs with different amperage at 300 kV are shown. The linear regression was calculated based on measured dosimetry (**a**). The conservation of RSV-F after LEEI was measured by an ELISA with the 18F12-antibody in the process control (0 kGy) and after inactivation with 25 kGy LEEI. Shown are the mean of each group and the fold reduction compared to 0 kGy (n = 4) (**b**).

The inactivation with LEEI was performed as described in the Methods section After the process, cells were infected with the LEEI-treated RSV-samples to verify inactivation. After the second passage, no CPE was detectable in the cells infected with RSV treated with 25 kGy of LEEI, thus being considered inactivated (Figure S1). This led to further testing of the material and analysis of the conservation of the F-glycoprotein. Hence, an ELISA was performed where the whole RSV-F-content was tested with 18F12 (Figure 1a). As a control, the processed material without irradiation (0 kGy) was tested and there was a non-statistically significant 1.4-fold reduction in the 25 kGy group (Figure 1b). To determine the RSV amount corresponding to the measured relative light unit (RLU) signal, a standard curve with a serial dilution of RSV was generated by ELISA (Figure S2). The amount of RSV was calculated for the processed material using this standard curve. For the process control (0 kGy), 30,276 RLU/s corresponded to 6.97×10^4 $TCID_{50}$ and for LEEI-RSV irradiated with 25 kGy, 21,392 RLU/s correlated with 4.58×10^4 TCID50 (Figure 1b).

3.1.2. Formulation of Inactivated Material

To apply inactivated RSV intranasally, the material was formulated in liposomes after inactivation. A solution of inactivated LEEI-RSV at a concentration of 2×10^5 TCID50/mL in PBS with 10% (w/v) sucrose was added to the unilamellar liposome formulation at a LEEI RSV:lipid ratio of 5:1. Zetasizer measurements showed a size distribution of 200 nm $\pm$ 49 nm. The rather large variation in liposomal size may be a result of the known variation in particle size of the RSV itself, which ranges between 150 and 250 nm [67]. The RSV content of the formulated liposomes was verified by ELISA.

3.2. Evaluation of Adverse and Immunogenic Effects of the New Vaccine Ex Vivo

An easy and fast procedure to test for adverse effects or the acute toxicity of drugs and vaccines on lung tissues is the human or rodent precision cut lung slice (PCLS) technology [68–70]. Freshly prepared living murine lung tissue was sliced (300 μm thickness) into cell culture dishes and incubated with the two compounds for 24 h. In addition to the toxicity, the secretion of cytokines was measured.

The non-formulated and formulated inactivated RSV material, LEEI-RSV and PC-LEEI-RSV, respectively, was tested on murine PCLS for adverse effects. The slices were incubated with tenfold serial dilutions of the material ranging from 10^3 to 10^6 TCID50/mL for 24 h. The LDH-release as a parameter for the cytotoxicity was measured in the supernatants. The background level of LDH-release in the medium control (TCID50 = 0) was around 30–45% of the Triton-lysed cells (Figure 2a). For both vaccines, up to a concentration of 10^5 TCID50/mL, no LDH release was detectable compared to the medium control (Figure 2a). In contrast, a dose of 10^6 TCID50/mL PC-LEEI-RSV induced a median LDH release of 80% compared to 45% in the medium control, while with LEEI-RSV, the induction was only to 39% compared to 31% (Figure 2a).

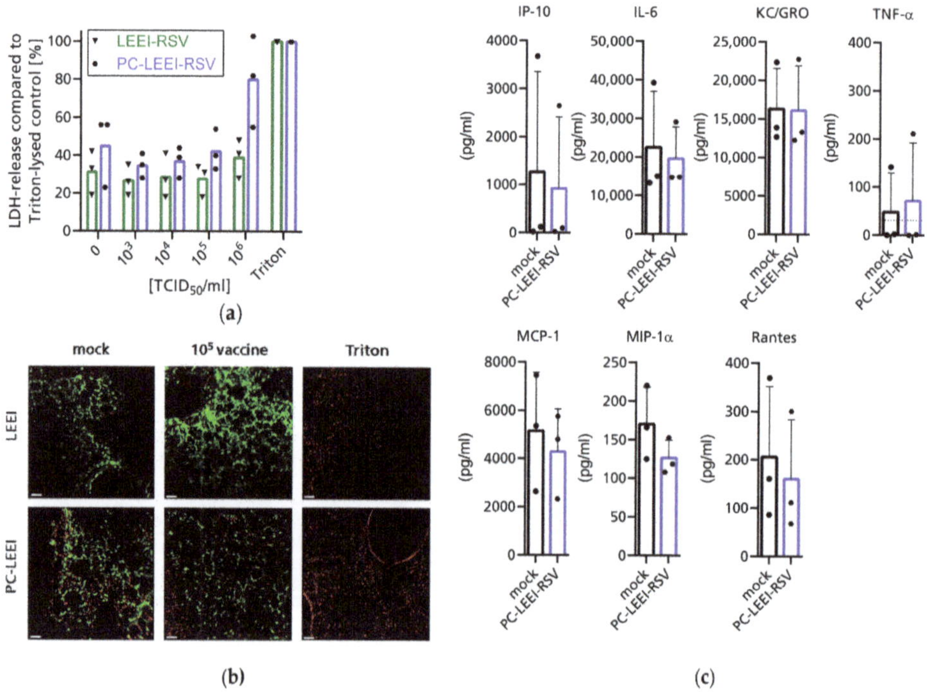

Figure 2. Evaluation of the adverse effects and toxicity of LEEI-RSV and PC-LEEI-RSV using precision cut lung slices (PCLSs). Murine PCLS were incubated for 24 h with different concentrations of LEEI-RSV or PC-LEEI-RSV medium (0 TCID50/mL or mock) or Triton (1% Triton X-100). LDH release after 24 h was measured and is shown as the percentage relative to Triton control (**a**). PCLSs were analyzed after 24 h incubation and stained with the fluorescence markers for living cells with calcein-AM (green) and for dead cells with Ethidium-homodimer (red) (**b**). Microscopic pictures of representative areas are shown (scale bar = 50 μm). After 24 h, the indicated cytokines were measured in culture supernatants at the concentration of 10^4 TCID50 per reaction (MSD U-Plex assay) (**c**). Single dots indicate an independent experiment and bars the mean of all experiments (n = 3) with (**c**) or without (**a**) standard deviations; dotted line indicates the limit of detection.

Looking at cell proliferation via the WST-conversion, a similar picture was visible: at a dose of 10^5 TCID50/mL, no changes compared to the medium control were visible (Figure S3). Subsequently, at a concentration of 10^6 TCID50/mL, both LEEI-RSV and PC-LEEI-RSV showed strongly reduced WST-conversion (Figure S3).

Analysis of the toxic effects in a live/dead staining supported the findings that a dose of 10^5 TCID50/mL PC-LEEI-RSV or LEEI-RSV showed no toxic effects on the lung tissue compared to only culture medium (mock) (Figure 2b).

Aside from the absence of adverse effects by the inactivated material, the secretion of cytokines was measured for PC-LEEI-RSV in a concentration of 10^4 TCID50/mL to ensure that the measured effects were not induced as by-products of toxicity but by the vaccine itself. The measured amounts of the indicated cytokines showed no significant differences between the PC LEEI-RSV in comparison to the medium control after 24 h (Figure 2c). Therefore, we concluded that PC-LEEI-RSV does not induce any of the measured cytokines in 24 h. In addition, IFN-alpha, -beta, and -gamma were measured but were under the limit of detection of 25 pg/mL, 1.5 pg/mL, and 0.9 pg/mL, respectively, in both conditions.

3.3. PC-LEEI-RSV Induces Immune Responses and Protection in Mice after Vaccination

To test the immunogenicity and protective efficacy of PC-LEEI-RSV, we used a well-established in vivo infection model of RSV. BALB/c mice were mucosally vaccinated in a homologous prime-boost regimen to analyze RSV specific humoral immune responses and protection against an RSV challenge. Routine follow-up of all animals showed no side effects at any timepoint after the applications of the different vaccines.

3.3.1. Humoral Systemic Immune Response after Vaccination

The induction of a humoral immune response was measured by analyzing serum samples from the blood of animals before vaccination (pre immune) and three weeks after prime and boost immunization, respectively.

Sera were used to analyze the systemic RSV neutralizing and RSV binding antibodies (Figure 3). RSV neutralizing antibodies in the unvaccinated control group were at the baseline level (Figure 3a). After prime and boost immunization, two out of five animals showed the induction of neutralizing antibodies after intranasal vaccination of PC-LEEI-RSV. The control group with intramuscular vaccination of Alhydrogel-adjuvanted LEEI-RSV showed a significant induction of RSV-neutralizing antibodies after boost vaccination (Figure 3a).

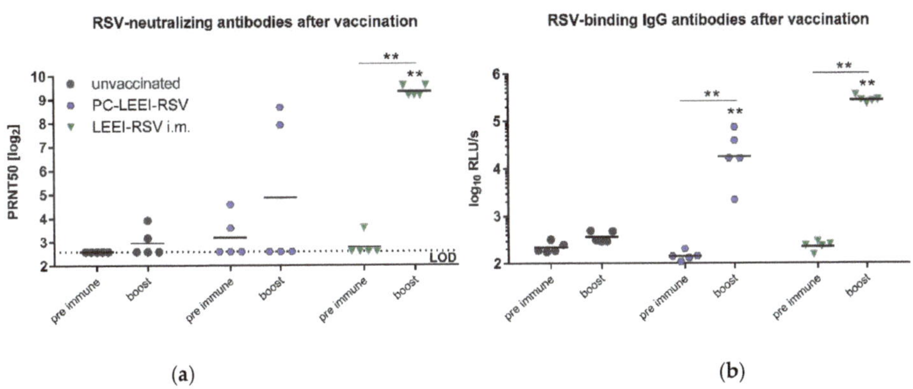

Figure 3. Systemic humoral immune response of the immunized animals. Mice were vaccinated in a homologous prime-boost regimen either with Alhydrogel adjuvanted LEEI-RSV intramuscularly (LEEI-RSV i.m.) or intranasally with PC-LEEI-RSV. Control animals were left unvaccinated. Before

(pre-immune) and three weeks after the prime and boost vaccination, blood samples were collected to monitor the systemic humoral immune responses. The 50% plaque neutralization titers (PRNT50) in sera were tested in a microneutralization assay (**a**) and RSV-binding serum IgG antibodies (**b**) by ELISA. Every dot represents the mean of two separate measurements in duplicates of one animal (**a**). In (**b**), every dot is the mean of the duplicate of one animal and shown is a representative experiment out of two. Statistical evaluation performed by the Mann–Whitney test, either in comparison to the respective unvaccinated animal (indicated above each group) or in comparison to the group against the different timepoints (line) (**: $p \leq 0.01$). (LOD = limit of detection for the virus neutralization titer at 1:6; $n = 5$). (relative light units per second = RLU/s).

All intranasally vaccinated PC-LEEI-RSV animals produced RSV-specific IgG antibodies (Figure 3b). The induction was statistically significant in comparison to the unvaccinated animals and compared to the pre-immune levels of these mice (Figure 3b). The LEEI-RSV i.m. group showed a 16-fold higher amount of IgG antibodies than the PC-LEEI-RSV group. In both vaccinated groups, the induction was statistically significant in comparison to the unvaccinated animals and to the respective pre-immune levels of the mice (Figure 3b). Untreated animals only had baseline levels of the antibodies (Figure 3b).

3.3.2. PC-ELLI-RSV Protects Mice after RSV Challenge

To test the induced protective efficacy, an RSV challenge experiment with 10^6 FFU per mouse was performed. Animals were scored daily, and no clinical symptoms were detectable after RSV infection. Five days after challenge, mice were euthanized and the viral load was determined in the lungs.

PC-LEEI-RSV, given intranasally by a homologous prime boost vaccination regimen containing 8×10^3 RSV particles per application, induced an immune response, which led to a 171-fold reduction in the viral load compared to the untreated animals (Figure 4). The viral load in LEEI-RSV i.m. mice was 966-fold lower than in the unvaccinated group, and therefore, the i.m. vaccination led to a 5.7-fold better reduction in the viral load compared to the intranasally applied PC-LEEI-RSV (Figure 4).

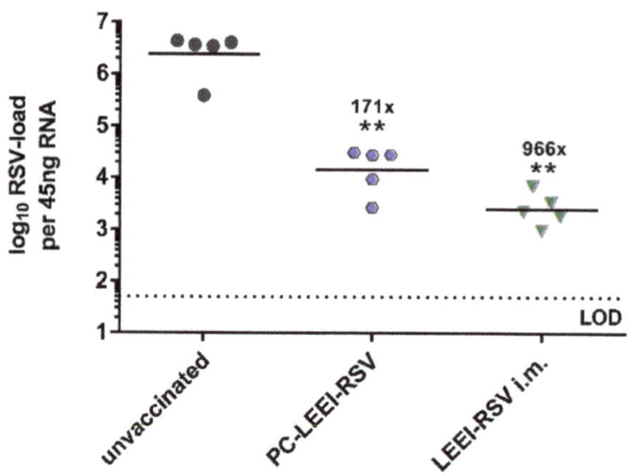

Figure 4. Viral load in the lungs after RSV challenge. BALB/c mice were vaccinated as described above. Four weeks after the boost immunization, animals were challenged with 10^6 FFU RSV per mouse. RSV-load was measured five days after challenge in the lung via qRT-PCR. Shown is the viral copy number of RSV of each animal measured in duplicate with the corresponding geometric mean of each group. Calculated viral load reduction to the untreated control is presented. Statistical evaluation of the data was performed by the Mann–Whitney test in comparison to the untreated animal (**: $p \leq 0.01$) (LOD = limit of detection at 50 FFU; $n = 5$).

In another animal trial, the vaccination of mice intranasally with LEEI-RSV, without adjuvantation, did not show the induction of a sufficient protective immune response (Figure S4). Low titers of RSV-binding IgG antibodies were induced (Figure S4a), and only a threefold reduction in viral load after RSV challenge was detectable (Figure S4b). This shows that the PC-formulation is critical for the immunogenicity.

Since we also could not detect any sign of enhanced disease severity shown by weight loss or disease development in mice, we can confirm that LEEI-RSV is a safe and highly efficacious vaccine. In addition, by formulation with PC, this vaccine candidate can be applied intranasally without any further adjuvants.

4. Discussion

Mucosal vaccines induce protection at the site of infection by activating local immune responses [29,30,71–73].

In this study, we showed, for the first time, protection with the intranasal application of a PC-formulated LEEI-RSV in mice against RSV. To exclude adverse effects on viability or immune activation in the lung tissue, we first showed in the PCLS model that no toxic effects were generated with our vaccines. In the analysis, neither the LEEI-RSV itself nor the formulated PC-LEEI-RSV induced adverse effects in the murine lung tissue. The viability staining showed good viability with the vaccine candidates, and the WST and LDH only showed adverse results at the highest concentration tested. Importantly, the PC-LEEI-RSV vaccine induced systemic RSV-binding antibodies and serum neutralizing antibodies after mucosal vaccination. The detected levels of neutralizing antibodies close to the baseline in the unvaccinated boost-sera and the PC-LEEI-RSV pre immune sera were more likely unspecific background. Upon RSV challenge, a statistically significant protection in viral load in the lungs of vaccinated animals compared to unvaccinated mice was observed. We could thus demonstrate that the intranasal vaccination provides a reduction in the viral load, which might be sufficient to protect against RSV infection if LRTIs are blocked [74].

These encouraging results demonstrate, to our knowledge, a proof-of-principle of a novel mucosal applied vaccine candidate that is inactivated by LEEI and formulated to improve immunogenicity and protective efficacy. It is worth mentioning that the liposome-based formulation and the mucosal application can be further improved. For intranasal application, sophisticated devices have been developed to produce size-specific droplets. However, the optimization of this mucosal vaccine candidate was beyond this proof-of-concept study. The application route of LEEI inactivated vaccines has so far been the intramuscular route, but in this proof-of-concept study, we observed that the intranasal route also induced protective efficacy based on the reduction in the viral load. Nevertheless, the mucosal route needs the liposomal based formulation for the efficient induction of a protective immune response. Liposomes, especially PC, are physiological substances that are known to be non-toxic and are already medically applied [59]. Liposomes are widely used for drug delivery across a variety of therapeutic areas [58–60]. Many studies have already proven the non-toxicity of liposomes [75,76]; especially PC, as a physiological substance should not interfere with the respiratory system as a high proportion of the pulmonary surfactant consists of lipids [61,77,78].

Mucosal vaccines are to date mostly based on live-attenuated viruses (LAAV) and viral vectors as these vaccine candidates can naturally infect mucosal cells and overcome the mucosal barriers. LAAVs activate multiple immune responses including the innate immune system, mimicking a normal virus infection [8,72]. However, an RSV-LAAV still induced unwanted disease symptoms after application because it was insufficiently attenuated [79]. In contrast, over-attenuated RSV-LAAV failed to induce sufficient protection [80].

Viral vectors present another promising approach for mucosal application as they infect mucosal cells and express an immune relevant transgene that induces an immune response. Adenoviral vectors have shown great potential as a mucosal immunization platform, as protection was induced compared to an intramuscular application after a vaccination with the same doses [31,34]. Furthermore, the adenoviral vector vaccine offers higher protection against RSV compared to natural infection in mice including better humoral and cellular immune response at the site of infection [30]. We have previously described that an intramuscular prime with a DNA-plasmid encoding RSV-F, followed by a mucosal adenoviral vector boost, induces high amounts of mucosal T-cells, systemic humoral responses, and protective efficacy upon RSV challenge [81]. With regard to intramuscular prime and mucosal boost, we have recently shown that an adenoviral vector against SARS-CoV-2 induces high protective efficacy and superior mucosal immune responses in comparison to a homologous vaccination using the formulated mRNA intramuscular vaccine [82]. Aside from adenoviral vectors, another vector platform based on the Modified Vaccinia Ankara virus (MVA) is currently in clinical trials against RSV using the intramuscular approach [8]. A similar MVA-based vaccine was recently tested using the intranasal route, showing good antibody induction and protection against RSV in mice [83]. One larger disadvantage of virus vector vaccines is the induced vector immunity [84,85]. In both live attenuated viruses and virus vectors, it is important that no immune response against the vaccines with IgA or cytotoxic T-cells is already established because it could block the vaccine [84–87]. In addition, new concerns against the adenoviral vector vaccines arose in the context of vaccines against SARS-CoV-2 due to vaccine-induced immune thrombotic thrombocytopenia [88,89].

The issues of vector immunity and safety concerns with LAAV or vector vaccines can be circumvented by using inactivated vaccines [90]. Inactivated vaccines are not able to replicate and are mostly applied intramuscularly with different adjuvants to enhance and prolong the immune response [91,92]. For respiratory viruses such as RSV, the site of infection is the respiratory tract, and a mucosal vaccination would induce higher mucosal immunity [28,30,34,57,93]. Furthermore, it has already been observed that the induction of immune responses via a mucosal application include broad mucosal immune response and systemic immunity [30,31,93]. The intranasal application of an adenoviral vector vaccine against RSV led to the induction of RSV-F specific CD8+ T cells, central memory CD8+ T cells, and most importantly, to tissue-resident memory CD8+ T cells in the lungs of vaccinated mice [30]. Similar observations were seen with an intranasal RSV-F Nanovaccine inducing tissue-resident memory CD4+ and CD8+ T cells besides neutralizing antibodies [93]. These cellular immune response mechanisms are important and necessary in a balanced and protective immunity against RSV [94]. However, the analysis of cellular immune responses was beyond the scope of this proof-of-principle study. Besides the beneficial immunological aspects, a needle-free application of a vaccine might also reduce vaccine hesitancy, as it is a non-traumatic application method [28]. Additionally, the application is more feasible than an injection and the immunization can likely be performed without medical assistance, enhancing the performance of global immunization [28].

For the effective induction of an immune response after mucosal application, the vaccine has to overcome the physical and biological barriers of the mucosa, namely the cell layer with tight junctions and the mucus with proteoglycans, lipids, or DNA [71]. To overcome elimination by the mucosal defense mechanism, resorption, and immune stimulatory reasons, adjuvanting or packaging of the vaccine is necessary, whereby the choice of substance can be essential. For example, in an intranasal influenza vaccine, the use of the mLT LTK3 adjuvant led to transient peripheral facial nerve palsy in some vaccines [71,95]. In contrast, PC is a lipid that is also present in the surfactant of the airways, and due to its physiological properties, no adverse effects are expected [60,96]. We showed that the formulation of the inactivated vaccine with PC is necessary, as shown in Figure S4 compared to Figure 4 [71]. It is worth mentioning that the vaccination of LEEI-RSV intramuscularly needed adjuvanting with Alhydrogel whereas the mucosal application only needed the formulation with PC.

In conclusion, this proof-of-principle study shows a novel potent method for the production of a mucosal inactivated whole virus vaccine formulated with PC. PC-LEEI-RSV protects against RSV by significantly reducing the viral load in vaccinated animals and presents a promising vaccine candidate. Further preclinical optimization and the clinical development of this vaccine candidate are still warranted.

Supplementary Materials: The following supporting information can be downloaded at: https://www.mdpi.com/article/10.3390/v15091846/s1, Figure S1: Inactivation of RSV by LEEI; Figure S2: Standard curve for RSV amount; Figure S3: Metabolic activity in PCLS incubation with inactivated material; Figure S4: Systemic humoral immune response after intranasal LEEI-RSV without formulation immunized animals and viral load after RSV infection.

Author Contributions: Conceptualization, T.G. and S.U.; Methodology, V.E., L.B., J.F., M.A., M.T., B.S., Y.M., T.G. and S.U.; Validation, V.E., J.F., M.A., J.K.B., B.S., M.T. and Y.M.; Formal analysis, V.E:, J.F., M.A. and J.K.B.; Investigation, V.E., L.B., J.F., M.A., J.K.B., B.S. and Y.M.; Resources, T.G., S.U., M.T., C.H., F.L., S.S. and J.D.; Data curation, V.E., J.F., L.B., J.K.B. and M.A.; Writing—original draft preparation, V.E. and T.G.; Writing—review and editing, V.E., T.G., S.U., L.B., J.F., C.H., J.K.B., S.S., M.T., Y.M. and J.D.; Visualization, V.E.; Supervision, T.G., S.U., M.T., C.H., F.L., S.S. and J.D.; Project administration, T.G., S.U., M.T., C.H., F.L., S.S. and J.D.; Funding acquisition, T.G. All authors have read and agreed to the published version of the manuscript.

Funding: This research was funded by the Cluster of Excellence of the Fraunhofer Gesellschaft, grant number 840277; MucoRSV.

Institutional Review Board Statement: The study was conducted in accordance with the Declaration of Helsinki and approved by the Ethics Committee of Landesdirektion Sachsen.

Informed Consent Statement: Not applicable.

Data Availability Statement: Data can be provided after request to the corresponding authors.

Acknowledgments: We would like to thank Fabian Pappe and André Poremba for their excellent technical assistance. We thank Jasmin Fertey for her scientific input and Jana Beckmann for her support in the development of the dosimetry.

Conflicts of Interest: S.U. and M.T. are authors of patents and patent applications covering LEEI to inactivate liquids: WO2018041953, DE102015224206B3, US10080795, DE 102017002645.9. All authors have no further conflict of interest.

References

1. Simoes, E.A. Respiratory syncytial virus infection. *Lancet* **1999**, *354*, 847–852. [CrossRef] [PubMed]
2. Rima, B.; Collins, P.; Easton, A.; Fouchier, R.; Kurath, G.; Lamb, R.A.; Lee, B.; Maisner, A.; Rota, P.; Wang, L.; et al. ICTV Virus Taxonomy Profile: Pneumoviridae. *J. Gen. Virol.* **2017**, *98*, 2912–2913. [CrossRef] [PubMed]
3. Bohmwald, K.; Espinoza, J.A.; Rey-Jurado, E.; Gómez, R.S.; González, P.A.; Bueno, S.M.; Riedel, C.A.; Kalergis, A.M. Human Respiratory Syncytial Virus: Infection and Pathology. *Semin. Respir. Crit. Care Med.* **2016**, *37*, 522–537. [CrossRef] [PubMed]
4. Li, Y.; Wang, X.; Blau, D.M.; Caballero, M.T.; Feikin, D.R.; Gill, C.J.; Madhi, S.A.; Omer, S.B.; Simões, E.A.F.; Campbell, H.; et al. Global, regional, and national disease burden estimates of acute lower respiratory infections due to respiratory syncytial virus in children younger than 5 years in 2019: A systematic analysis. *Lancet* **2022**, *399*, 2047–2064. [CrossRef]
5. Ackerson, B.; Tseng, H.F.; Sy, L.S.; Solano, Z.; Slezak, J.; Luo, Y.; Fischetti, C.A.; Shinde, V. Severe Morbidity and Mortality Associated With Respiratory Syncytial Virus Versus Influenza Infection in Hospitalized Older Adults. *Clin. Infect. Dis.* **2019**, *69*, 197–203. [CrossRef] [PubMed]
6. Savic, M.; Penders, Y.; Shi, T.; Branche, A.; Pirçon, J.-Y. Respiratory syncytial virus disease burden in adults aged 60 years and older in high-income countries: A systematic literature review and meta-analysis. Influenza Other Respir. *Viruses* **2023**, *17*, e13031. [CrossRef]
7. Falsey, A.R.; Hennessey, P.A.; Formica, M.A.; Cox, C.; Walsh, E.E. Respiratory syncytial virus infection in elderly and high-risk adults. *N. Engl. J. Med.* **2005**, *352*, 1749–1759. [CrossRef]
8. Mazur, N.I.; Terstappen, J.; Baral, R.; Bardají, A.; Beutels, P.; Buchholz, U.J.; Cohen, C.; Crowe, J.E.; Cutland, C.L.; Eckert, L.; et al. Respiratory syncytial virus prevention within reach: The vaccine and monoclonal antibody landscape. *Lancet Infect. Dis.* **2023**, *23*, e2–e21. [CrossRef]
9. Kim, H.W.; Canchola, J.G.; Brandt, C.D.; Pyles, G.; Chanock, R.M.; Jensen, K.; Parrott, R.H. Respiratory syncytial virus disease in infants despite prior administration of antigenic inactivated vaccine. *Am. J. Epidemiol.* **1969**, *89*, 422–434. [CrossRef]
10. Castilow, E.M.; Olson, M.R.; Varga, S.M. Understanding respiratory syncytial virus (RSV) vaccine-enhanced disease. *Immunol. Res.* **2007**, *39*, 225–239. [CrossRef]
11. BLOUNT, R.E.; MORRIS, J.A.; SAVAGE, R.E. Recovery of cytopathogenic agent from chimpanzees with coryza. *Proc. Soc. Exp. Biol. Med.* **1956**, *92*, 544–549. [CrossRef] [PubMed]

12. Anderson, E.J.; Carosone-Link, P.; Yogev, R.; Yi, J.; Simões, E.A.F. Effectiveness of Palivizumab in High-risk Infants and Children: A Propensity Score Weighted Regression Analysis. *Pediatr. Infect. Dis. J.* **2017**, *36*, 699–704. [CrossRef] [PubMed]
13. CHANOCK, R.; ROIZMAN, B.; MYERS, R. Recovery from infants with respiratory illness of a virus related to chimpanzee coryza agent (CCA). I. Isolation, properties and characterization. *Am. J. Hyg.* **1957**, *66*, 281–290. [CrossRef]
14. Hammitt, L.L.; Dagan, R.; Yuan, Y.; Baca Cots, M.; Bosheva, M.; Madhi, S.A.; Muller, W.J.; Zar, H.J.; Brooks, D.; Grenham, A.; et al. Nirsevimab for Prevention of RSV in Healthy Late-Preterm and Term Infants. *N. Engl. J. Med.* **2022**, *386*, 837–846. [CrossRef]
15. Guendoul, S.; Bain, S.; Berland, E.; Conway, K. Press Release: FDA Accepts Nirsevimab Application as First Protective Option against RSV Disease for All Infants—Sanofi. Available online: https://www.sanofi.com/en/media-room/press-releases/2023/2023-01-05-07-00-00-2583365 (accessed on 9 May 2023).
16. Vidal Valero, M. 'A good day': FDA approves world's first RSV vaccine. *Nature* 2023. [CrossRef] [PubMed]
17. GSK. US FDA Approves GSK's Arexvy, the World's First Respiratory Syncytial Virus (RSV) Vaccine for Older Adults | GSK. Available online: https://www.gsk.com/en-gb/media/press-releases/us-fda-approves-gsk-s-arexvy-the-world-s-first-respiratory-syncytial-virus-rsv-vaccine-for-older-adults/ (accessed on 9 May 2023).
18. Papi, A.; Ison, M.G.; Langley, J.M.; Lee, D.-G.; Leroux-Roels, I.; Martinon-Torres, F.; Schwarz, T.F.; van Zyl-Smit, R.N.; Campora, L.; Dezutter, N.; et al. Respiratory Syncytial Virus Prefusion F Protein Vaccine in Older Adults. *N. Engl. J. Med.* **2023**, *388*, 595–608. [CrossRef]
19. GSK. GSK's Respiratory Syncytial Virus Older Adult Vaccine Candidate Gains Positive European Medicines Agency CHMP Opinion | GSK. Available online: https://www.gsk.com/en-gb/media/press-releases/gsk-s-respiratory-syncytial-virus-older-adult-vaccine-candidate-gains-positive-european-medicines-agency-chmp-opinion/ (accessed on 9 May 2023).
20. Walsh, E.E.; Pérez Marc, G.; Zareba, A.M.; Falsey, A.R.; Jiang, Q.; Patton, M.; Polack, F.P.; Llapur, C.; Doreski, P.A.; Ilangovan, K.; et al. Efficacy and Safety of a Bivalent RSV Prefusion F Vaccine in Older Adults. *N. Engl. J. Med.* **2023**, *388*, 1465–1477. [CrossRef]
21. Pfizer Announces Positive Top-Line Data from Phase 3 Trial of Older Adults for its Bivalent Respiratory Syncytial Virus (RSV) Vaccine Candidate | Pfizer. Available online: https://www.pfizer.com/news/press-release/press-release-detail/pfizer-announces-positive-top-line-data-phase-3-trial-older (accessed on 9 May 2023).
22. Pfizer Receives Positive FDA Advisory Committee Votes Supporting Potential Approval for Vaccine Candidate to Help Combat RSV in Older Adults | Pfizer. Available online: https://www.pfizer.com/news/press-release/press-release-detail/pfizer-receives-positive-fda-advisory-committee-votes (accessed on 9 May 2023).
23. Pfizer Announces Positive Top-Line Data of Phase 3 Global Maternal Immunization Trial for its Bivalent Respiratory Syncytial Virus (RSV) Vaccine Candidate | Pfizer. Available online: https://www.pfizer.com/news/press-release/press-release-detail/pfizer-announces-positive-top-line-data-phase-3-global (accessed on 9 May 2023).
24. U.S. FDA Accepts Biologics License Application for Pfizer's Respiratory Syncytial Virus Maternal Vaccine Candidate for Priority Review | Pfizer. Available online: https://www.pfizer.com/news/press-release/press-release-detail/us-fda-accepts-biologics-license-application-pfizers (accessed on 9 May 2023).
25. White, K. FDA approved Pfizer's RSV vaccine for older adults. Morning Brew [Online]. 6 May 2023. Available online: https://www.healthcare-brew.com/stories/2023/06/05/fda-approved-pfizer-s-rsv-vaccine-for-older-adults (accessed on 28 June 2023).
26. Inc, M. Moderna Announces mRNA-1345, an Investigational Respiratory Syncytial Virus (RSV) Vaccine, Has Met Primary. ACCESSWIRE.com [Online]. 17 January 2023. Available online: https://www.accesswire.com/735567/Moderna-Announces-mRNA-1345-an-Investigational-Respiratory-Syncytial-Virus-RSV-Vaccine-Has-Met-Primary-Efficacy-Endpoints-in-Phase-3-Trial-in-Older-Adults (accessed on 9 May 2023).
27. Carvalho, T. mRNA vaccine effective against RSV respiratory disease. *Nat. Med.* **2023**, *29*, 755–756. [CrossRef]
28. Skwarczynski, M.; Toth, I. Non-invasive mucosal vaccine delivery: Advantages, challenges and the future. *Expert. Opin. Drug Deliv.* **2020**, *17*, 435–437. [CrossRef]
29. Lycke, N. Recent progress in mucosal vaccine development: Potential and limitations. *Nat. Rev. Immunol.* **2012**, *12*, 592–605. [CrossRef]
30. Maier, C.; Fuchs, J.; Irrgang, P.; Wißing, M.H.; Beyerlein, J.; Tenbusch, M.; Lapuente, D. Mucosal immunization with an adenoviral vector vaccine confers superior protection against RSV compared to natural immunity. *Front. Immunol.* **2022**, *13*, 920256. [CrossRef]
31. Pierantoni, A.; Esposito, M.L.; Ammendola, V.; Napolitano, F.; Grazioli, F.; Abbate, A.; Del Sorbo, M.; Siani, L.; D'Alise, A.M.; Taglioni, A.; et al. Mucosal delivery of a vectored RSV vaccine is safe and elicits protective immunity in rodents and nonhuman primates. *Mol. Ther. Methods Clin. Dev.* **2015**, *2*, 15018. [CrossRef] [PubMed]
32. Lavelle, E.C.; Ward, R.W. Mucosal vaccines—Fortifying the frontiers. *Nat. Rev. Immunol.* **2022**, *22*, 236–250. [CrossRef]
33. Mueller, S.; Le Nouen, C.; Buchholz, U.J.; Kalkeri, R.; Koide, F.; Collins, P.; Coleman, J.R. 2777. Live-Attenuated Vaccine Against RSV Generates Robust Cellular and Humoral Immune Responses. *Open Forum Infect. Dis.* **2019**, *6*, S980. [CrossRef]
34. Kohlmann, R.; Schwannecke, S.; Tippler, B.; Ternette, N.; Temchura, V.V.; Tenbusch, M.; Uberla, K.; Grunwald, T. Protective efficacy and immunogenicity of an adenoviral vector vaccine encoding the codon-optimized F protein of respiratory syncytial virus. *J. Virol.* **2009**, *83*, 12601–12610. [CrossRef]
35. World Health Organization. Polio vaccines: WHO position paper, January 2014. *Wkly. Epidemiol. Rec.* **2014**, *89*, 73–92.

36. Jorba, J.; Diop, O.M.; Iber, J.; Henderson, E.; Zhao, K.; Sutter, R.W.; Wassilak, S.G.F.; Burns, C.C. Update on Vaccine-Derived Polioviruses—Worldwide, January 2017-June 2018. MMWR Morb. *Mortal. Wkly. Rep.* **2018**, *67*, 1189–1194. [CrossRef] [PubMed]
37. Lemiale, F.; Kong, W.; Akyürek, L.M.; Ling, X.; Huang, Y.; Chakrabarti, B.K.; Eckhaus, M.; Nabel, G.J. Enhanced mucosal immunoglobulin A response of intranasal adenoviral vector human immunodeficiency virus vaccine and localization in the central nervous system. *J. Virol.* **2003**, *77*, 10078–10087. [CrossRef]
38. Furuya, Y. Return of inactivated whole-virus vaccine for superior efficacy. *Immunol. Cell Biol.* **2012**, *90*, 571–578. [CrossRef]
39. Nichol, K.L. The efficacy, effectiveness and cost-effectiveness of inactivated influenza virus vaccines. *Vaccine* **2003**, *21*, 1769–1775. [CrossRef]
40. Sanders, B.; Koldijk, M.; Schuitemaker, H. Inactivated Viral Vaccines. In *Vaccine Analysis: Strategies, Principles, and Control*; Nunnally, B.K., Turula, V.E., Sitrin, R.D., Eds.; Springer: Berlin/Heidelberg, Germany, 2015; pp. 45–80. ISBN 978-3-662-45023-9.
41. Fertey, J.; Thoma, M.; Beckmann, J.; Bayer, L.; Finkensieper, J.; Reißhauer, S.; Berneck, B.S.; Issmail, L.; Schönfelder, J.; Casado, J.P.; et al. Automated application of low energy electron irradiation enables inactivation of pathogen- and cell-containing liquids in biomedical research and production facilities. *Sci. Rep.* **2020**, *10*, 12786. [CrossRef] [PubMed]
42. Fertey, J.; Standfest, B.; Beckmann, J.; Thoma, M.; Grunwald, T.; Ulbert, S. Low-Energy Electron Irradiation (LEEI) for the Generation of Inactivated Bacterial Vaccines. *Methods Mol. Biol.* **2022**, *2414*, 97–113. [CrossRef] [PubMed]
43. Fertey, J.; Bayer, L.; Grunwald, T.; Pohl, A.; Beckmann, J.; Gotzmann, G.; Casado, J.P.; Schönfelder, J.; Rögner, F.-H.; Wetzel, C.; et al. Pathogens Inactivated by Low-Energy-Electron Irradiation Maintain Antigenic Properties and Induce Protective Immune Responses. *Viruses* **2016**, *8*, 319. [CrossRef] [PubMed]
44. Bayer, L.; Fertey, J.; Ulbert, S.; Grunwald, T. Immunization with an adjuvanted low-energy electron irradiation inactivated respiratory syncytial virus vaccine shows immunoprotective activity in mice. *Vaccine* **2018**, *36*, 1561–1569. [CrossRef]
45. Finkensieper, J.; Issmail, L.; Fertey, J.; Rockstroh, A.; Schopf, S.; Standfest, B.; Thoma, M.; Grunwald, T.; Ulbert, S. Low-Energy Electron Irradiation of Tick-Borne Encephalitis Virus Provides a Protective Inactivated Vaccine. *Front. Immunol.* **2022**, *13*, 825702. [CrossRef]
46. Finkensieper, J.; Mayerle, F.; Rentería-Solís, Z.; Fertey, J.; Makert, G.R.; Lange, F.; Besecke, J.; Schopf, S.; Poremba, A.; König, U.; et al. Apicomplexan parasites are attenuated by low-energy electron irradiation in an automated microfluidic system and protect against infection with Toxoplasma gondii. *Parasitol. Res.* **2023**, *122*, 1819–1832. [CrossRef]
47. Seo, H.S. Application of radiation technology in vaccines development. *Clin. Exp. Vaccine Res.* **2015**, *4*, 145–158. [CrossRef]
48. Gotzmann, G.; Portillo, J.; Wronski, S.; Kohl, Y.; Gorjup, E.; Schuck, H.; Rögner, F.H.; Müller, M.; Chaberny, I.F.; Schönfelder, J.; et al. Low-energy electron-beam treatment as alternative for on-site sterilization of highly functionalized medical products—A feasibility study. *Radiat. Phys. Chem.* **2018**, *150*, 9–19. [CrossRef]
49. Wetzel, C.; Schönfelder, J.; Schwarz, W.; Funk, R. Surface modification of polyurethane and silicone for therapeutic medical technics by means of electron beam. *Surf. Coat. Technol.* **2010**, *205*, 1618–1623. [CrossRef]
50. Delrue, I.; Delputte, P.L.; Nauwynck, H.J. Assessing the functionality of viral entry-associated domains of porcine reproductive and respiratory syndrome virus during inactivation procedures, a potential tool to optimize inactivated vaccines. *Vet. Res.* **2009**, *40*, 62. [CrossRef]
51. Zhao, Y.; Ma, C.; Yang, J.; Zou, X.; Pan, Z. Dynamic Host Immune and Transcriptomic Responses to Respiratory Syncytial Virus Infection in a Vaccination-Challenge Mouse Model. *Virol. Sin.* **2021**, *36*, 1327–1340. [CrossRef]
52. Killikelly, A.M.; Kanekiyo, M.; Graham, B.S. Pre-fusion F is absent on the surface of formalin-inactivated respiratory syncytial virus. *Sci. Rep.* **2016**, *6*, 34108. [CrossRef]
53. Murphy, B.R.; Walsh, E.E. Formalin-inactivated respiratory syncytial virus vaccine induces antibodies to the fusion glycoprotein that are deficient in fusion-inhibiting activity. *J. Clin. Microbiol.* **1988**, *26*, 1595–1597. [CrossRef]
54. Moghaddam, A.; Olszewska, W.; Wang, B.; Tregoning, J.S.; Helson, R.; Sattentau, Q.J.; Openshaw, P.J.M. A potential molecular mechanism for hypersensitivity caused by formalin-inactivated vaccines. *Nat. Med.* **2006**, *12*, 905–907. [CrossRef]
55. Openshaw, P.J.; Culley, F.J.; Olszewska, W. Immunopathogenesis of vaccine-enhanced RSV disease. *Vaccine* **2001**, *20* (Suppl. 1), S27–S31. [CrossRef]
56. Polack, F.P.; Teng, M.N.; Collins, P.L.; Prince, G.A.; Exner, M.; Regele, H.; Lirman, D.D.; Rabold, R.; Hoffman, S.J.; Karp, C.L.; et al. A role for immune complexes in enhanced respiratory syncytial virus disease. *J. Exp. Med.* **2002**, *196*, 859–865. [CrossRef] [PubMed]
57. Kim, S.-H.; Jang, Y.-S. The development of mucosal vaccines for both mucosal and systemic immune induction and the roles played by adjuvants. *Clin. Exp. Vaccine Res.* **2017**, *6*, 15–21. [CrossRef]
58. Watson, D.S.; Endsley, A.N.; Huang, L. Design considerations for liposomal vaccines: Influence of formulation parameters on antibody and cell-mediated immune responses to liposome associated antigens. *Vaccine* **2012**, *30*, 2256–2272. [CrossRef] [PubMed]
59. Torchilin, V.P. Recent advances with liposomes as pharmaceutical carriers. *Nat. Rev. Drug Discov.* **2005**, *4*, 145–160. [CrossRef] [PubMed]
60. Schwendener, R.A. Liposomes as vaccine delivery systems: A review of the recent advances. *Ther. Adv. Vaccines* **2014**, *2*, 159–182. [CrossRef] [PubMed]
61. Joshi, S.; Bawage, S.; Tiwari, P.; Kirby, D.; Perrie, Y.; Dennis, V.; Singh, S.R. Liposomes: A promising carrier for respiratory syncytial virus therapeutics. Expert Opin. *Drug Deliv.* **2019**, *16*, 969–980. [CrossRef]

62. Ternette, N.; Tippler, B.; Uberla, K.; Grunwald, T. Immunogenicity and efficacy of codon optimized DNA vaccines encoding the F-protein of respiratory syncytial virus. *Vaccine* **2007**, *25*, 7271–7279. [CrossRef] [PubMed]
63. Issmail, L.; Ramsbeck, D.; Jäger, C.; Henning, T.; Kleinschmidt, M.; Buchholz, M.; Grunwald, T. Identification and evaluation of a novel tribenzamide derivative as an inhibitor targeting the entry of the respiratory syncytial virus. *Antiviral Res.* **2023**, *211*, 105547. [CrossRef]
64. Reed, L.J.; Muench, H. A simple method of estimating fifty per cent endpoints. *Am. J. Epidemiol.* **1938**, *27*, 493–497. [CrossRef]
65. Schopf, S.; Gotzmann, G.; Dietze, M.; Gerschke, S.; Kenner, L.; König, U. Investigations Into the Suitability of Bacterial Suspensions as Biological Indicators for Low-Energy Electron Irradiation. *Front. Immunol.* **2022**, *13*, 814767. [CrossRef]
66. Virology Culture Guide. Available online: https://www.atcc.org/resources/culture-guides/virology-culture-guide (accessed on 2 August 2023).
67. Utley, T.J.; Ducharme, N.A.; Varthakavi, V.; Shepherd, B.E.; Santangelo, P.J.; Lindquist, M.E.; Goldenring, J.R.; Crowe, J.E. Respiratory syncytial virus uses a Vps4-independent budding mechanism controlled by Rab11-FIP2. *Proc. Natl. Acad. Sci. USA* **2008**, *105*, 10209–10214. [CrossRef] [PubMed]
68. Pearson, H.; Todd, E.J.A.A.; Ahrends, M.; Hover, S.E.; Whitehouse, A.; Stacey, M.; Lippiat, J.D.; Wilkens, L.; Fieguth, H.-G.; Danov, O.; et al. TMEM16A/ANO1 calcium-activated chloride channel as a novel target for the treatment of human respiratory syncytial virus infection. *Thorax* **2021**, *76*, 64–72. [CrossRef]
69. Hess, A.; Wang-Lauenstein, L.; Braun, A.; Kolle, S.N.; Landsiedel, R.; Liebsch, M.; Ma-Hock, L.; Pirow, R.; Schneider, X.; Steinfath, M.; et al. Prevalidation of the ex-vivo model PCLS for prediction of respiratory toxicity. *Toxicol. Vit.* **2016**, *32*, 347–361. [CrossRef]
70. Lauenstein, L.; Switalla, S.; Prenzler, F.; Seehase, S.; Pfennig, O.; Förster, C.; Fieguth, H.; Braun, A.; Sewald, K. Assessment of immunotoxicity induced by chemicals in human precision-cut lung slices (PCLS). *Toxicol. Vit.* **2014**, *28*, 588–599. [CrossRef]
71. Baker, J.R.; Farazuddin, M.; Wong, P.T.; O'Konek, J.J. The unfulfilled potential of mucosal immunization. *J. Allergy Clin. Immunol.* **2022**, *150*, 1–11. [CrossRef]
72. Neutra, M.R.; Kozlowski, P.A. Mucosal vaccines: The promise and the challenge. *Nat. Rev. Immunol.* **2006**, *6*, 148–158. [CrossRef] [PubMed]
73. Miquel-Clopés, A.; Bentley, E.G.; Stewart, J.P.; Carding, S.R. Mucosal vaccines and technology. *Clin. Exp. Immunol.* **2019**, *196*, 205–214. [CrossRef]
74. Modjarrad, K.; Giersing, B.; Kaslow, D.C.; Smith, P.G.; Moorthy, V.S. WHO consultation on Respiratory Syncytial Virus Vaccine Development Report from a World Health Organization Meeting held on 23–24 March 2015. *Vaccine* **2016**, *34*, 190–197. [CrossRef] [PubMed]
75. Perrie, Y.; Mohammed, A.R.; Kirby, D.J.; McNeil, S.E.; Bramwell, V.W. Vaccine adjuvant systems: Enhancing the efficacy of sub-unit protein antigens. *Int. J. Pharm.* **2008**, *364*, 272–280. [CrossRef]
76. Joshi, S.; Chaudhari, A.A.; Dennis, V.; Kirby, D.J.; Perrie, Y.; Singh, S.R. Anti-RSV Peptide-Loaded Liposomes for the Inhibition of Respiratory Syncytial Virus. *Bioengineering* **2018**, *5*, 37. [CrossRef] [PubMed]
77. Numata, M.; Chu, H.W.; Dakhama, A.; Voelker, D.R. Pulmonary surfactant phosphatidylglycerol inhibits respiratory syncytial virus-induced inflammation and infection. *Proc. Natl. Acad. Sci. USA* **2010**, *107*, 320–325. [CrossRef] [PubMed]
78. Numata, M.; Nagashima, Y.; Moore, M.L.; Berry, K.Z.; Chan, M.; Kandasamy, P.; Peebles, R.S.; Murphy, R.C.; Voelker, D.R. Phosphatidylglycerol provides short-term prophylaxis against respiratory syncytial virus infection. *J. Lipid Res.* **2013**, *54*, 2133–2143. [CrossRef]
79. Rostad, C.A.; Stobart, C.C.; Todd, S.O.; Molina, S.A.; Lee, S.; Blanco, J.C.G.; Moore, M.L. Enhancing the Thermostability and Immunogenicity of a Respiratory Syncytial Virus (RSV) Live-Attenuated Vaccine by Incorporating Unique RSV Line19F Protein Residues. *J. Virol.* **2018**, *92*. [CrossRef]
80. Karron, R.A.; Wright, P.F.; Belshe, R.B.; Thumar, B.; Casey, R.; Newman, F.; Polack, F.P.; Randolph, V.B.; Deatly, A.; Hackell, J.; et al. Identification of a recombinant live attenuated respiratory syncytial virus vaccine candidate that is highly attenuated in infants. *J. Infect. Dis.* **2005**, *191*, 1093–1104. [CrossRef]
81. Grunwald, T.; Tenbusch, M.; Schulte, R.; Raue, K.; Wolf, H.; Hannaman, D.; de Swart, R.L.; Uberla, K.; Stahl-Hennig, C. Novel vaccine regimen elicits strong airway immune responses and control of respiratory syncytial virus in nonhuman primates. *J. Virol.* **2014**, *88*, 3997–4007. [CrossRef]
82. Lapuente, D.; Fuchs, J.; Willar, J.; Vieira Antão, A.; Eberlein, V.; Uhlig, N.; Issmail, L.; Schmidt, A.; Oltmanns, F.; Peter, A.S.; et al. Protective mucosal immunity against SARS-CoV-2 after heterologous systemic prime-mucosal boost immunization. *Nat. Commun.* **2021**, *12*, 6871. [CrossRef] [PubMed]
83. Endt, K.; Wollmann, Y.; Haug, J.; Bernig, C.; Feigl, M.; Heiseke, A.; Kalla, M.; Hochrein, H.; Suter, M.; Chaplin, P.; et al. A Recombinant MVA-Based RSV Vaccine Induces T-Cell and Antibody Responses That Cooperate in the Protection Against RSV Infection. *Front. Immunol.* **2022**, *13*, 841471. [CrossRef] [PubMed]
84. Shirley, J.L.; de Jong, Y.P.; Terhorst, C.; Herzog, R.W. Immune Responses to Viral Gene Therapy Vectors. *Mol. Ther.* **2020**, *28*, 709–722. [CrossRef] [PubMed]
85. Ahi, Y.S.; Bangari, D.S.; Mittal, S.K. Adenoviral vector immunity: Its implications and circumvention strategies. *Curr. Gene Ther.* **2011**, *11*, 307–320. [CrossRef]
86. Monto, A.S.; Ohmit, S.E.; Petrie, J.G.; Johnson, E.; Truscon, R.; Teich, E.; Rotthoff, J.; Boulton, M.; Victor, J.C. Comparative efficacy of inactivated and live attenuated influenza vaccines. *N. Engl. J. Med.* **2009**, *361*, 1260–1267. [CrossRef]

87. Chang, J. Adenovirus Vectors: Excellent Tools for Vaccine Development. *Immune Netw.* **2021**, *21*, e6. [CrossRef]
88. Klok, F.A.; Pai, M.; Huisman, M.V.; Makris, M. Vaccine-induced immune thrombotic thrombocytopenia. *Lancet Haematol.* **2022**, *9*, e73–e80. [CrossRef]
89. Gordon, S.F.; Clothier, H.J.; Morgan, H.; Buttery, J.P.; Phuong, L.K.; Monagle, P.; Chunilal, S.; Wood, E.M.; Tran, H.; Szer, J.; et al. Immune thrombocytopenia following immunisation with Vaxzevria ChadOx1-S (AstraZeneca) vaccine, Victoria, Australia. *Vaccine* **2021**, *39*, 7052–7057. [CrossRef]
90. Zhang, M.-X.; Zhang, T.-T.; Shi, G.-F.; Cheng, F.-M.; Zheng, Y.-M.; Tung, T.-H.; Chen, H.-X. Safety of an inactivated SARS-CoV-2 vaccine among healthcare workers in China. Expert Rev. *Vaccines* **2021**, *20*, 891–898. [CrossRef]
91. Lee, S.; Nguyen, M.T. Recent advances of vaccine adjuvants for infectious diseases. *Immune Netw.* **2015**, *15*, 51–57. [CrossRef]
92. Facciolà, A.; Visalli, G.; Laganà, A.; Di Pietro, A. An Overview of Vaccine Adjuvants: Current Evidence and Future Perspectives. *Vaccines* **2022**, 10. [CrossRef] [PubMed]
93. Stephens, L.M.; Ross, K.A.; Waldstein, K.A.; Legge, K.L.; McLellan, J.S.; Narasimhan, B.; Varga, S.M. Prefusion F-Based Polyanhydride Nanovaccine Induces Both Humoral and Cell-Mediated Immunity Resulting in Long-Lasting Protection against Respiratory Syncytial Virus. *J. Immunol.* **2021**, *206*, 2122–2134. [CrossRef] [PubMed]
94. Openshaw, P.J.; Chiu, C. Protective and dysregulated T cell immunity in RSV infection. *Curr. Opin. Virol.* **2013**, *3*, 468–474. [CrossRef] [PubMed]
95. Mutsch, M.; Zhou, W.; Rhodes, P.; Bopp, M.; Chen, R.T.; Linder, T.; Spyr, C.; Steffen, R. Use of the inactivated intranasal influenza vaccine and the risk of Bell's palsy in Switzerland. *N. Engl. J. Med.* **2004**, *350*, 896–903. [CrossRef] [PubMed]
96. Huhn, C. Die inhalative Substitution von Phosphatidylcholin. Neue Wege bei Surfactantstörungen des respiratorischen Systems. *Atemwegs Lungenkrankh* **2012**, *38*, 1–9. [CrossRef]

viruses

MDPI

Article

The Insertion of an Evolutionary Lost Four-Amino-Acid Cytoplasmic Tail Peptide into a Syncytin-1 Vaccine Increases T- and B-Cell Responses in Mice

Isabella Skandorff [1,2], Jasmin Gille [3], Emeline Ragonnaud [2,4], Anne-Marie Andersson [2], Silke Schrödel [5], Christian Thirion [5], Ralf Wagner [3] and Peter Johannes Holst [2,4,*]

1 Department of Immunology and Microbiology, University of Copenhagen, Blegdamsvej 3B, 2200 Copenhagen, Denmark; isa@inprother.com
2 InProTher, COBIS, Ole Maaloesvej 3, 2200 Copenhagen, Denmark; era@inprother.com (E.R.); aca@inprother.com (A.-M.A.)
3 Institute of Medical Microbiology and Hygiene, Molecular Microbiology, University of Regensburg Germany, 93053 Regensburg, Germany; jasmin.gille@klinik.uni-regensburg.de (J.G.); ralf.wagner@klinik.uni-regensburg.de (R.W.)
4 Department of Biomedical Sciences, University of Copenhagen, Blegdamsvej 3B, 2200 Copenhagen, Denmark
5 Sirion Biotech GmbH, Am Haag 6, 82166 Graefelfing, Germany; schroedel@sirion-biotech.de (S.S.); thirion@sirion-biotech.de (C.T.)
* Correspondence: pjh@inprother.com

Abstract: Human endogenous retrovirus type W (HERV-W) is expressed in various cancers. We previously developed an adenovirus-vectored cancer vaccine targeting HERV-W by encoding an assembled HERV-W group-specific antigen sequence and the HERV-W envelope sequence Syncytin-1. Syncytin-1 is constitutively fusogenic and forms large multinucleated cell fusions when overexpressed. Consequently, immunising humans with a vaccine encoding Syncytin-1 can lead to the formation of extensive syncytia, which is undesirable and poses a potential safety issue. Here, we show experiments in cell lines that restoring an evolutionary lost cleavage site of the fusion inhibitory R-peptide of Syncytin-1 inhibit cell fusion. Interestingly, this modification of the HERV-W vaccine's fusogenicity increased the expression of the vaccine antigens in vitro. It also enhanced Syncytin-1-specific antibody responses and CD8$^+$-mediated T-cell responses compared to the wildtype vaccine in vaccinated mice, with a notable enhancement in responses to subdominant T-cell epitopes but equal responses to dominant epitopes and similar rates of survival following a tumour challenge. The impairment of cell–cell fusion and the enhanced immunogenicity profile of this HERV-W vaccine strengthens the prospects of obtaining a meaningful immune response against HERV-W in patients with HERV-W-overexpressing cancers.

Keywords: adenoviral vector; cell fusion; human endogenous retrovirus type W (HERV-W); R-peptide; Syncytin-1

Citation: Skandorff, I.; Gille, J.; Ragonnaud, E.; Andersson, A.-M.; Schrödel, S.; Thirion, C.; Wagner, R.; Holst, P.J. The Insertion of an Evolutionary Lost Four-Amino-Acid Cytoplasmic Tail Peptide into a Syncytin-1 Vaccine Increases T- and B-Cell Responses in Mice. *Viruses* **2023**, *15*, 1686. https://doi.org/10.3390/v15081686

Academic Editors: Ronald N. Harty, Pietro Hiram Guzzi, Marianna Milano and Jayanta Kumar Das

Received: 6 June 2023
Revised: 30 July 2023
Accepted: 31 July 2023
Published: 3 August 2023

1. Introduction

Human endogenous retrovirus (HERV) genes originate from now-extinct exogenous retroviruses that integrated into the genome of our ancestors. HERV type W (HERV-W) proviruses were acquired approximately 30–40 million years ago; subsequently, HERV-W genes spread throughout the genome via reinfections and retro-transpositions [1]. Since then, accumulations of mutations and truncations have compromised the coding capacity of most HERV genes, including HERV-W [1,2]. However, some HERV genes are still coding-competent and express functional proteins. For example, the *ERVWE1* provirus encodes the envelope (Env) protein Syncytin-1, which has been co-opted to play crucial roles in placentation [3–5]. Syncytin-1 facilitates the fusion of cytotrophoblasts, leading to the formation of the placental syncytiotrophoblast cell layer [3,4,6]; furthermore, Syncytin-1 is believed to be involved in maintaining foeto-maternal tolerance [7,8]. The expression of Syncytin-1 is strictly regulated by epigenetic mechanisms [9,10], and the aberrant expression of Syncytin-1 is associated with various pathologies, among others, cancer [11,12]. There is a growing number of reports on the expression of Syncytin-1 and gene products of other HERV-W loci in human cancer cells and tissues at either mRNA or protein levels (reviewed in [11,13–16]). The implication of HERV-W/Syncytin-1 expression in cancer is not fully understood, but the expression of HERV-W in cancer makes these antigens potential targets for cancer immunotherapy.

In a recent study, we aimed to make a HERV-W-targeting cancer vaccine based on the virus-like-vaccine concept [17]. In a replication-deficient adenovirus vector, an assembled HERV-W group-specific antigen (Gag) sequence and the HERV-W Env sequence of Syncytin-1 were encoded [18]. We evaluated the vaccine-induced immune responses and anti-cancer efficacy in inbred mice, and we found that the HERV-W vaccine elicited both T-cell responses to different domains of the Env protein and the antibody recognition of the native form of Syncytin-1, which was expressed on the surfaces of the mouse cancer cells [18].

The HERV-W Env Syncytin-1 is, however, an atypical vaccine antigen as it is constitutively fusion-competent. When expressed on a cell surface, Syncytin-1 can induce cell–cell fusion via interaction with either of the identified receptors, human ASCT1 and ASCT2, and when Syncytin-1 is overexpressed, these interactions can form large, multinucleated cells. The immunisation of humans with a vaccine encoding the constitutively fusogenic Syncytin-1 would not only result in an immune attack of the HERV-W vaccine-infected cells but also adjacently fused non-transduced cells, worsening the local vaccine-induced tissue damage. To avoid this effect, we were interested in developing an immunogenic but non-fusogenic HERV-W vaccine.

In the functional characterisation of Syncytin-1, mutagenesis studies found that Syncytin-1 fusion can be abolished by different mutations in both the extracellular and intracellular parts of Syncytin-1 while still enabling its surface expression [19–21]. One study showed that the substitution of cysteine with alanine at the third cysteine in a CX_6CC disulfide motif, which is located in the extracellular part of the transmembrane subunit (TU) of Syncytin-1, impaired disulfide binding to a CXXC motif in its surface subunit (SU). This mutation inhibited cell–cell fusion but maintained Syncytin-1 surface expression [20]. Another study compared the cytoplasmic tail (CT) of Syncytin-1 with paralogous and orthologous HERV-W Env sequences from humans and different apes. It revealed that Syncytin-1 is constitutively fusogenic because it has lost the four amino acids that originally constituted the cleavage site for the fusion-inhibitory R-peptide in the CT [21]. When re-introducing a four-amino-acid-long consensus sequence, LQMV, Syncytin-1 was still surface-expressed, but the cell fusion activity was inhibited. This is because the human genome no longer holds any functional proteases from the HERV-W Pro-Pol loci. Thus, the LQMV-containing Syncytin-1 mutant is locked in a non-fusogenic conformation [21,22].

In the present study, we aimed to construct a non-fusogenic HERV-W vaccine without compromising the immunogenicity of the vaccine. We tested CX_6CC motif-mutated and LQMV-reconstituted Syncytin-1 Env proteins as possible non-fusogenic HERV-W Env candidates in our vaccine constructs, and we explored the fusion capacity and surface expression of these constructs in human cell lines. Furthermore, we encoded the LQMV mutant in an adenovirus-vectored vaccine and examined its immunogenicity compared to the wt HERV-W vaccine in mice [18]. We show that the non-fusogenic LQMV mutant vaccine increased the cell surface expression of HERV-W Env in both human and murine cells. Additionally, this vaccine increased higher $CD8^+$ T-cell responses towards subdominant antigens and increased antibody responses towards cancer cells expressing the native Syncytin-1 protein compared to the corresponding fusion-competent vaccine. The fusogenic and non-fusogenic HERV-W vaccines increased the survival of tumour-challenged mice to the same extent, possibly reflecting similar dominant epitope-specific responses. Thus, these results imply that we can avoid vaccine-induced cell–cell fusions while obtaining quantitatively higher antigen-specific immune responses by inserting LQMV into the CT of Syncytin-1.

2. Materials and Methods

2.1. Cell Lines

The human T24 urinary bladder carcinoma cell line (HTB-4; ATCC, Manassas, VA, USA) and the human HEK293 epithelial kidney cell line (CRL-1573; ATCC, Manassas, VA, USA) were cultured in DMEM GlutaMAX. The murine RenCa renal cortical adenocarcinoma cell line from a male BALB/c mouse (CRL-2947, ATCC, Manassas, VA, USA) and the human A549 lung carcinoma cell line (CCL-185, ATCC, Manassas, VA, USA) were cultured in RPMI 1640 GlutaMAX and Ham's F12 Nutrient Mix GlutaMAX media (31765035, Thermo Scientific™, Waltham, MA, USA), respectively. All media were supplemented with 10% FBS, 100 units/mL of penicillin–streptomycin (pen/strep) (15140122; Thermo Scientific™, Waltham, MA, USA), and 1 mM sodium pyruvate (11360070; Thermo Scientific™, Waltham, MA, USA). All cells were maintained at 37 °C and 5% CO_2.

2.2. Antigen and Viral Vector Design

The sequences of HERV-W Gag and the wt HERV-W Env, Syncytin-1, were described in our recent study [18]. The HERV-W Gag sequence was obtained from an assembled sequence derived from the viral particles of the multiple sclerosis-associated retrovirus of the HERV-W family in a study by Komurian-Pradel et al. [23]. Based on the virus-like-vaccine principle, we included the assembled HERV-W Gag sequence in the vaccine construct to obtain in situ-formed HERV-W Gag particles presenting the HERV-W Env Syncytin-1 in a highly immunogenic fashion [17]. In the present study, three HERV-W vaccine plasmids were constructed: one encoding the wt Syncytin-1 and Gag (HERV-W_{WT}), one encoding Syncytin-1 with four amino acids inserted into the CT region and HERV-W Gag (HERV-W_{LQMV}), and one encoding Syncytin-1 with a cysteine-to-alanine exchange in the ectodomain and HERV-W Gag (HERV-$W_{C>A}$).

The Syncytin-1 sequence, incl. the above-mentioned mutated variants in bold and/or underlined, is as follows: MALPYHIFLFTVLLPSFTLTAPPPCRCMTSSSPYQEFLWRMQRPG-NIDAPSYRSLSKGTPTFTAHTHMPRNCYHSATLCMHANTHYWTGKMINPSCPGGLGVT-VCWTYFTQTGMSDGGGVQDQAREKHVKEVISQLTRVHGTSSPYKGLDLSKLHETLRTH-TRLVSLFNTTLTGLHEVSAQNPTNCWICLPLNFRPYVSIPVPEQWNNFSTEINTTSVLVGP-LVSNLEITHTSNLTCVKFSNTTYTTNSQCIRWVTPPTQIVCLPSGIFFVCGTSAYRCLNGSS-ESMCFLSFLVPPMTIYTEQDLYSYVISKPRNKRVPILPFVIGAGVLGALGTGIGGITTSTQFY-YKLSQELNGDMERVADSLVTLQDQLNSLAAVVLQNRRALDLLTAERGGTCLFLGEEC**C**Y-YVNQSGIVTEKVKEIRDRIQRRAEELRNTGPWGLLSQWMPWILPFLGPLAAIILLLLFGPC-IFNLLVNFVSSRIEAVKLQMVLQMEPKMQSKTKIYRRPLDRPASPRSDVNDIKGTPPEEIS-AAQPLLRPNSAGSS*. In all three plasmids, the sequences coding for the Env and Gag proteins were separated by a self-cleavable P2 A peptide sequence, and a CMV promoter controlled the antigen expression. A control plasmid for transfection encoded a copGFP sequence. GenScript Biotech (Piscataway, NJ, USA) synthesised all plasmids.

The HERV-W_{WT} and HERV-W_{LQMV} constructs were further cloned into a shuttle vector in *E. coli* and hereafter into a BAC vector containing the backbone of a replication-deficient human adenoviral vector type 19a/64 (hAd19a/64) (lacking E1 and E3 genes) [24]. A negative control vaccine (Neg. ctrl vaccine) contained the same vector but did not encode any antigens.

2.3. Adenoviral Vector Production

The hAd19a/64 vaccines were produced by Sirion Biotech, following the procedure described in [25]. In brief, after the cloning of the HERV-W antigens into the hAd19a/64 backbone in BAC cells, the DNA was purified, linearized, and then transfected into a modified HEK293 production cell line. In the HEK293 cells, the viral constructs were amplified to a large-scale lysate, and from here, the viruses were purified. The purified viruses were tittered in parallel via the immunohistochemical staining of the adenoviral hexon protein. Virus-derived DNA was isolated and sequenced for quality control.

2.4. Surface Expression of Transfected HEK293 and T24 Cells and Transduced A549 Cells

HEK293 and T24 cells were transfected with either HERV-W_{WT}, HERV-W_{LQMV}, or HERV-$W_{C>A}$ plasmids using PEI and Opti-MEM (11058021; Thermo Scientific™, Waltham, MA, USA) in complete DMEM media without pen/strep. Transfection with PEI and Opti-MEM was carried out in the ratios of DNA to PEI, 1:3, and DNA to Opti-MEM, 1:100, meaning 3 µg of DNA to 9 µL (1 mg/mL) of PEI to 300 µL of Opti-MEM. Each condition was performed in duplicates. The cells were incubated for 24 h prior to cell surface staining.

Human A549 cells were transduced with the hAd19a/64 HERV-W_{WT}, HERV-W_{LQMV}, or an empty vaccine (Neg. ctrl vaccine) at a multiplicity of infection (MOI) of 10. Each condition was performed in triplicate, and the cells were stained after 24 h of incubation.

The following antibody staining of the transfected and transduced cells was carried out in a FACS buffer consisting of PBS with 1% BSA and 0.1% NaN_3. For the cell surface staining of the HERV-W Env surface subunit, the cells were incubated with 15 µg/mL of primary rabbit anti-human HERV polyclonal antibody (PA5-22819; Invitrogen™, Waltham, MA, USA) for 1 h at 4 °C. Following this, the cells were stained with secondary PE donkey anti-rabbit IgG antibody (406421; BioLegend®, San Diego, CA, USA; 1:100) and eBioscience™ Fixable Viability Dye eFlour™ 780 (65-0865; Invitrogen™, Waltham, MA, USA; 1:1000) for 30 min at 4 °C. Next, the cells were fixated in 1% paraformaldehyde (PFA) for 15 min at 4 °C, and flow cytometry was performed using either the LSRFortessa™ 3-laser or 5-laser cell analyser (BD Biosciences, Franklin Lakes, NJ, USA). The flow cytometry data were analysed with FlowJo™ v10 analysis software and GraphPad Prism 9.

2.5. Visualisation of Cells Using Transmission Electron Microscopy (TEM) and Light Microscopy

T24 and A549 cells were seeded on Thermanox coverslips (150067; Thermo Scientific™, Waltham, MA, USA) in 24-well plates and hereafter transfected or transduced as described above except in the case of the A549 cells, which were transduced with 50MOI. Prior to cell seeding, the coverslips for the A549 cells were pre-coated with poly-L-lysine. Following 24 h of incubation, the cells were fixed with 2% glutaraldehyde in a 0.05 M sodium phosphate buffer (pH 7.2). As a control, T24 cells were transfected with a plasmid encoding a copGFP sequence and incubated for 48 h. TEM was performed by the Core Facility for Integrated Microscopy at the University of Copenhagen (see detailed description in [18,25]).

Using a ZOE Cell Imager (BioRad, Hercules, CA, USA), light microscopy pictures of unstained and non-fixed T24 cells 24 h after transduction with 25MOI of the hAd19a/64 HERV-W_{WT}, HERV-W_{LQMV}, or Neg. ctrl vaccine were obtained.

2.6. Evaluation of Gag Expression via Western Blotting

HERV-W Gag expression was evaluated via Western blotting, as described previously in [18]. In brief, A549 cells were lysed 24 h after transduction with 50MOI of either the HERV-W_{WT} vaccine or the HERV-W_{LQMV} vaccine. Denatured samples were run on a NuPAGE™ Bis-Tris Mini Gel (NP0321; Invitrogen™, Waltham, MA, USA) under reducing conditions and transferred to a nitrocellulose membrane (IB230002; Invitrogen™, Waltham, MA, USA). The membrane was incubated overnight at 4 °C with the primary antibodies: the anti-T2A-antibody (Crb200569d; CRB discoveries, Cleveland, UK; 1:2000) to detect the P2A peptide on HERV-W Gag and the housekeeping control protein anti-GAPDH antibody (ab181602; Abcam, Cambridge, UK; 1:8000). The bound primary antibody was detected after 1 h of incubation with the secondary polyclonal goat anti-rabbit IgG antibody (P0448; Dako, Glostrup, Denmark), using LumiGlo Chemiluminescent (5430; KPL, LGC group, Teddington, UK) or SuperSignal West Femto Maximum Sensitivity Substrate (34095; Thermo Scientific™, Waltham, MA, USA). The relative expression differences in the protein bands were analysed with iBright analysis software, V5.1.0.

2.7. Maturation, Transduction, and Staining of Murine Bone Marrow-Derived Dendritic Cells (BMDCs) and Measurement of Pro-Inflammatory Biomarkers

BMDCs derived from the femurs of two BALB/c mice were isolated and matured based on a study published by Jin et al. [26]. Detailed descriptions of the maturation conditions, transduction, and staining are provided in [18]. Twenty-four hours after the transduction of the murine BMDCs, the supernatant was collected. The concentrations of four different proinflammatory biomarkers were determined using a customised V-PLEX mouse proinflammatory cytokine panel 1 kit (K15048D; Mesoscale, Rockville, MD, USA). The supernatants were diluted 1:5, and the samples were assessed in duplicates. The biomarker concentrations were analysed with a MESO QuickPlex SQ 120 MM instrument (Mesoscale, Rockville, MD, USA).

2.8. Animal Procedures and Serum Isolation

Female BALB/c mice from Envigo were obtained at 6–8 weeks of age and housed at the Panum Institute, University of Copenhagen. The mice were acclimatised for at least one week prior to any experiments, and all the experiments were performed in accordance with the national guidelines. The experimental procedures were approved by the National Animal Experiments Inspectorate (Dyreforsøgstilsynet, license no. 2019-15-0201-00203).

Prior to subcutaneous (s.c.) vaccinations, the mice were anaesthetised with isoflurane. The mice were vaccinated s.c. in the lower right limb with 30 µL of the relevant vaccine, which contained 2×10^7 infectious units (IFU) diluted in $1 \times$ PBS. At the end of each experiment, the mice were euthanised via cervical dislocation.

Blood samples were collected prior to and on the last day of the vaccination studies. Serum was isolated from the blood samples via two centrifugations at $800\times g$ for 8 min at 8 °C.

Intravenous (i.v.) tumour inoculation was performed as described previously in Skandorff et al. [18]. Each mouse was challenged with 0.5×10^6 RenCa cells modified to express the HERV-W Env Syncytin-1 (see previous study [18]). These cells were diluted in $1 \times$ PBS, and 100 µL of cell suspension was i.v. injected into the tail vein. The mice were randomised, and four days after the tumour challenge, the mice were divided into three groups; each group was vaccinated with either 2×10^7 IFU of the HERV-W_{WT} vaccine ($n = 15$), HERV-W_{LQMV} vaccine ($n = 15$), or the Neg. ctrl vaccine ($n = 10$). The mice were continuously monitored, and when humane endpoints were reached (starey coat, bent over position, or reduced mobility) the mice were euthanised via cervical dislocation. The lungs were collected and evaluated for the presence of tumour nodules. One mouse in the HERV-W_{WT} vaccinated group and two in the HERV-W_{LQMV} group showed no signs of sickness throughout the experiment. When evaluating the presence of tumour nodules in the lungs of these remaining mice, the HERV-W_{WT}-vaccinated mouse showed one minor nodule. In contrast, no nodules were visible in the HERV-W_{LQMV}-vaccinated mice. The probability of survival was calculated between groups two-and-two using a log-rank Mantel-Cox test in GraphPad Prism 9 with a statistical significance level defined as * = $p < 0.05$.

2.9. Evaluation of HERV-W Env-Specific Antibody Responses

Antibodies, from serum isolated from BALB/c mice prior to and at the end of the vaccination, were evaluated via flow cytometry for binding to RenCa cells modified to stably express HERV-W Env. Pre- and end-bleed sera were diluted 1:20 and added to the HERV-W Env^+ Renca cells for 1 h at 4 °C. Following this, the cells were stained with PE goat anti-mouse IgG (405307; BioLegend®, San Diego, CA, USA; 1:100) and eBioscience™ Fixable Viability Dye eFlour™ 780 (65-0865; Invitrogen™, Waltham, MA, USA; 1:1000) for 30 min at 4 °C. Finally, the cells were fixated in 1% PFA, and flow cytometry was performed on an LSRFortessa™ 3-laser flow cytometer (BD Biosciences, Franklin Lakes, NJ, USA). FlowJo™ v10 and GraphPad Prism 9 were used to analyse the antibody responses. Statistical significance levels were calculated using the Mann–Whitney t-test (* = $p < 0.05$ and ** = $p < 0.01$).

2.10. Evaluation of HERV-W Gag and Env T-Cell Responses

T-cell responses were evaluated using splenocytes isolated from the spleens of the vaccinated mice, as described previously [18]. The splenocytes were stimulated for 5 h with pools of 16-mer peptides, which overlapped by 11 amino acid and that together spanned either the HERV-W Env Syncytin-1 cell surface subunit, the ectodomain of Syncytin-1's transmembrane subunit, the Syncytin-1 transmembrane domain and the cytoplasmic tail of the transmembrane subunit, or the assembled HERV-W Gag (see Figure 3B). The splenocytes were also stimulated with two previously identified 9-mer peptides: peptide 28 (p28, FGPCIFNLL), which originated from the overlapping sequence between the transmembrane domain and the cytoplasmic tail of the transmembrane subunit of Syncytin-1, and peptide 34 (p34, CYYVNQSGI), which originated from the ectodomain of the transmembrane subunit of Syncytin-1 [18].

Following peptide stimulation, the splenocytes were stained with the following cell surface fluorophore-conjugated antibodies: BV421 rat anti-mouse CD8b antibody (126629; BioLegend®, San Diego, CA, USA), PE-Cy7 rat anti-mouse CD4 (561099; BD Biosciences, Franklin Lakes, NJ, USA), PerCP-Cy5.5 rat anti-mouse CD45R/B220 (552771; BD Biosciences, Franklin Lakes, NJ, USA), and FITC rat anti-mouse CD44 (553133; BD Biosciences, Franklin Lakes, NJ, USA). The splenocytes were fixated in 1% PFA and permeabilised with saponin before they were stained with two intracellular fluorophore-conjugated antibodies: APC rat anti-mouse IFNγ (554413; BD Biosciences, Franklin Lakes, NJ, USA) and PE rat anti-mouse TNFα (554419; BD Biosciences, Franklin Lakes, NJ, USA). Finally, flow cytometry was carried out on an LSRFortessa-3 flow cytometer (BD Biosciences, Franklin Lakes, NJ, USA), and the data were analysed using FlowJo™ v10 and GraphPad Prism 9.

The percentage of the responses measured in the unstimulated samples for each mouse (background) was subtracted from the percentage of responses measured in the stimulated samples of the corresponding mouse. As a lower limit of positive geometric mean fluorescent intensity (MFI) IFNγ T-cell responses, the minimum response was defined as a response above the mean plus two times the standard deviation of the MFI of the splenocytes that were not stimulated with peptides (Neg. ctrl) (as seen in Figure 3G–J). However, when calculating response differences between the two vaccine groups, all responses were included. Response differences were calculated using the non-parametric Mann–Whitney test, with statistical significance levels defined as * = $p < 0.05$ and ** = $p < 0.01$. Values below zero were excluded from the graphs.

3. Results

3.1. Characterisation of Non-Fusogenic HERV-W Envs

To generate a non-fusogenic Syncytin-1 antigen, we first considered the evolutionary adaptation that rendered Syncytin-1 constitutively fusogenic. The C-terminal R-peptide of gamma-retroviruses inhibits Env fusion activity, but in the domestication of the HERV-W Env Syncytin-1, this protein acquired a four-amino-acid deletion mutation in the R-peptide cleavage site, making Syncytin-1 constitutively fusogenic. Based on human HERV-W Env paralogous sequences, Bonnaud et al. deduced a four-amino-acid sequence, LQMV [21]. Thus, in our first mutant HERV-W vaccine, we inserted the "lost" fusion-inhibitory R-peptide cleavage site into the Env CT (amino acids LQMV) and encoded this Env sequence in a plasmid together with an assembled sequence of HERV-W Gag (HERV-W_{LQMV}) (Figure 1A). Other functional and structural studies of Syncytin-1 have shown that certain insertion and deletion mutations can maintain Env surface expression while abolishing fusion activity [19,20]. For our second HERV-W mutant vaccine, we exchanged the last cysteine in the CX_6CC motif of the Env TU ectodomain with alanine ($CX_6CC \rightarrow CX_6CA$), which was previously shown to abrogate the disulfide bond between the TU and SU [20]. Again, we encoded this Env into a plasmid together with HERV-W Gag (HERV-$W_{C>A}$) (Figure 1A). The expression of the antigens was controlled using a CMV promoter, and the Gag and Env sequences were separated using a self-cleavable P2A peptide sequence.

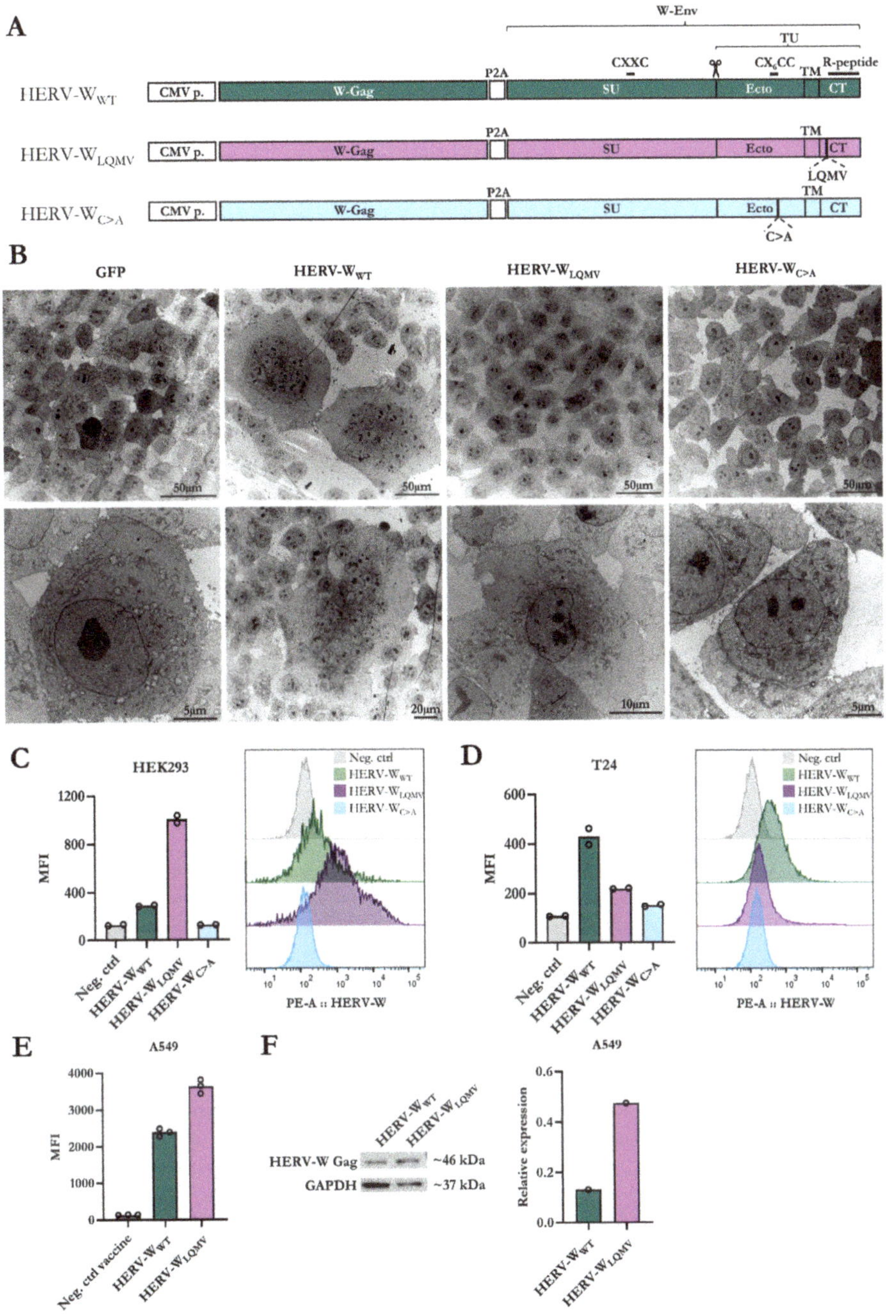

Figure 1. Characterisation of non-fusogenic mutants of HERV-W Env. (**A**) Schematic overview of plasmids encoding an assembled HERV-W Gag (W-Gag) sequence and three different versions of the HERV-W envelope (W-Env), Syncytin-1, sequences: wildtype Env (green, HERV-W_{WT}), Env with LQMV insertion (purple, HERV-W_{LQMV}), and Env with a C>A amino acid substitution (blue, HERV-$W_{C>A}$). The Gag and Env antigens are separated by a self-cleavable P2A peptide sequence, and the expression of the antigens is controlled by a CMV promoter (CMV p.). The illustration also

depicts the location of the R-peptide, the CX_6CC and CXXC disulfide motifs, and the different domains of the Env: the surface subunit (SU) and the transmembrane subunit (TU), where the TU contains the ectodomain (Ecto), the transmembrane domain (TM), and the cytoplasmic tail (CT). The scissor illustrates the cleavage site between SU and TU. (**B**) Transmission electron microscopy (TEM) pictures of human T24 cells transfected with plasmids encoding GFP, HERV-W_{WT}, HERV-W_{LQMV}, or HERV-$W_{C>A}$. The top row shows an overview of several cells, and the bottom row shows individual cells at a higher magnification. (**C**) (**right**) Graph showing the geometric mean fluorescent intensity (MFI) of the HERV-W Env surface expression of two technical repeats of transfected HEK293 cells, as indicated by bullets, and (**left**) histograms showing the distribution of one representative sample of each of the HERV-W plasmids and a negative control without transfection (Neg. ctrl). (**D**) The same as (**C**) but depicting results for T24 cells. (**E**) The MFI of the HERV-W Env surface expression of A549 cells 24 h after transduction with the hAd19a/64 vaccines encoding either HERV-W_{WT}, HERV-W_{LQMV}, or the empty expression cassette (Neg. ctrl vaccine). Bullets illustrate the three technical replicates per condition. (**F**) The Western blot shows HERV-W Gag expression in A549 cells 24 h after transduction with either the HERV-W_{WT} or HERV-W_{LQMV} vaccine. The right graph shows the difference in the expression of the HERV-W Gag in the Western blot between the two vaccines, relative to the expression of the housekeeping protein GAPDH.

We assessed the cell morphology of the transfected human T24 cells via transmission electron microscopy (TEM). T24 cells are highly fusion-permissive when expressing Syncytin-1, which results in large, multinuclear cells, as seen when transfected with plasmids containing the wt Syncytin-1 together with the assembled HERV-W Gag sequence (HERV-W_{WT}) (Figure 1B, Supplementary Figure S1A). In contrast, the T24 cells transfected with plasmids of HERV-W_{LQMV} and HERV-$W_{C>A}$ were mono-nuclear, like the GFP control (Figure 1B).

We further assessed the surface expression of the Env SU region of the easily transfected human HEK293 (Figure 1C) and T24 (Figure 1D) cells via flow cytometry. Consistent with the original study by Cheynet et al., the surface expression of the Env of the HERV-$W_{C>A}$ mutant was lower than for HERV-W_{WT} in the T24 cells and almost absent in the HEK293 cells [20] (Figure 1C,D). Oppositely, the geometric mean fluorescent intensity (MFI) of the surface expression of Env was higher for the HERV-W_{LQMV} mutant than the HERV-W_{WT} in transfected HEK293 cells (Figure 1C) but lower than for HERV-W_{WT} in transfected T24 cells (Figure 1D). The removal of the disulfide bond connecting the TU and SU of the Env of the HERV-$W_{C>A}$ mutant possibly affects the structure of the surface-expressed Env via the introduction of novel antigen structures that do not mimic the native Syncytin-1 protein. Considering this, and that the HERV-$W_{C>A}$ mutant's level of the cell surface expression of Env was low, we decided to continue with the HERV-W_{LQMV} mutant for the generation of a virus-vectored HERV-W vaccine.

To generate a vaccine candidate for in vivo testing, we produced replication-deficient human adenoviral vector type 19 a/64 (hAd19a/64 vector) vaccines encoding the antigens from the HERV-W_{WT} and HERV-W_{LQMV} plasmids (as seen in Figure 1A). Following the transduction of human A549 cells, we evaluated the surface expression levels of the Env protein via flow cytometry (Figure 1E; gating strategy in Supplementary Figure S1B) and the intracellular expression of the Gag protein Western blotting (Figure 1F), respectively. In line with the transfection experiments with HEK293 cells, the surface expression level of Env was higher for the A549 cells transduced with the HERV-W_{LQMV} vaccine than the HERV-W_{WT} vaccine (Figure 1E, Supplementary Figure S1C). Similarly, the Gag protein concentration was higher in the HERV-W_{LQMV} transduced cells than in the HERV-W_{WT} transduced cells (Figure 1F). By examining the transduced A549 cells using light microscopy, we could again confirm the presence of multinucleated cells among the HERV-W_{WT} transduced cells, but none of these were present in the cells transduced with HERV-W_{LQMV} or an empty vectored vaccine (Neg. ctrl vaccine) (Supplementary Figure S1D).

In a recent study of the HERV-W_{WT} vaccine, we further explored whether encoding HERV-W Gag and Env could give rise to Env-covered extracellular virus-like particles (VLPs) [18]. However, in accordance with our previous study, we could not observe any extracellular VLPs via TEM with either the HERV-W_{WT} or HERV-W_{LQMV} vaccine (Supplementary Figure S1E).

3.2. *Effects of Syncytin-1 in Transduced Bone Marrow-Derived Dendritic Cells*

Previously, we observed the association of vaccine immunogenicity with differences in the expression of activation markers on the surfaces of matured murine bone marrow-derived dendritic cells (BMDCs) [18,25]. In this study, we measured the surface expression of Syncytin-1 (from here on HERV-W Env) 24 h after transduction with either HERV-W_{WT} or HERV-W_{LQMV} and compared it to a lack of transduction (Neg. ctrl) or an empty vectored hAd19a/64 vaccine (Neg. ctrl vaccine) (Supplementary Figure S2A). Regarding its expression in human HEK293 with plasmids and the use of viruses in A549 cells, the Env cell surface expression on murine BMDCs was higher for HERV-W_{LQMV} than HERV-W_{WT} (Figure 2A), while the cell viability was similar (Figure 2B). No cell–cell fusions were observed with any of the vaccines; therefore, we did not anticipate any differences in the expression of the activation markers. Indeed, the percentages and MFI of HERV-W Env^+ cells expressing MHC class II (MHC-II) and CD86 on their surfaces were similar between the two vaccines (Figure 2B, Supplementary Figure S2B). However, the MFI of CD40 on HERV-W Env^+ cells appeared slightly lower for HERV-W_{LQMV} than HERV-W_{WT} (Figure 2C). We also measured the concentrations of four different secreted proinflammatory biomarkers in the supernatant 24 h after transduction but found no differences except for a modest increase in the concentration of TNFα with the HERV-W_{LQMV} vaccine when compared to HERV-W_{WT} and the Neg. ctrl vaccine (Supplementary Figure S2C). Overall, from these experiments, it is not possible to claim any clear differences in innate immunogenicity within the ex vivo-transduced BMDCs apart from the greater levels of cell surface Env expression of the BMDCs transduced with the HERV-W_{LQMV} vaccine.

3.3. *The LQMV Mutant Vaccine Exhibit Increased T-Cell Immunogenicity*

To explore whether the LQMV mutant vaccine was equally immunogenic to the wt vaccine in vivo, we vaccinate BALB/c mice subcutaneously (s.c.) and evaluated their cellular and humoral immune responses after 14 days (the expected peak of T-cell responses) and 28 days (early memory responses) (Figure 3A). Using flow cytometry, we measured the IFNγ and TNFα responses of $CD8^+$ and $CD4^+$ T-cells via intracellular staining after peptide stimulation with two identified HERV-W Env-responding 9-mer peptides (peptide 28 and peptide 34), as well as three pools of 16-mer peptides overlapping by 11 amino acids (for the gating strategy, see Supplementary Figure S3A). The peptide pools were made to span three different sections of the Env protein: SU, the ectodomain of the TU, and the transmembrane (TM) and CT regions of the TU (Figure 3B), and these and the single peptides were identified and evaluated in our recent study [18]. In this present study, we also used a pool of overlapping 16-mer peptides spanning the HERV-W Gag protein sequence (Gag) to measure potential T-cell responses to HERV-W Gag (Figure 3B).

Interestingly, the T-cell responses to the HERV-W_{LQMV} vaccine were generally higher than the responses to the HERV-W_{WT} vaccine. The percentage of IFNγ and TNFα $CD8^+$ T-cell responses to peptide 34 from the TU ectodomain and the SU peptide pool were significantly higher for the HERV-W_{LQMV} vaccine group compared to the HERV-W_{WT} vaccine group on day 14 (Figure 3C). The same tendency was observed on day 28 (Figure 3D) and for IFNγ-producing $CD8^+$ T-cell responses on day 14 (Supplementary Figure S3B), and with a reduced magnitude on day 28 (Supplementary Figure S3C). Consistent with our previous HERV-W vaccine study, the double-positive ($IFN\gamma^+TNF\alpha^+$) and single-positive ($IFN\gamma^+$) $CD8^+$ T-cell responses elicited to peptide 28 were generally the highest of all the peptide conditions [18]. However, in this current study, the two HERV-W vaccine groups responded similarly to this major epitope.

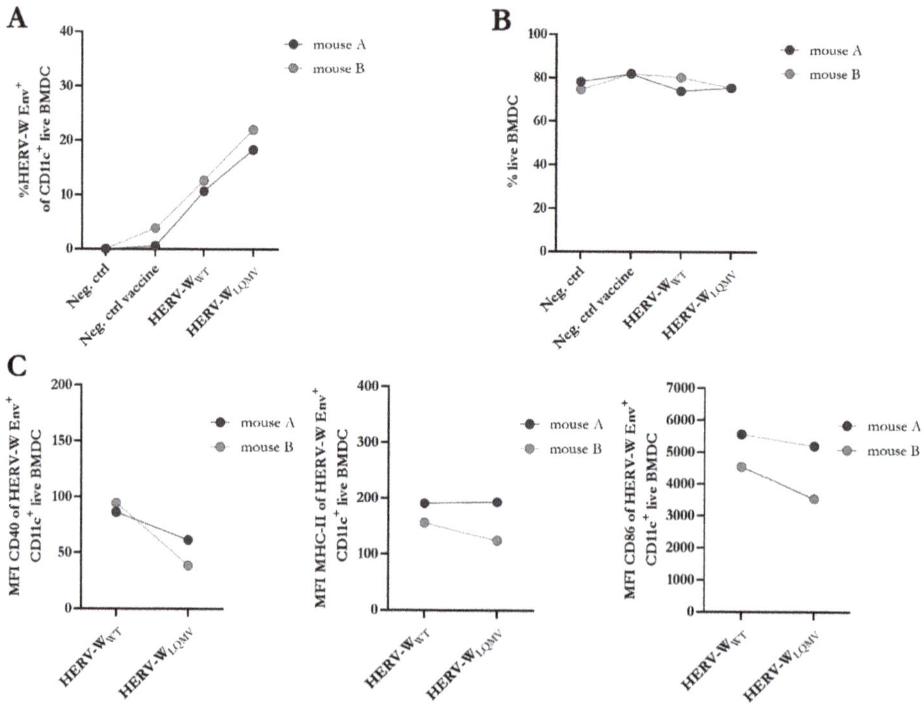

Figure 2. Evaluation of the effect of the non-fusogenic HERV-W$_{LQMV}$ vaccine on activated BMDCs. (**A**) Percentage of surface expression of HERV-W Env of CD11^{+} live bone-marrow-derived dendritic cells (BMDCs) from two BALB/c mice, analysed via flow cytometry. (**B**) Percentage of live BMDCs. (**C**) Geometric mean fluorescent intensity (MFI) of the activation markers CD40 (**left**), MHC-II (**middle**), and CD86 (**right**) on the cell surfaces of the BMDCs expressing CD11 c and HERV-W Env. All bullets depict the mean of two technical repeats from each of the two mice.

Furthermore, and in agreement with the previous study [18], vaccine-specific CD4^{+} T-cell responses were generally quite low in the BALB/c mice. However, we could still observe IFNγ^{+}TNFα^{+}-producing CD4^{+} T-cells responding to the peptide pools of the SU region and the ectodomain on day 14 and day 28; in most cases, these responses were the strongest for the HERV-W$_{LQMV}$ vaccine group, but the difference between the groups did not achieve statistical significance (Figure 3E,F). The percentages of IFNγ^{+} CD4^{+} T-cell responses were almost indistinguishable on day 14 (Supplementary Figure S3D), but on day 28, the responses to the ectodomain were significantly higher for the HERV-W$_{LQMV}$ group than for the HERV-W$_{WT}$ group (Supplementary Figure S3E).

In contrast to all the other results, the MFI of IFNγ^{+} CD8^{+} T-cell responses to peptide 28 of the TM-CT region was higher for the HERV-W$_{WT}$ vaccine group than the HERV-W$_{LQMV}$ vaccine group on day 14 (Figure 3G), but then the opposite occurred on day 28 (Figure 3H). We speculate whether this observation can reflect a still-maturing immune response on day 14 which settled on day 28. At later stages of the response, both vaccine groups are expected to be at the resting stage, and we would imagine that a more immunogenic vaccine would provide both the quantitatively and qualitatively strongest responses, which is what we observed for the HERV-W$_{LQMV}$ vaccine compared to the HERV-W$_{WT}$ vaccine on day 28.

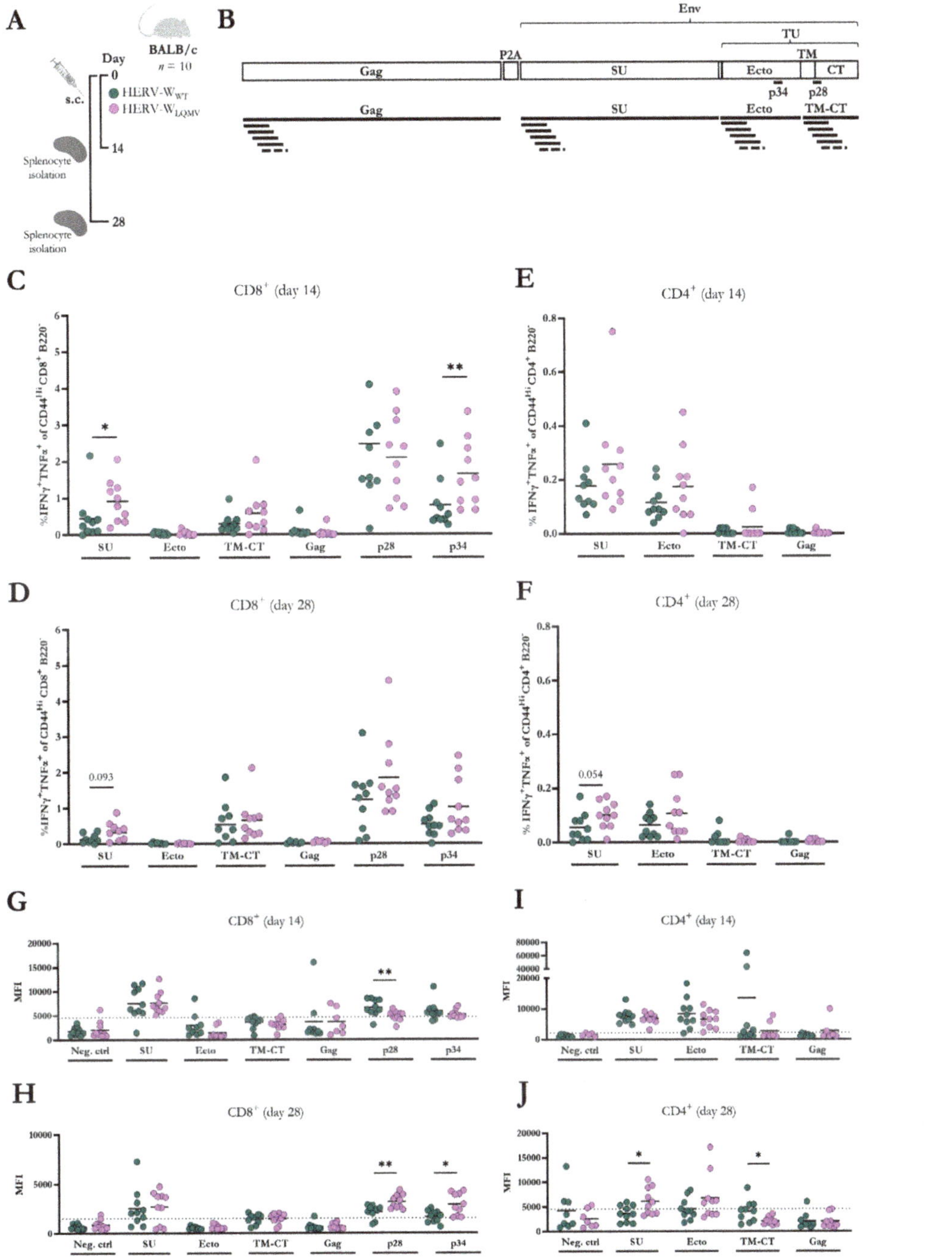

Figure 3. CD8$^+$ and CD4$^+$ T-cell responses in BALB/c mice vaccinated with the two HERV-W vaccines. (**A**) Schematic representation of the immunisation study with the HERV-W$_{WT}$ and HERV-W$_{LQMV}$ vaccines. BALB/c mice were immunised subcutaneously (s.c.) in the lower limb on day zero, and

splenocytes were harvested either 14 days or 28 days later. For each of the two experiments, each vaccine group included ten mice. (**B**) Illustration of the vaccine-encoded antigens, visualising the four different 16-mer peptide pools spanning HERV-W Gag, HERV-W Env SU, or the two HERV-W TU regions: the ectodomain (Ecto) or the TM-CT. The figure also illustrates the origin of the two 9-mer peptides: peptide 28 (p28, TM-CT) and peptide 34 (p34, Ecto). Graphs show the percentages of IFNγ- and TNFα-expressing $CD8^+$ T-cells 14 (**C**) and 28 (**D**) days after immunisation or $CD4^+$ T-cells 14 (**E**) and 28 (**F**) days after immunisation. Geometric mean fluorescent intensity (MFI) of IFNγ-expressing $CD8^+$ T-cells on day 14 (**G**) and day 28 (**H**), as well as $CD4^+$ T-cells on day 14 (**I**) and day 28 (**J**) after vaccination. The gating strategy is illustrated in supplementary Figure S3A. The individual bullet depicts the response from one mouse, and the horizontal line indicates the mean of the group. Statistically significant differences (non-parametric, two-tailed Mann–Whitney test) between T-cell responses are marked by * = $p < 0.05$ and ** = $p < 0.01$. The dotted lines indicate a lower limit for a positive response, and they are based on the mean plus two times the standard deviation of the mean of the responses for T-cells without peptide stimulation (Neg. ctrl).

The MFI of the IFNγ $CD4^+$ T-cell responses on day 14 and day 28 were similar for most epitopes, but at 28 days, we observed a significantly higher MFI for the T-cells responding to the SU peptide pool in the HERV-W_{LQMV}-immunised mice (Figure 3I,J). We oppositely observed a higher MFI for the TM-CT peptide pool in the HERV-W_{WT}-immunised mice, but this was a mean responsiveness calculated from a very low-responding cell population (comparing Figure 3E,J).

In this study, we also aimed to measure T-cell responses to HERV-W Gag by using a pool of 16-mer peptides that spanned the protein sequence. However, the responses to this region were very sparse and low. We only observed relevant responses in a few cases (Supplementary Figure S3F), but not to an extent to which we could observe differences between the two vaccine groups.

Overall, we conclude that the HERV-W_{LQMV} vaccine increased T-cell responses, particularly for minor epitopes at early time points, and increased T-cell functionality with increasing consistency over time.

3.4. *HERV-W_{LQMV} Vaccine Induces Higher Antibody Responses but Similar Tumour Survival as the HERV-W_{WT} Vaccine*

From the same mice that were used to measured T-cell responses, blood samples were also obtained prior to vaccination (pre-bleed) and on the day of the splenocyte isolation (end-bleed): day 14 or day 28 after vaccination (Figure 4A). We isolated serum, and via flow cytometry, we tested the binding of serum IgG to a murine RenCa cancer cell line that we had, in a previous study, modified to stably express the native HERV-W Env protein on the surface (Supplementary Figure S4A) [18]. The total serum IgG binding from HERV-W_{LQMV}-vaccinated mice was significantly greater compared to the HERV-W_{WT} mice on day 14, and this difference was further increased on day 28 after vaccination (Figure 4B, Supplementary Figure S4B).

Since both T-cell and IgG responses were higher for the HERV-W_{LQMV}-vaccinated mice than the HERV-W_{WT}-vaccinated mice, we were curious whether this could also reflect improved tumour protection against HERV-W Env-expressing tumours. The BALB/c mice were injected intravenously (i.v.) in the tail with the HERV-W Env-expressing murine RenCa tumour cells. The mice were randomised, and four days after the tumour challenge, the mice were vaccinated with the HERV-W_{WT} ($n = 15$), the HERV-W_{LQMV} ($n = 15$), or the Neg. ctrl vaccine ($n = 10$) (Figure 4C). Together (Figure 4D) and individually (Figure 4E,F), the two HERV-W vaccines significantly increased the probability of survival compared to the control group. However, there was no difference in the overall survival when comparing the two HERV-W vaccinated groups with each other (Figure 4G).

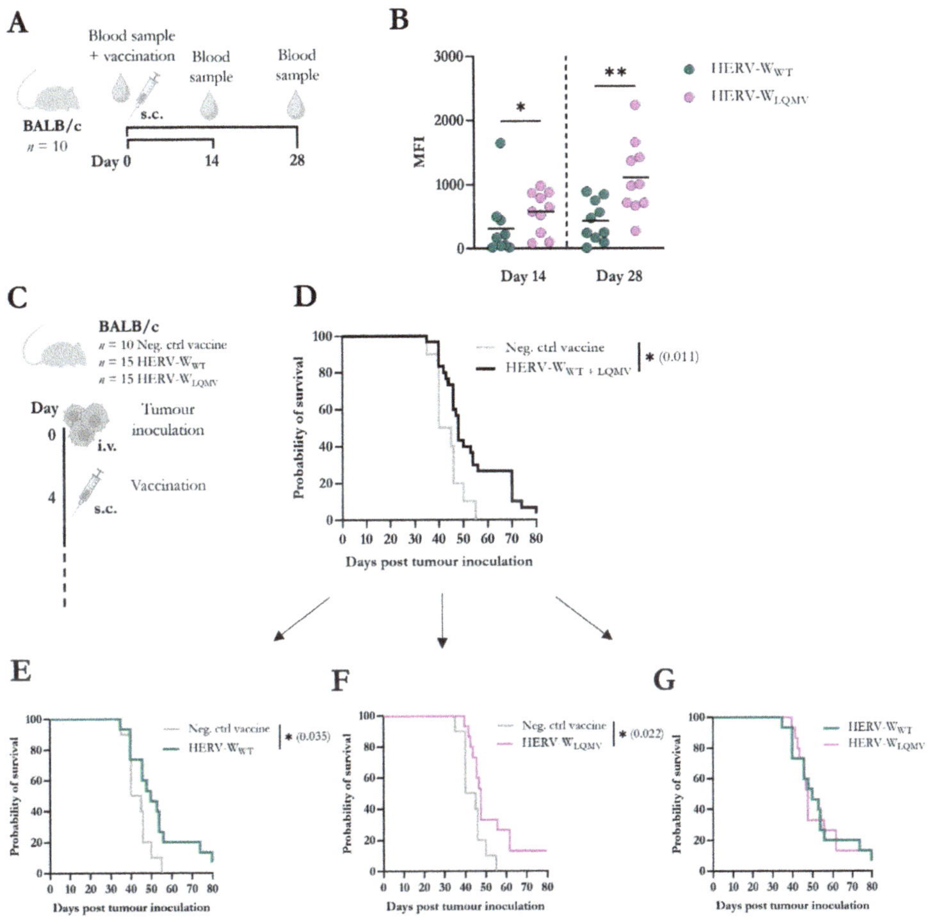

Figure 4. Evaluation of antibody responses and tumour efficacy in mice vaccinated with HERV-W_{LQMV}. (**A**) Schematic representation of the vaccination study using BALB/c mice immunised with the HERV-W_{WT} or HERV-W_{LQMV} vaccine (n = 10 per group) (also illustrated in Figure 3A). Mice were vaccinated subcutaneously (s.c.) on day zero. Blood samples were taken prior to vaccination and at the end of the study, either on day 14 or day 28. (**B**) The graph shows the geometric mean fluorescent intensity (MFI) of serum IgG binding to the surface of HERV-W Env-expressing live RenCa cells, measured via flow cytometry. Each bullet represents one mouse, and the bold line shows the mean of the responses of the group. Response differences between the two vaccine groups were calculated using the Mann–Whitney test, and statistical significance is defined as: * = $p < 0.05$, ** = $p < 0.01$. (**C**) Overview of tumour study with HERV-W Env^+ RenCa cells injected intravenously (i.v.) into the tail veins of BALB/c mice on day zero. On day 4, the mice were vaccinated s.c. with either of the two HERV-W vaccines or the empty control vaccine (Neg. ctrl vaccine). The mice were monitored until humane endpoints were reached. (**D**) Graph showing the survival of mice challenged with HERV-W Env^+ RenCa tumours and immunised with the Neg. ctrl vaccine compared to the two HERV-W vaccines together. (**E**–**G**) Survival curves show the three vaccine groups compared two-and-two. The statistical survival difference between the groups was calculated using the log-rank test (Mantel–Cox), * = $p < 0.05$.

4. Discussion

In a recent study, we generated a potential cancer vaccine targeting HERV-W which encoded the sequence for the fusogen Syncytin-1 and an assembled HERV-W Gag sequence [18]. Following this study, we were interested in developing a non-fusogenic HERV-W vaccine without compromising the vaccine's immunogenicity. Therefore, in this present study, we tested two previously reported non-fusogenic HERV-W Env/Syncytin-1 mutants: one mutant containing a neutral amino acid substitution (C > A) in the disulfide bond motif of the Env TU ectodomain, and one mutant with four amino acids (LQMV) inserted into the Env CT, restoring an evolutionary lost cleavage site of the fusion inhibitory R-peptide. Neither of the HERV-W mutants induced fusion, but only the HERV-W_{LQMV} mutant increased Env cell surface expression in HEK293 cells and was clearly expressed on the cell surfaces of T24 cells. Furthermore, mutating the disulfide bond might affect protein folding; therefore, we proceed with the HERV-W_{LQMV} variant. Interestingly, when encoded in an hAd19a/64 vector, the HERV-W_{LQMV} mutant also induced higher levels of Env cell surface expression than the HERV-W_{WT} vaccine in both murine BMDCs and human A549 cells.

Next, we assessed potential differences in innate immunogenicity between the HERV-W_{WT} and HERV-W_{LQMV} vaccines by measuring the expression of cell surface activation markers and secreted proinflammatory biomarkers of transduced murine BMDCs. Most metrics were similar between the two vaccines, but we observed a reduction in CD40 cell surface expression concomitant with an increase in TNFα in the supernatant of HERV-W_{LQMV} transduced BMDCs. It was not possible to pinpoint a more or less activated phenotype of these transduced BMDCs, but as previously mentioned, the BMDCs confirmed the HERV-W_{LQMV} vaccine-induced increase in Env protein cell surface expression.

Increased antigen expression has previously been found to increase T-cell responses to adenovirus-vector-encoded antigens [27]. Interestingly, prior studies have also shown that dramatically increased antigen presentation or diminished epitope competition can yield similar modest effects on the responses to dominant epitopes and primarily improve subdominant epitope-specific responses [28,29]. This agrees with what we observed in our study. The vaccination of inbred BALB/c mice with the HERV-W_{LQMV} vaccine resulted in significantly higher $CD8^+$ T-cell responses to the HERV-W_{LQMV} vaccine compared to HERV-W_{WT} vaccine at 14 and 28 days after vaccination for the less-responsive epitopes and peptide pools, such as peptide 34 and the SU peptide pool. In contrast, responses to the most immunogenic peptide, peptide 28, were unchanged.

Furthermore, we observed significantly higher humoral IgG responses to the HERV-W_{LQMV} vaccine on day 14 and day 28 after vaccination compared to the HERV-W_{WT} vaccine. Both vaccines improved the overall probability of survival of the mice subjected to distantly injected lung tumours expressing HERV-W Env. However, there was no difference between the two vaccines, likely reflecting similar responses to major epitopes, such as peptide 28.

To our knowledge, no other study has characterised the immunogenicity of the LQMV mutant of Syncytin-1 nor assessed its expression in multiple cell lines and primary cells. We were consequently surprised by the findings of higher levels of cell surface expression in most cells, as well as higher and broader Env-specific $CD8^+$ T-cell responses and antibody responses to the HERV-W_{LQMV} vaccine compared to the HERV-W_{WT} vaccine. While it may seem likely that increased antigen expression can contribute to increased immunogenicity, we cannot be sure that this is the main reason for the observed differences. Further, we have no explanation for why we observe this, nor why this profile was inverted in T24 cells.

We speculate that the increase in HERV-W_{LQMV} Env cell surface expression via transfection in HEK293 cells and transduced murine BMDCs, as well as the increase in the expression levels of both Env and Gag of transduced A549 cells, could be attributed to changes in protein homeostasis, such as increased production, increased surface accumulation, and decreased breakdown. Alternatively, cell stress can induce increased transactivation of the CMV promoter through which the expression of the HERV-W antigens is determined [30]. The increased CMV promoter activation could increase antigen transcription if the HERV-W_{LQMV} mutant induced more cell stress than the HERV-W_{WT}. This could explain why we see higher levels of expression of HERV-W Gag and Env in HERV-W_{LQMV}-transduced A549 cells compared to HERV-W_{WT}. Furthermore, the increased transcription and expression of antigens caused by cell stress would be expected to affect viability. However, potential differences in cell stress were not reflected in cell viability, which was equal between the groups in both murine BMDCs (Figure 2B) and A549 cells (data not shown).

Oppositely, we did not observe an increased level of Env expression in transfected T24 cells. While A549 and HEK293 cells expressing Syncytin-1 fused modestly, Syncytin-1 expression in T24 cells induced the formation of large, multinucleated cells. Syncytin-1-expressing TE671 cells are also highly fusion-permissive (Figure 2 of [3]). When these cells were transfected with the LQMV mutant in the original study by Bonnaud et al., the Env surface expression of the LQMV mutants was similarly slightly lower than for the wt Syncytin-1 (Figure 1C of [21]). Thus, differences in surface expression of the LQMV mutant may reflect cell-dependent differences in receptor engagement.

Two fusion receptors for Syncytin-1 have been identified in humans: the sodium-dependent transporters ASCT1 and ASCT2 [3,31]. Mice express ortholog versions of these transporters (murine ASCT1 and ASCT2), which are quite diverse from the human ones. However, early studies showed that pseudo-typed Syncytin-1 virions could infect via the murine ASCT1 or the de-glycosylated version of murine ASCT2 [31]. A more recent study reported the cell–cell fusion of murine B16F10 cells expressing Syncytin-1 [32]. However, both studies indicated that the cell–cell fusion induced by Syncytin-1 in murine cells was less than in human cells.

We did not observe cell–cell fusions of HERV-W_{WT}-transduced murine BMDCs (data not shown) in this study, nor in a similar experimental setup with the transduction of human PBMC-derived dendritic cells (data not shown). This could indicate that cell–cell fusion is highly regulated in antigen-presenting cells. However, in vivo vaccination studies in mice involve many different cell types for which cell–cell fusion induced by Syncytin-1 is possible. Still, cell–cell fusion induced by Syncytin-1 would most certainly be more pronounced in humans. Therefore, we might observe greater differences in cellular and humoral responses between the two HERV-W vaccines if tested in humans. Accordingly, the HERV-W_{LQMV} would not only be a more desired vaccine with respect to fusion-related side effects but it would also have a higher potential to break immunological tolerance in humans. These differences in the mechanism of action in humans versus mice gain further importance when considering that HERV-W is a self-antigen in humans and it is thus likely that there will be a reduced frequency of immune precursors recognising HERV-W with high affinity. Here, it is highly encouraging to observe an enhancement of subdominant antigen T-cell responses by the HERV-W_{LQMV} vaccine, demonstrated by a broadening of the anti-HERV-W Env T-cell response. Indeed, vaccines designed for increased MHC-I- and MHC-II-restricted antigen presentation have demonstrated a wider efficacy gap to non-modified antigens when targeting cancers through their dominant epitopes in tolerant models [33]. Therefore, the HERV-W_{LQMV} vaccine may outcompete the HERV-W_{WT} vaccine in the treatment of human HERV-W-expressing cancers, despite eliciting similar levels of dominant epitope immunogenicity and anti-cancer efficacy against HERV-W-expressing cancers in the highly immunogenic murine system. Additionally, the immune responses to the HERV-W_{LQMV} vaccine would, at the very least, be less sensitive to immune escape.

Whether or not the improvements in HERV-W_{LQMV} immunity will translate into anti-cancer efficacy in humans relies, first of all, on breaking the immunological tolerance to the endogenous (self-antigen) target in the tumour, here HERV-W, without inducing autoimmune reactions to other tissues potentially expressing HERV-W antigens at low levels. Breaking tolerance and avoiding autoimmunity remain major challenges in cancer immunotherapy. Innate and adaptive immune responses to HERV-W in pathological conditions, such as multiple sclerosis and type 1 diabetes, indicate an incomplete tolerance which can be broken in inflamed conditions [34–37]. Whether a vaccine-induced response can break tolerance in humans and cause autoimmune reactions cannot be answered before tests in humans, but it is encouraging that having multiple sclerosis does not necessarily imply acquiring type 1 diabetes and vice versa.

To test the claim that subtle differences in antigen display and immunogenicity translate into wide differences in a tolerant system, we explored vaccines against a murine endogenous melanoma-associated retrovirus (MelARV). The Env of MelARV is a murine self-antigen overexpressed in many cancer models [38]. In a previous study, we showed that mice were tolerant to DNA plasmids expressing wt MelARV Gag and Env. Still, we could break the immunological tolerance to MelARV through the use of viral vectors [39]. Additionally, via the delivery of a vaccine with immunogenic replication-deficient adenoviral vectors in a prime-boost regimen, in combination with relief of the immunosuppressive activity of the intrinsic Env immunosuppressive domain (ISD) via two point-mutations, we achieved considerably stronger T-cell responses of up to 10% of the $CD8^+$ T-cells [25]. These data illustrate that gradually augmenting immunogenicity by improving the delivery vehicle and antigen used allows for an intrinsically tolerant antigen to induce responses of the magnitudes that are typically associated with live viral infections. Furthermore, the MelARV study showed that the adenovirus-vectored vaccine encoding the MelARV Gag and the mutated Env, in combination with the immune checkpoint inhibitor, anti-PD-1 antibody, could eradicate established colorectal tumours in mice [25].

Our current HERV-W vaccines are also encoded in adenoviral vectors, but in contrast to the MelARV study, a mutation of the ISD is unnecessary because the ISD of the HERV-W Env Syncytin-1 is natively non-immunosuppressive [18,40]. Notably, the non-immunosuppressive ISD of Syncytin-1 might help explain the reported spontaneous break of tolerance in humans [34–36]. While that presents an obstacle to further improving the immunogenicity of Syncytin-1 through changes to its ISD [25,41], the HERV-W_{LQMV} mutant here represents a fortunate and surprising strategy for increasing the HERV-W vaccine immunogenicity.

5. Conclusions

In this study we generated a non-fusogenic HERV-W cancer vaccine capable of inducing cellular and humoral immune responses without the risk of causing undesired and potentially dangerous vaccinate-induced cell–cell fusions. We found that the HERV-W_{LQMV} vaccine significantly improved survival in mice with HERV-W Env-positive cancers compared to the control vaccine. Relative to the wt HERV-W vaccine, the HERV-W_{LQMV} vaccine increased specific immunogenicity, which enhances the prospects for developing immunotherapies for patients with HERV-W-overexpressing cancers.

Supplementary Materials: The following supporting information can be downloaded at: https://www.mdpi.com/article/10.3390/v15081686/s1.

Author Contributions: Conceptualization: P.J.H., R.W. and C.T.; funding acquisition: P.J.H., R.W. and C.T.; project administration: I.S., J.G., R.W., P.J.H. and C.T.; methodology: E.R., I.S. and J.G.; investigation: A.-M.A., E.R., I.S. and J.G.; data curation, visualisation, and formal analysis: I.S.; supervision: E.R., P.J.H., R.W. and C.T.; resources: S.S. and C.T.; writing—original draft: I.S.; review and editing: All authors. All authors have read and agreed to the published version of the manuscript.

Funding: The Eurostars-2 programme, TREATCANCERV (113556), InProTher ApS, and Sirion Biotech funded this study.

Institutional Review Board Statement: All animal experiments were performed according to national guidelines, and experimental procedures were approved by the National Animal Experiments Inspectorate (Dyreforsøgstilsynet, license no. 2019-15-0201-00203).

Data Availability Statement: Data is contained within the article or Supplementary Material. The raw data is available upon reasonable request from the corresponding author.

Acknowledgments: We want to thank the core facilities for flow cytometry and integrated microscopy at Panum, the University of Copenhagen, for their assistance. We would also like to thank the VAR2 group at the Institute for Immunology and Microbiology at the University of Copenhagen for kindly providing the T24 cells. Finally, we would also like to thank our colleague L. Neukirch for assistance with the in vivo tumour experiments and L. Piilgaard Petersen for reviewing the manuscript.

Conflicts of Interest: The project was sponsored by InProTher ApS and Sirion Biotech, where the latter provided the used vaccines. P.J.H. and C.T. are co-inventors of the immunotherapeutic cancer technology called virus-like-vaccine. P.J.H. is the current CSO of InProTher ApS. P.J.H. and C.T. are share- and warrant-holders and board members of InProTher ApS. I.S., E.R. and A.-M.A. are current employees and warrant-holders of InProTher ApS. Despite this, the aforementioned facts had no influence on the design of the experiments, nor the data or representation of these.

Abbreviations

BMDC	Bone marrow-derived dendritic cells
CT	Cytoplasmic domain
Env	Envelope
Gag	Group-specific antigen
hAd19a/64	Human adenovirus type 19a/64
HERV	Human endogenous retrovirus
HERV-$W_{C>A}$	HERV-W Env with C to A mutation, HERV-W Gag
HERV-W_{LQMV}	HERV-W Env with LQMV insertion, HERV-W Gag
HERV-W_{WT}	HERV-W Env wt, HERV-W Gag
IFU	Infectious units
ISD	Immunosuppressive domain
i.v.	Intravenous
MelARV	Melanoma associated retrovirus
MFI	Geometric mean fluorescent intensity
MHC-I/MHC-II	MHC class I/MHC class II
MOI	Multiplicity of infection
Neg. ctrl	Negative control
Pen/strep	Penicillin-streptomycin
PFA	Paraformaldehyde
P28	Peptide 28
P34	Peptide 34
s.c.	Subcutaneous
SU	Surface subunit
TEM	Transmission electron microscopy
TM	Transmembrane domain
TU	Transmembrane subunit
VLP	Virus-like particle
wt	Wildtype

References

1. Grandi, N.; Cadeddu, M.; Blomberg, J.; Mayer, J.; Tramontano, E. HERV-W group evolutionary history in non-human primates: Characterization of ERV-W orthologs in Catarrhini and related ERV groups in Platyrrhini. *BMC Evol. Biol.* **2018**, *18*, 6. [CrossRef]
2. Villesen, P.; Aagaard, L.; Wiuf, C.; Pedersen, F.S. Identification of endogenous retroviral reading frames in the human genome. *Retrovirology* **2004**, *1*, 32. [CrossRef] [PubMed]
3. Blond, J.-L.; Lavillette, D.; Cheynet, V.; Bouton, O.; Oriol, G.; Chapel-Fernandes, S.; Mandrand, B.; Mallet, F.; Cosset, F.-L. An Envelope Glycoprotein of the Human Endogenous Retrovirus HERV-W Is Expressed in the Human Placenta and Fuses Cells Expressing the Type D Mammalian Retrovirus Receptor. *J. Virol.* **2000**, *74*, 3321–3329. [CrossRef]
4. Mi, S.; Lee, X.; Li, X.-P.; Veldman, G.M.; Finnerty, H.; Racie, L.; LaVallie, E.; Tang, X.-Y.; Edouard, P.; Howes, S.; et al. Syncytin is a captive retroviral envelope protein involved in human placental morphogenesis. *Nature* **2000**, *403*, 785–789. [CrossRef]
5. Denner, J. Expression and function of endogenous retroviruses in the placenta. *APMIS* **2016**, *124*, 31–43. [CrossRef]
6. Frendo, J.-L.; Olivier, D.; Cheynet, V.; Blond, J.-L.; Bouton, O.; Vidaud, M.; Rabreau, M.; Evain-Brion, D.; Mallet, F. Direct Involvement of HERV-W Env Glycoprotein in Human Trophoblast Cell Fusion and Differentiation. *Mol. Cell. Biol.* **2003**, *23*, 3566–3574. [CrossRef]

7. Holder, B.S.; Tower, C.L.; Forbes, K.; Mulla, M.J.; Aplin, J.D.; Abrahams, V.M. Immune cell activation by trophoblast-derived microvesicles is mediated by syncytin 1. *Immunology* **2012**, *136*, 184–191. [CrossRef]
8. Tolosa, J.M.; Schjenken, J.; Clifton, V.; Vargas, A.; Barbeau, B.; Lowry, P.; Maiti, K.; Smith, R. The endogenous retroviral envelope protein syncytin-1 inhibits LPS/PHA-stimulated cytokine responses in human blood and is sorted into placental exosomes. *Placenta* **2012**, *33*, 933–941. [CrossRef] [PubMed]
9. Matoušková, M.; Blažková, J.; Pajer, P.; Pavlíček, A.; Hejnar, J. CpG methylation suppresses transcriptional activity of human syncytin-1 in non-placental tissues. *Exp. Cell Res.* **2006**, *312*, 1011–1020. [CrossRef] [PubMed]
10. Gimenez, J.; Montgiraud, C.; Oriol, G.; Pichon, J.-P.; Ruel, K.; Tsatsaris, V.; Gerbaud, P.; Frendo, J.-L.; Evain-Brion, D.; Mallet, F. Comparative methylation of ERVWE1/syncytin-1 and other human endogenous retrovirus LTRs in placenta tissues. *DNA Res.* **2009**, *16*, 195–211. [CrossRef] [PubMed]
11. Wang, Q.; Shi, Y.; Bian, Q.; Zhang, N.; Wang, M.; Wang, J.; Li, X.; Lai, L.; Zhao, Z.; Yu, H. Molecular mechanisms of syncytin-1 in tumors and placental development related diseases. *Discov. Oncol.* **2023**, *14*, 104. [CrossRef]
12. Grandi, N.; Tramontano, E. HERV envelope proteins: Physiological role and pathogenic potential in cancer and autoimmunity. *Front. Microbiol.* **2018**, *9*, 462. [CrossRef]
13. Müller, M.D.; Holst, P.J.; Nielsen, K.N. A Systematic Review of Expression and Immunogenicity of Human Endogenous Retroviral Proteins in Cancer and Discussion of Therapeutic Approaches. *Int. J. Mol. Sci.* **2022**, *23*, 1330. [CrossRef]
14. Grandi, N.; Tramontano, E. Type W human endogenous retrovirus (HERV-W) integrations and their mobilization by L1 machinery: Contribution to the human transcriptome and impact on the host physiopathology. *Viruses* **2017**, *9*, 162. [CrossRef] [PubMed]
15. Dittmar, T.; Weiler, J.; Luo, T.; Hass, R. Cell-Cell Fusion Mediated by Viruses and HERV-Derived Fusogens in Cancer Initiation and Progression. *Cancers* **2021**, *13*, 5363. [CrossRef]
16. Bastida-Ruiz, D.; Van Hoesen, K.; Cohen, M. The dark side of cell fusion. *Int. J. Mol. Sci.* **2016**, *17*, 638. [CrossRef]
17. Andersson, A.-M.C.; Schwerdtfeger, M.; Holst, P.J. Virus-Like-Vaccines against HIV. *Vaccines* **2018**, *6*, 10. [CrossRef] [PubMed]
18. Skandorff, I.; Ragonnaud, E.; Gille, J.; Andersson, A.-M.; Schrödel, S.; Duvnjak, L.; Turner, L.; Thirion, C.; Wagner, R.; Holst, P.J. Human Ad19a/64 HERV-W Vaccines Uncover Immunosuppression Domain-Dependent T-Cell Response Differences in Inbred Mice. *Int. J. Mol. Sci.* **2023**, *24*, 9972. [CrossRef] [PubMed]
19. Chang, C.; Chen, P.T.; Chang, G.D.; Huang, C.J.; Chen, H. Functional characterization of the placental fusogenic membrane protein syncytin. *Biol. Reprod.* **2004**, *71*, 1956–1962. [CrossRef]
20. Cheynet, V.; Ruggieri, A.; Oriol, G.; Blond, J.-L.; Boson, B.; Vachot, L.; Verrier, B.; Cosset, F.-L.; Mallet, F. Synthesis, Assembly, and Processing of the Env ERVWE1/Syncytin Human Endogenous Retroviral Envelope. *J. Virol.* **2005**, *79*, 5585–5593. [CrossRef]
21. Bonnaud, B.; Bouton, O.; Oriol, G.; Cheynet, V.; Duret, L.; Mallet, F. Evidence of selection on the domesticated ERVWE1 env retroviral element involved in placentation. *Mol. Biol. Evol.* **2004**, *21*, 1895–1901. [CrossRef] [PubMed]
22. Bouton, O.; Bedin, F.; Duret, L.; Mandrand, B.; Voisset, C.; Blancher, A.; Perron, H.; Mallet, F.; Paranhos-Baccala, G.; Kim, H.-S.; et al. Chromosomal Distribution and Coding Capacity of the Human Endogenous Retrovirus HERV-W Family. *AIDS Res. Hum. Retroviruses* **2000**, *16*, 731–740. [CrossRef]
23. Komurian-Pradel, F.; Paranhos-Baccala, G.; Bedin, F.; Ounanian-Paraz, A.; Sodoyer, M.; Ott, C.; Rajoharison, A.; Garcia, E.; Mallet, F.; Mandrand, B.; et al. Molecular Cloning and Characterization of MSRV-Related Sequences Associated with Retrovirus-like Particles. *Virology* **1999**, *260*, 1–9. [CrossRef] [PubMed]
24. Ruzsics, Z.; Lemnitzer, F.; Thirion, C. Engineering adenovirus genome by Bacterial Artificial Chromosome (BAC) technology. *Methods Mol. Biol.* **2014**, *1089*, 143–158. [CrossRef]
25. Daradoumis, J.; Ragonnaud, E.; Skandorff, I.; Nielsen, K.N.; Bermejo, A.V.; Andersson, A.-M.; Schroedel, S.; Thirion, C.; Neukirch, L.; Holst, P.J. An Endogenous Retrovirus Vaccine Encoding an Envelope with a Mutated Immunosuppressive Domain in Combination with Anti-PD1 Treatment Eradicates Established Tumours in Mice. *Viruses* **2023**, *15*, 926. [CrossRef]
26. Jin, D.; Sprent, J. GM-CSF Culture Revisited: Preparation of Bulk Populations of Highly Pure Dendritic Cells from Mouse Bone Marrow. *J. Immunol.* **2018**, *201*, 3129–3139. [CrossRef] [PubMed]
27. Sridhar, S.; Reyes-Sandoval, A.; Draper, S.J.; Moore, A.C.; Gilbert, S.C.; Gao, G.P.; Wilson, J.M.; Hill, A.V.S. Single-Dose Protection against Plasmodium berghei by a Simian Adenovirus Vector Using a Human Cytomegalovirus Promoter Containing Intron A. *J. Virol.* **2008**, *82*, 3822–3833. [CrossRef]
28. Holst, P.J.; Sorensen, M.R.; Jensen, C.M.M.; Orskov, C.; Thomsen, A.R.; Christensen, J.P. MHC Class II-Associated Invariant Chain Linkage of Antigen Dramatically Improves Cell-Mediated Immunity Induced by Adenovirus Vaccines 1. *J. Immunol.* **2008**, *180*, 3339–3346. [CrossRef]
29. Holst, P.J.; Jensen, B.A.H.; Ragonnaud, E.; Thomsen, A.R.; Christensen, J.P. Targeting of non-dominant antigens as a vaccine strategy to broaden T-cell responses during chronic viral infection. *PLoS ONE* **2015**, *10*, e0117242. [CrossRef]
30. Bruening, W.; Giasson, B.; Mushynski, W.; Durham, H.D. Activation of stress-activated MAP protein kinases up-regulates expression of transgenes driven by the cytomegalovirus immediate/early promoter. *Nucleic Acids Res.* **1998**, *26*, 486–489. [CrossRef]
31. Lavillette, D.; Marin, M.; Ruggieri, A.; Mallet, F.; Cosset, F.-L.; Kabat, D. The Envelope Glycoprotein of Human Endogenous Retrovirus Type W Uses a Divergent Family of Amino Acid Transporters/Cell Surface Receptors. *J. Virol.* **2002**, *76*, 6442–6452. [CrossRef] [PubMed]

32. Mo, H.; Ouyang, D.; Xu, L.; Gao, Q.; He, X. Human endogenous retroviral syncytin exerts inhibitory effect on invasive phenotype of B16F10 melanoma cells. *Chin. J. Cancer Res.* **2013**, *25*, 556–564. [PubMed]
33. Sorensen, M.R.; Holst, P.J.; Pircher, H.; Christensen, J.P.; Thomsen, A.R. Vaccination with an adenoviral vector encoding the tumor antigen directly linked to invariant chain induces potent $CD4^+$ T-cell-independent $CD8^+$ T-cell-mediated tumor control. *Eur. J. Immunol.* **2009**, *39*, 2725–2736. [CrossRef] [PubMed]
34. Arru, G.; Sechi, E.; Mariotto, S.; Farinazzo, A.; Mancinelli, C.; Alberti, D.; Ferrari, S.; Gajofatto, A.; Capra, R.; Monaco, S.; et al. Antibody response against HERV-W env surface peptides differentiates multiple sclerosis and neuromyelitis optica spectrum disorder. *Mult. Scler. J. Exp. Transl. Clin.* **2017**, *3*, 2055217317742425. [CrossRef]
35. Mameli, G.; Cossu, D.; Cocco, E.; Frau, J.; Marrosu, M.G.; Niegowska, M.; Sechi, L.A. Epitopes of HERV-Wenv induce antigen-specific humoral immunity in multiple sclerosis patients. *J. Neuroimmunol.* **2015**, *280*, 66–68. [CrossRef]
36. Brudek, T.; Christensen, T.; Aagaard, L.; Petersen, T.; Hansen, H.J.; Møller-Larsen, A. B cells and monocytes from patients with active multiple sclerosis exhibit increased surface expression of both HERV-H Env and HERV-W Env, accompanied by increased seroreactivity. *Retrovirology* **2009**, *6*, 104. [CrossRef]
37. Niegowska, M.; Wajda-Cuszlag, M.; Stępień-Ptak, G.; Trojanek, J.; Michałkiewicz, J.; Szalecki, M.; Sechi, L.A. Anti-HERV-W Env antibodies are correlated with seroreactivity against Mycobacterium avium subsp. paratuberculosis in children and youths at T1D risk. *Sci. Rep.* **2019**, *9*, 6282. [CrossRef]
38. Scrimieri, F.; Askew, D.; Corn, D.J.; Eid, S.; Bobanga, I.D.; Bjelac, J.A.; Tsao, M.L.; Allen, F.; Othman, Y.S.; Wang, S.-C.G.; et al. Murine leukemia virus envelope gp70 is a shared biomarker for the high-sensitivity quantification of murine tumor burden. *Oncoimmunology* **2013**, *2*, e26889. [CrossRef]
39. Neukirch, L.; Nielsen, T.K.; Laursen, H.; Daradoumis, J.; Thirion, C.; Holst, P.J. Adenovirus based virus-like-vaccines targeting endogenous retroviruses can eliminate growing colorectal cancers in mice. *Oncotarget* **2019**, *10*, 1458–1472. [CrossRef]
40. Mangeney, M.; Renard, M.; Schlecht-Louf, G.; Bouallaga, I.; Heidmann, O.; Letzelter, C.; Richaud, A.; Ducos, B.; Heidmann, T. Placental syncytins: Genetic disjunction between the fusogenic and immunosuppressive activity of retroviral envelope proteins. *Proc. Natl. Acad. Sci. USA* **2007**, *104*, 20534–20539. [CrossRef]
41. Holst, P.; Thirion, C.; Neukirch, L. A Vaccine For Use in The Prophylaxis and/or Treatment of a Disease. U.S. Patent WO2019043127, 30 August 2019.

viruses

MDPI

Article

Effect of Pre-Treatment with a Recombinant Chicken Interleukin-17A on Vaccine Induced Immunity against a Very Virulent Marek's Disease Virus

Nitish Boodhoo, Ayumi Matsuyama-Kato, Sugandha Raj, Fatemeh Fazel, Myles St-Denis and Shayan Sharif *

Department of Pathobiology, Ontario Veterinary College, University of Guelph, Guelph, ON N1G 2W1, Canada; boodhoon@uoguelph.ca (N.B.); matsuyam@uoguelph.ca (A.M.-K.); rajs@uoguelph.ca (S.R.); ffazel@uoguelph.ca (F.F.); stdenis@uoguelph.ca (M.S.-D.)
* Correspondence: shayan@uoguelph.ca

Abstract: The host response to pathogenic microbes can lead to expression of interleukin (IL)-17, which has antimicrobial and anti-viral activity. However, relatively little is known about the basic biological role of chicken IL-17A against avian viruses, particularly against Marek's disease virus (MDV). We demonstrate that, following MDV infection, upregulation of IL-17A mRNA and an increase in the frequency of IL-17A+ T cells in the spleen occur compared to control chickens. To elaborate on the role of chIL-17A in MD, the full-length chIL-17A coding sequence was cloned into a pCDNA3.1-V5/HIS TOPO plasmid. The effect of treatment with pcDNA:chIL-17A plasmid in combination with a vaccine (HVT) and very virulent(vv)MDV challenge or vvMDV infection was assessed. In combination with HVT vaccination, chickens that were inoculated with the pcDNA:chIL-17A plasmid had reduced tumor incidence compared to chickens that received the empty vector control or that were vaccinated only (66.6% in the HVT + empty vector group and 73.33% in HVT group versus 53.3% in the HVT + pcDNA:chIL-17A). Further analysis demonstrated that the chickens that received the HVT vaccine and/or plasmid expressing IL-17A had lower MDV-*Meq* transcripts in the spleen. In conclusion, chIL-17A can influence the immunity conferred by HVT vaccination against MDV infection in chickens.

Keywords: MDV; chickens; Th17 cells; IL-17A; interferon-gamma and adaptive immunity

Citation: Boodhoo, N.; Matsuyama-Kato, A.; Raj, S.; Fazel, F.; St-Denis, M.; Sharif, S. Effect of Pre-Treatment with a Recombinant Chicken Interleukin-17A on Vaccine Induced Immunity against a Very Virulent Marek's Disease Virus. *Viruses* **2023**, *15*, 1633. https://doi.org/10.3390/v15081633

Academic Editors: Pietro Hiram Guzzi, Marianna Milano and Jayanta Kumar Das

Received: 22 June 2023
Revised: 23 July 2023
Accepted: 24 July 2023
Published: 27 July 2023

1. Introduction

Since the discovery of murine and human T helper (Th) 17 cells, much progress has been made in defining their developmental origin, lineage, relative biological function, and respective expression of the IL-17 cytokine family members (IL-17A to IL-17F) [1,2]. It is clear that, in mice and humans, IL-17 has a role in immune responses associated with allergenic disease, autoimmune disease, malignancy, transplantation rejection, and modulation of host defense against both viral and bacterial microbes. While the expression of the avian IL-17 family of cytokines has been identified using intestinal intraepithelial lymphocytes (IELs), the functional role of these cytokines beyond differential gene expression has not been described [3].

IL-17 signaling is mediated through the IL-17A/F heterodimeric receptor complex (IL-17RA). The latter was demonstrated to have heterogeneous tissue distribution, which may serve to allow tissue-specific signal transduction [4]. As in mammalian hosts, in chickens, the IL-17RA molecule constitutes a distinct family of transmembrane proteins that are expressed by various immune system cells and fibroblasts [5,6]. IL-17A/IL-17RA interaction results in the induction of various cytokines ((IL-6, tumor necrosis factor (TNF)-α, and granulocyte monocyte colony stimulator factor (GM-CSF)), chemokines, antimicrobial peptides, and tissue-remodeling molecules. While Th17 cells are the best characterized source of IL-17, various innate immune system cells and T cell subsets can also express various IL-17 cytokines [7–9]. IL-17A expression may be preferentially induced in response to bacterial and fungal pathogens [4]. In fact, the functional role of IL-17A is much more understood in bacterial infections, while little or much less is known about its role in viral infection. In viral infection models, IL-17A overexpression modulates Th1/Th2 responses, leading to exacerbation of vaccinia virus virulence in infected mice [10]. In contrast, no direct anti-viral effects against herpes simplex virus 2 (HSV-2) were observed. However, its expression can stimulate CD4+ T cells to respond to HSV-2 reactivation in peripheral neurons [11,12]. Most of these studies were centered on non-avian hosts, and relatively little is known about the immune-protective function of chicken IL-17A against avian viruses [13].

Marek's disease virus (MDV), also known as gallid herpesvirus 2 (GaHV-2), is the causative agent of Marek's disease (MD) in chickens [14,15]. The clinical manifestation of MD is associated with transient paralysis, immunosuppression, metabolic dysregulation, and CD4+ T cell lymphoma formation in infected chickens [15–17]. Because of the cell-associated nature of MDV, T cell-mediated immunity is believed to be crucial in the control of MD [18]. The administration of chicken IFN-γ with MDV vaccine has been shown to increase vaccine efficacy, suggesting that IFN-γ may play an important role in boosting protection against MD [19]. IL-17 and IFN-γ play diverse roles and can modulate differentiation of the distinct Th1 and Th17 lineages, respectively [20]. That MDV infection modulates the host immune system suggests a potential underlying dueling role between IL-17 and IFN-γ in MD. The latter could indicate a relationship between pro-inflammatory cytokine expression and MDV infection [21,22]. Therefore, the aim of the present study was to identify and assess the potential role of IL-17A in MDV-infected chickens. To that end, cytokine expression from MDV-infected chickens, as well as tumor lesion severity, was assessed following treatment of chickens with recombinant IL-17A.

2. Materials and Methods

2.1. Plasmid and Cloning

The chicken IL-17A gene (accession number: NM_204460.2) was amplified from PMA/ION stimulated splenocytes by high-fidelity PCR (GoTaq polymerase; Promega, WI, USA) per the manufacturer's recommendation and using designer primers (Table 1) to generate an amplicon (approximately 499 bp) with respective 5′-*Hin*dIII and 3′-*Eco*RV restriction sites. The gel purified amplicon (Gel purification Kit; Qiagen, Toronto, ON, Canada) was ligated into a pDRIVE vector (TA cloning; Qiagen, ON, Canada) and subcloned (*Hin*dIII and *Eco*RV restriction digestion; New England Bioscience, ON, Canada) into the pCDNA3.1/V5-HIS TOPO plasmid (Life Technologies, Mississauga, ON, Canada). DH5α competent cells were transformed and utilized to propagate the respective plasmids for purification with a midi-prep kit (Qiagen, ON, Canada). All plasmids (pCDNA3.1/V5-HIS TOPO and pCDNA3.1/rchIL-17A-V5-HIS TOPO) were linearized by *Bgl*II (New England Bioscience, Whitby, ON, Canada) digestion for in vitro transfection.

Table 1. Primer sequences used for high-fidelity PCR.

Target		Primer Sequences	Accession No.
Full length chIL-17A	FWD	**AAGCTT**ATGTCTCCGATCCCTTG	NM_204460.2
	REV	**GATATC**AGCCTGGTGCTGGATCAGTGGG	

2.2. Cell Culture

(i) Cells: HEK 293-T cells (ATCC CRL-3216) were cultured and maintained (37 °C and 5% CO_2) in minimum essential medium (MEM; Life Technologies, Mississauga, ON, Canada) supplemented with 10% fetal bovine serum (FBS; Gibco, Life Technologies, Canada) and 0.1% penicillin–streptomycin (Gibco, Life Technologies, Canada).

(ii) Transfection/stably expressing cell lines: One million HEK 293-T cells were seeded per well in a 6 well plate. Overnight cells (>80% confluent) were transfected, according to

the manufacturer's recommendation, using Lipofectamine stem reagent (Life Technologies, Canada) with up to 3 µg of linearized pCDNA3.1/V5-HIS-TOPO empty vector or pCDNA3.1/rchIL-17A-V5-HIS TOPO. After 48 h (hrs), cells were passaged (0.5% trypsin; Life Technologies, Canada) into T25 flasks and treated with 800 µg/mL G418 in MEM complete medium (10% FBS and 0.1% penicillin–streptomycin). The medium was changed every 48 hrs until proliferating islands of cells were observed. After 2–3 weeks of antibiotic selection, stably expressing cells were cultured without G418 for the purpose of collecting both cells and supernatant to confirm mRNA expression, as well as extracellular secretion of recombinant chicken IL-17A (rchIL-17A), and the cells were stored at −80 °C until required.

2.3. Virus Preparation

The very virulent MDV-RB1B strain (vvMDV), was provided courtesy of Dr. K.A. Schat (Cornell University, Ithaca, NY, USA) [17]. vvMDV-RB1B virus titers were calculated on primary chicken kidney cells obtained from 2- to 3-week-old specific pathogen free (SPF) chickens to establish infectious doses of inoculums, as well as stocks (liquid nitrogen storage) [23]. The HVT vaccine strain (MD-Vac-CFL; Fort Dodge Animal Health, Fort Dodge, IA, USA) was diluted in the recommended diluent and stored on ice prior to use.

2.4. Experimental Design and Sampling

(i) Experimental animals: Two hundred ninety-eight-day-old SPF chickens (layers) were purchased from the Animal Disease Research Institute, Canadian Food Inspection Agency (Ottawa, ON, Canada), and were accommodated in the isolation unit at the University of Guelph. For the duration of the experiments, all chickens were given ad libitum access to food and water. All experiments were approved by the Animal Care Committee of the University of Guelph and were conducted according to their guidelines.

(ii) TLR ligands and MDV vaccine: Synthetic class B CpG oligodeoxynucleotides 2007 (CPG ODN 2007; Invivogen, San Diego, CA, USA) and polyinosinic:polycytidylic acid (Poly I:C; Sigma-Aldrich, Oakville, ON, Canada) were re-suspended in phosphate-buffered saline (PBS, pH 7.4) and stored at −20 °C. The cell-free HVT vaccine (MD-Vac-CFL; Animal Health section, Boehringer-Ingelheim Canada, Burlington, ON, Canada) was resuspended in the recommended diluent and stored on ice until required for same-day inoculations.

(iii) Experimental outline: In the first experiment, 2-week-old chicks (n = 36) were injected intramuscularly (I/M) in the pectoral muscle with 100 µL of either CpG ODN 2007 (10 µg; n = 12) or Poly I:C (400 µg; n = 12). The control group (n = 12) received 100 µL of PBS (negative control). At 8 and 16 h post-inoculation, 6 chicks from each group were euthanized for sample collection.

In the second experiment, 1-day-old chicks (n = 108) were administered the HVT vaccine at one-half recommended dosage via a subcutaneous route. A reduced vaccine dose was used to observe the full potential of recombinant chicken IL-17A when used on its own or as a vaccine adjuvant. On day 3 post-vaccination, two groups of chicks were either inoculated intramuscular with 10 µg/chick of pCDNA3.1/V5-HIS-TOPO empty vector (n = 72) or pCDNA3.1/rchIL-17A-V5-His TOPO (n = 72). On day 4 post-vaccination, 198 chicks were infected with 250 plaque-forming units of the vvMDV-RB1B strain or sham/diluent treatment via the intraabdominal route.

(iv) Sampling: At various time points, tissue samples were collected for various analytical processes. Chicks (n = 6/group) from the first experiment were euthanized at 8 and 18 h post-TLR treatment, after which whole spleens were collected and stored (−20 °C) in RNA (Invitrogen, Toronto, ON, Canada). In the second experiment, whole spleens (n = 12/group) were collected in PBS containing penicillin (10 U/mL), and streptomycin (10 µg/mL) at 4, 10, and 21 days post infection (DPI), and parts of the whole spleen were either stored in RNA later for RNA extraction or were frozen in OCT embedding medium for immunostaining. Whole spleen samples were also collected from 5 control (non-infected chickens) and vvMDV-RB1B infected chickens at 4, 10, and 21 DPI and stored on ice for mononuclear cell isolation.

2.5. Spleen Mononuclear Cell Preparation and Stimulation

(i) Mononuclear cells: Whole spleens collected aseptically at various time points were applied onto 40-μm BD cell strainers (BD Biosciences, Mississauga, ON, Canada), crushed through using the rubber end of a 10-mL syringe plunger, and washed in RPMI 1640 with penicillin (10 U/mL) and streptomycin (10 μg/mL) [24]. Gradient suspensions were prepared by layering cells (2:1) onto histopaque 1077 (Millipore-Sigma, Oakville, ON, Canada) and centrifugation at 2100 rpm (600× *g* with no brakes) for 20 min to allow for the separation of mononuclear cells. Aspirated buffy coats were washed (2×) at 1500 rpm (400× *g*) for 5 min in RPMI 1640 with penicillin (10 U/mL), and streptomycin (10 μg/mL). Mononuclear cells were suspended in complete RPMI cell culture medium; RPMI 1640 medium contains 10% fetal bovine serum (Millipore-Sigma, Canada), penicillin (10 U/mL), and streptomycin (10 μg/mL). Cell numbers and viability were calculated using a hemocytometer and trypan the blue exclusion method. Mononuclear cells were suspended in complete RPMI cell culture medium at a density of 5×10^6 cells/mL and kept on ice.

(ii) In vitro stimulation assays: Spleen mononuclear cells were seeded in 96-well U-bottom plates at a density of 1.0×10^6 cells/200 μL and stimulated first with vehicle and then either CpG ODN 2007 (5 ug/mL) or lipopolysaccharides (LPS; 10 μg/mL). Cells were subsequently incubated (41 °C and 5% CO_2) overnight (18 h), and supernatants were collected for anti-IL-17A enzyme-linked immunosorbent assay (ELISA). All stimulation assays were performed at least three times in triplicate on different days. Second, cells were co-incubated with supernatant from pCDNA3.1/V5-HIS-TOPO empty vector or pCDNA3.1/rchIL-17A-V5-HIS TOPO stably expressing cell lines on ice (15 min in the dark) for FACS analysis.

2.6. RNA Extraction and Reverse Transcription

RNA was extracted from in vitro mononuclear cell cultures, as well as spleens, using Trizol following the manufacturer's instructions and treated with DNA-free DNase (ThermoFisher Scientific, Mississauga, ON, Canada) as previously described [25]. The quality, as well as the quantity, of RNA was estimated using NanoDrop ND-1000 spectrophotometry (NanoDrop Technologies, Wilmington, DE, USA). Subsequently, 1 μg of purified RNA was reverse transcribed to cDNA using Oligo(dT)12–18 primers (SuperScript II First-Strand Synthesis System; Invitrogen Life Technologies, Carlsbad, Ottawa, ON, Canada) according to the manufacturer's recommended protocol. The resulting cDNA was diluted 1:9 in diethyl pyrocarbonate-treated (DEPC) water.

2.7. Real-Time Polymerase Chain Reaction (RT-PCR)

Quantitative real-time PCR with SYBR green was performed on diluted cDNA using a LightCycler 480 II (Roche Diagnostics GmbH, Mannheim, GER) according to the manufacturer's recommendation. In brief, quantitative real-time PCR was performed with the following conditions: initial denaturation was performed at 95 °C for 5 min, followed by 40 cycles of denaturation at 95 °C for 10 s (s), primer annealing listed in Table 2 for 15 s, and extension at 72 °C for 20 s, with end-point melt-curve analysis. The relative fold change of target genes was calculated by $2-\Delta\Delta CT$ method. The Ct value for each sample was normalized against the GAPDH housekeeping gene for respective samples. Data represent the means from six biological replicates, using primers outlined in Table 2.

Table 2. Primer sequences used for real-time PCR.

Target		Primer Sequences	Reference
IL-1β	FWD	GTGAGGCTCAACATTGCGCTGTA	[23]
	REV	TGTCCAGGCGGTAGAAGATGAAG	

Table 2. *Cont.*

Target		Primer Sequences	Reference
TGF-β	FWD	CGGCCGACGATGAGTGGCTC	[26]
	REV	CGGGGCCCATCTCACAGGGA	
COX-2	FWD	CTGCTCCCTCCCATGTCAGA	[16]
	REV	CACGTGAAGAATTCCGGTGTT	
IL-2	FWD	TGCAGTGTTACCTGGGAGAAGTGGT	[27]
	REV	ACTTCCGGTGTGATTTAGACCCGT	
IL-6	FWD	CGTGTGCGAGAACAGCATGGAGA	[28]
	REV	TCAGGCATTTCTCCTCGTCGAAGC	
IL-10	FWD	TTTGGCTGCCAGTCTGTGTC	[29]
	REV	CTCATCCATCTTCTCGAACGTC	
IL-12p40	FWD	CCAAGACCTGGAGCACACCGAAG	[28]
	REV	CGATCCCTGGCCTGCACAGAGA	
IL-17A	FWD	TATCAGCAAACGCTCACTGG	[30]
	REV	AGTTCACGCACCTGGAATG	
IFN-γ	FWD	ACACTGACAAGTCAAAGCCGCACA	[26]
	REV	AGTCGTTCATCGGGAGCTTGGC	
MDV-*Meq*	FWD	GTCCCCCCTCGATCTTTCTC	[23]
	REV	CGTCTGCTTCCTGCGTCTTC	
β-actin	FWD	CAACACAGTGCTGTCTGGTGGTA	[31]
	REV	ATCGTACTCCTGCTTGCTGATCC	

2.8. Flow Cytometry

Following a wash in FACS staining buffer (PBS with 0.5% Bovine Serum Albumin; BSA), 5.0×10^5 spleen mononuclear cells were counter-stained in two assays to demonstrate IL-17A functional interaction, as well as intracellular IL-17A cytokine staining. Antibodies used for the monocyte/macrophage panel (mouse anti-chicken Kul01-fluorescein isothiocyanate: FITC and mouse anti-chicken major histocompatibility complex (MHC) II-phycoerythrin: PE were all purchased from Southern Biotech, Birmingham, AL, USA) and T cell panel (mouse anti-chicken CD3ε-PB, mouse anti-chicken CD4-PE-CY7, mouse anti-chicken CD8α-FITC and mouse anti-chicken γδTCR-PE were all purchased from Southern Biotech, Birmingham, AL, USA).

(i) rchIL-17A surface binding: Mononuclear cells were incubated (4 °C for 15 min) in the presence of rchIL-17A. The cells were subsequently washed in FACS staining buffer (400× *g* for 5 min) and incubated (15 min at 4 °C) in the presence of unpurified mouse anti-chIL-17A antibody (IgG2a isotype), kindly provided by Dr. Thomas W. Göbel (Tierärztliche Fakultät, LMU München, Veterinärstrasse 13, 80539 München, Germany) [5]. The cells were washed again and incubated (15 min at 4 °C) in the presence of mouse anti-IgG2a-APC antibody prior to counter-staining per the APC or T cell panels listed. Following a final wash, mononuclear cells were counterstained with the live/dead marker 7-AAD (BDTM Pharmingen, Mississauga, ON, Canada) and acquired for analysis using a BD FACS Canto II.

(ii) Intracellular cytokine staining: IL-17 intracellular staining: Spleen mononuclear cells were washed in FACS staining buffer and subsequently counterstained (15 min at 4 °C) with the T cell panel antibodies. Following a wash, mononuclear cells were counterstained with the live/dead marker 7-AAD (BDTM Pharmingen, Canada). After washing twice with FACS staining buffer, the cells were incubated in 200 μL of Cytofix/Cytoperm solution

(Beckton Dickinson Biosciences, Mississauga, ON, Canda) for 40 min at room temperature (RT), followed by washing three times with Perm/Wash solution (Beckton Dickinson Biosciences). The cells were stained with mouse anti-chIL-17A antibody (IgG2a isotype) for 15 min at RT in the dark and subsequently washed twice with Perm/Wash solution. Following final incubation (15 min at RT) in the presence of mouse anti-IgG2a-APC antibody, the cells washed in Perm/Wash solution and stored in FACS staining buffer.

IFN-γ intracellular staining: One million splenocytes were cultured with Golgi Plug and Dnase I in a 96-well round-bottom plate for 4 h at 41 °C in 5% CO_2. The cell staining protocol for flow cytometry was described elsewhere [32].

All samples were acquired on a BD FACS Canto II. Data were processed by FlowJo software, version V10.

2.9. Confocal Microscopy

(i) Sample preparation: OCT-embedded frozen spleen tissue samples were cut using a cryostat at approximately 10 μm in thickness. Samples were mounted onto slides and stored (−80 °C) for immunostaining. The slides were subsequently fixed (1 h in the dark at RT) in 4% paraformaldehyde (PFA). They were washed (PBS), permeabilized with 0.1% triton-X buffer solution (10 min in the dark at RT), and blocked (1 h in the dark at RT) with 1% bovine serum albumin (BSA-PBS). The slides were then washed 3× (PBS) and incubated (1 h in the dark at RT) with the mouse anti-chIL-17A antibody (IgG2a; 500 μL in 0.5% BSA–PBS). Following another wash (3× with PBS), the slides were again incubated overnight (4 °C in the dark at RT) with a goat anti-mouse IgG2a-488 (from ThermoFisher Scientific, ON, Canada) antibody (0.5 μg/500 μL in 0.5% BSA–PBS). The next day, the slides were washed (3× with PBS), and nuclei were labeled with Vectashield containing 4′,6-diamidino-2-phenylindole (DAPI from Vector laboratories/Cedarlane, ON, Canada). Tissue samples were sealed with round coverslips for fluorescence imaging.

(ii) Visualization: Tissue samples were viewed using a Leica SP5 laser scanning confocal microscope, and optical sections were recorded using either 405 or 488 nM with a numerical aperture of 1.4 or 1.25, respectively. All data were collected sequentially to minimize cross talk between fluorescence signals. The data are presented as maximum projections of each channel and overlaid for analysis with LAS AF software.

2.10. ELISA

Stably expressing rchIL-17A or vector control HEK 293-T cell culture supernatants, as well as diluent control-, CpG ODN 2007- and LPS-stimulated spleen mononuclear cell culture supernatants (100 μL), were incubated overnight at 4 °C with ELISA coating buffer. The next day, wells were washed (PBS) and blocked (5% BSA-PBS) for 1 h at RT prior to incubation (1 h at RT) with the mouse anti-chIL-17A antibody (IgG2a; 500 μL in 0.5% BSA–PBS). Following another wash (3× with 0.01% tween-PBS), the wells were again incubated (1 h at RT) with a goat anti-mouse IgG H/L-horse-radish peroxidase (HRP) antibody (0.5 μg/500 μL in 0.5% BSA–PBS from Southern Biotech, Al, USA). Following another wash (3× with 0.01% Tween-PBS), the wells were developed with the substrate (ABTS, peroxidase substrate system; Kirkegaard and Perry Laboratories, Gaithersburg, MD, USA). After 20−30 min of incubation, the optical densities were measured at 450 nm using an ELISA plate reader (Bio-Tek Instruments, Inc., Winooski, VT, USA).

2.11. Statistical Analysis

Graph-Pad Prism software, version 8 for Windows, was utilized to generate graphs and perform statistical analysis. All data are presented as means + standard deviations (SDs) and analyzed by Tukey's post hoc test, the Kruskal–Wallis non-parametric test, Wilcoxon's test (Mann–Whitney,) or Fisher's exact test with the results shown as the mean ± standard deviation by Graph-Pad Prism software, version 8. The results were considered statistically significant at $p < 0.05$ (*).

3. Results

3.1. Expression of IL-17A, COX-2, TGF-β, and IFN-γ in vvMDV-RB1B Infected Chickens

The expression of IL-17A in the spleens of vvMDV-RB1B- and non-infected chickens was probed using confocal microscopy and real-time PCR (Figure 1). Our results demonstrated the detection of IL-17A in the spleens of vvMDV-RB1B-infected and not in non-infected chickens at 4 and 10 dpi (Figure 1A,B). IL-17A could be detected in the spleens of non-infected chickens at 21 dpi (Figure 1C). Based on punctate fluorescence, expression of IL-17A (488 nm) at the protein level was the highest at 21 days post infection (dpi) (Figure 1C) in vvMDV-RB1B-infected chicken when compared to 4 (Figure 1A) and 10 dpi (Figure 1B). The expression of IL-17A was also confirmed at the mRNA level based on real-time PCR (Figure 1D). The results demonstrated that IL-17A expression was significantly higher ($p < 0.005$) at 4 dpi compared to non-infected chickens. No difference was observed at 10 dpi. At 21 dpi, the expression of IL-17A was significantly lower ($p < 0.005$) compared to non-infected chickens.

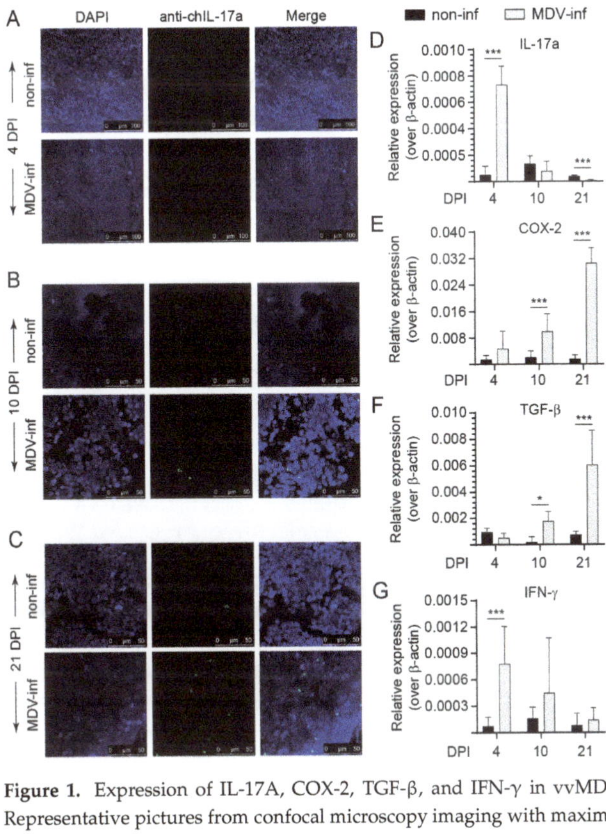

Figure 1. Expression of IL-17A, COX-2, TGF-β, and IFN-γ in vvMDV-RB1B-infected chickens. Representative pictures from confocal microscopy imaging with maximum projections of z-stacks for each channel demonstrating the localization of IL-17A (anti-IL-17A antibody; 488nm) in vvMDV-RB1B- and non-infected chickens at (**A**) 4, (**B**) 10, and (**C**) 21 dpi. Images are presented as maximum projections of all combined channels (merge). Real-time PCR analysis of (**D**) IL-17A, (**E**) COX-2, (**F**) TGF-β, and (**G**) IFN-γ gene expression at 4, 10, and 21 dpi in whole spleens. Target and reference gene expression was quantified by real-time-PCR, and expression is presented relative to β-actin expression. The results are based on at least 5–6 biological replicates in each group. Wilcoxon's non-parametric test (Mann–Whitney) was used to test significance with the results shown as means ± standard deviations. * ($p \leq 0.05$) and *** ($p \leq 0.005$) indicates a significant difference.

The expression of COX-2 (Figure 1E), TGF-β (Figure 1F), and IFN-γ (Figure 1G) was also assessed. The results demonstrated upregulation of both COX-2 (Figure 1E) and TGF-β (Figure 1F) transcripts during the later stages of MDV-RB1B infection. COX-2 transcripts were significantly higher at 10 ($p < 0.005$) and 21 dpi ($p < 0.005$) compared to non-infected chickens (Figure 1E). Furthermore, TGF-β transcripts were significantly higher at 10 ($p < 0.05$) and 21 dpi ($p < 0.005$) compared to non-infected chickens (Figure 1F). During the early stages of MDV-RB1B infection, IFN-γ transcripts were increased (Figure 1G). At 4 dpi, IFN-γ gene expression was significantly higher ($p < 0.005$) in MDV-RB1B infected chickens but not at 10 or 21 dpi compared to non-infected chickens (Figure 1G).

3.2. Detection of IL-17A+ and IFN-γ+ T Cells Post vvMDV Infection

Spleen mononuclear cells were probed to define the IL-17A-producing cells at 4, 10, and 21 dpi (Figure 2). The gating strategy per FACS analysis defining intracellular IL-17A+ αβ or γδ T cells is shown (Figure 2A). Intracellular IL-17A cytokine staining was confirmed in spleen mononuclear cells of both control and vvMDV-RB1B-infected chickens at 4, 10, and 21 dpi (Figure 2B–D). The results demonstrated an increase in the frequency of CD3ε+CD8α+IL-17A+ γδ (Figure 2B), CD3ε+CD4+IL-17A+ (Figure 2C), and CD3ε+CD8α+IL-17A+ αβ T cells (Figure 2D). Specifically, the frequency of CD8α+IL-17A+ γδ T cells was significantly increased at 10 ($p < 0.05$) and 21 dpi ($p < 0.01$) compared to non-infected chickens (Figure 2B). αβ T cells were subdivided into CD3ε+ CD4+/CD8α+ αβ T cells. Within CD4+ αβ T cells, the frequency of CD3ε+CD4+IL-17A+ αβ T cells was significantly increased at 21 dpi ($p < 0.05$) and not at 4 or 10 dpi compared to non-infected chickens (Figure 2C). Within CD8α+ αβ T cells, the frequency of CD8α+IL-17A+ αβ T cells was significantly increased at 10 ($p < 0.05$), and 21 dpi ($p < 0.01$) compared to non-infected chickens (Figure 2D).

Splenocytes derived from vvMDV-RB1B- and non-infected chickens were cultured and IFN-γ production in various T cell subsets (αβ or γδ T cells) were analyzed by flow cytometry. We have previously shown, based on intracellular cytokine staining, that there is no significant difference in the frequency of CD8α+IFN-γ+ γδ T at 4, 10, and 21 dpi between vvMDV-RB1B- and non-infected chickens [32]. The gating strategy to define CD4+IFN-γ+ αβ T cells and CD8α+IFN-γ+ αβ T cells is shown (Figure 2A). The frequency of CD4+IFN-γ+ αβ T cells was significantly higher in vvMDV-RB1B infected chickens at 10 ($p < 0.05$) and 21 dpi ($p < 0.01$) than that in the control group (Figure 2E). IFN-γ production was significantly increased in CD8α+ αβ T cells derived from spleens of MDV-challenged chickens compared to control chickens at 4 ($p < 0.05$), 10 ($p < 0.01$), and 21 dpi ($p < 0.05$) (Figure 2F).

3.3. Cloning, Expression and Evaluation of the rchIL-17A Bioactivity Based on Receptor Binding

The full-length coding sequence of chIL-17A with signal peptide was subcloned into a pCDNA3.1/V5-HIS-TOPO plasmid by digestion/ligation reactions from a pDRIVE-IL-17A vector (TA cloning) (Figure 3A,B). HEK 293 T cells were transfected with the linearized pCDNA3.1/V5-HIS-TOPO and pCDNA3.1/rchIL-17A-V5-HIS TOPO to generate supernatant containing the recombinant chIL-17a from stably expressing cell lines. Expression of rchIL-17A in the culture medium was confirmed by an anti-V5 ELISA (Figure 3C). To demonstrate that a biologically active rchIL-17A protein (Figure 3B) is expressed by a eukaryotic expression system, chicken spleen mononuclear cells were co-incubated with supernatants from pCDNA3.1/V5-HIS-TOPO (control) or pCDNA3.1/rchIL-17A-V5-HIS-TOPO (IL-17A+)-transfected HEK-293 T cells. Spleen mononuclear cells were further phenotypically differentiated as either APCs (KUL01+MHCII+) or T cells (CD3+ αβ or γδ) per FACS analysis to demonstrate specific interactions (Figure 3D). Surface binding of rchIL-17A to its biological receptor, IL-17RA, was probed, and the gating strategy demonstrated detection of the rchIL-17A on the cell surfaces of both APCs and various T cell subsets following incubation with both control and IL-17A+ supernatant (Figure 3D). The results demonstrated that rchIL-17A can bind to a cell surface receptor found on both

APCs and CD3+ T cells (αβ and γδ subsets). Co-incubation with the IL-17A+ supernatant delineated the frequency of APCs (Figure 3E), γδ (Figure 3F), and αβ (Figure 3G,H) CD3+ T cell subsets that express the cell surface IL-17RA molecule. Significantly higher ($p < 0.05$) frequencies of both CD4+ (Figure 3G) and CD8α+ (Figure 3H) αβ T cells were observed as expressing the IL-17RA complex.

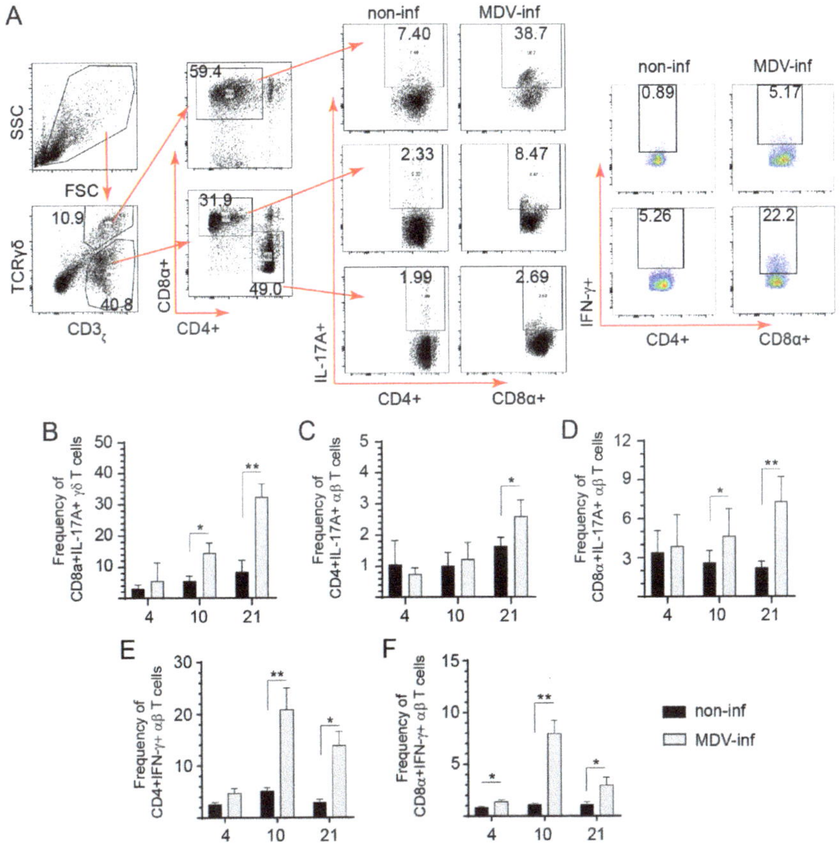

Figure 2. Frequency of IL-17A+ and IFN-γ+ T cells in spleens of vvMDV-RB1B-infected chickens at 4, 10, and 21, dpi. Representative FACS plots showing the frequency of IL-17A+ and IFN-γ+ T cells in the spleens of vvMDV-RB1B-infected chickens compared to non-infected chickens. (**A**) Dot plots demonstrate the gating strategy for identification of various CD3ε+ T cell subsets from vvMDV-RB1B- and non-infected chickens at 4, 10, and 21 dpi. The percentage of intracellular IL-17A+ (**B**) CD3ε+CD8α+ γδ T cells, (**C**) CD3ε+CD4+, and (**D**) CD3ε+CD8α+ αβ T cells in spleens from vvMDV-RB1B- and non-infected chickens are shown. Percentages of (**E**) CD4+IFN-γ+ αβ T cells within CD4+ αβ T cells and (**F**) CD8α+IFN-γ+ αβ T cells within CD8α^{+} αβ T cells. Wilcoxon's non-parametric test (Mann–Whitney) was used to test significance. Data represent means ± standard deviations (n = 5) at each time point. * ($p < 0.05$) and ** ($p < 0.01$) indicates a significant difference.

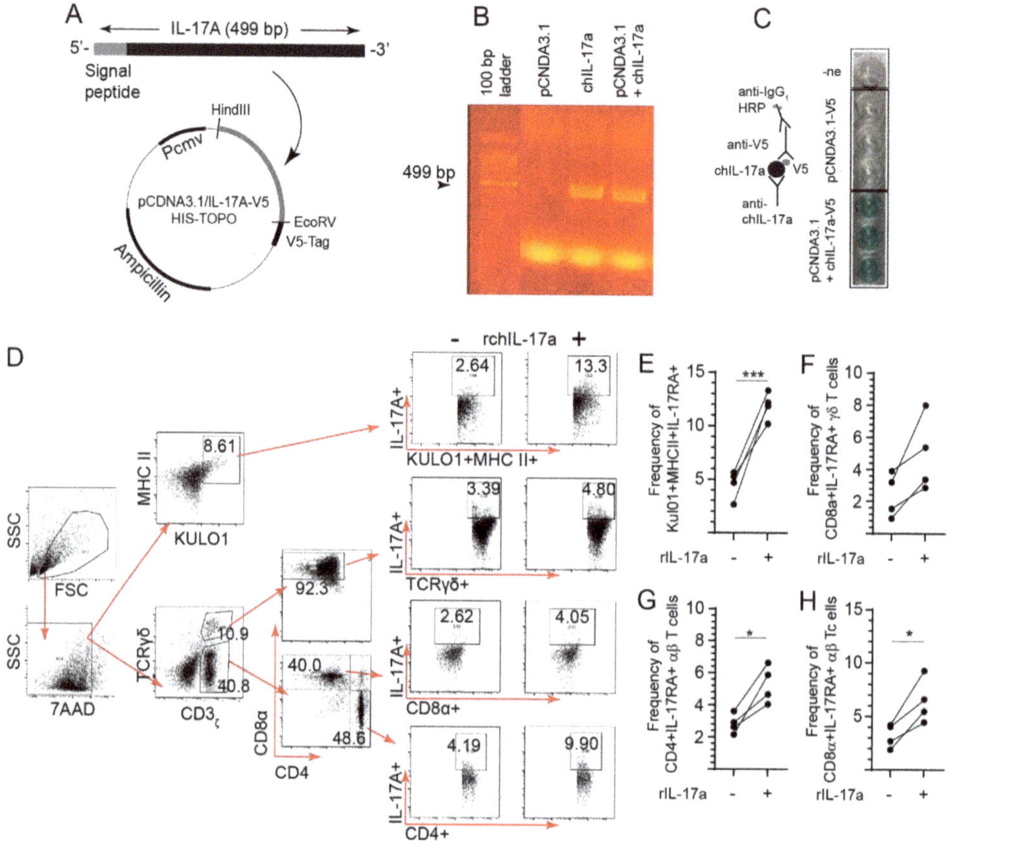

Figure 3. Overexpression of recombinant chicken IL-17A (rchIL-17A). Cloning of complete coding sequence of chicken IL-17A from PMA/ION-stimulated splenocytes into a (**A**) pCDNA3.1/V5-HIS-TOPO plasmid at the HindIII and EcoRV ligation sites. HEK-293 T cell were either mock-transfected (with empty vector) or with the recombinant plasmid (pCDNA3.1/V5-HIS-TOPO-rchIL-17A) to generate cell lines stably expressing IL-17A. Both cells and culture supernatant were collected 24 and 72 hrs post-transfection, respectively, to confirm expression of rchIL-17A at both (**B**) RNA level and (**C**) protein level. (**B**) Data shown are representative gel electrophoresis for visualization of rchIL-17A amplicon following standard PCR. (**C**) Indirect ELISA demonstrating extracellular expression of rchIL-17A. (**D**) FACS dot plot showing gating strategy to confirm bioactivity by functional interaction with the IL-17RA on (**E**) KUL01+MHCII+ (antigen presenting cells), (**F**) CD3ε+CD8α+ γδ T cell, and (**G**) CD3ε+CD4+ and (**H**) CD3ε+CD8α+ αβ T cells in live mononuclear cells by 7-AAD exclusion. The results are based on at least 5–6 biological replicates in each group. Wilcoxon's non-parametric test (Mann–Whitney) was used to test significance with the results shown as means ± standard deviations. * ($p \leq 0.05$) and *** ($p \leq 0.001$) indicates a significant difference.

3.4. Neither CpG ODN 2007 Nor Poly:IC Treatment Induce IL-17A Expression

The class B CpG ODN 2007 (5 µg/mL) and LPS (10 µg/mL) were used for in vitro stimulation of spleen mononuclear cells (Figure 4). After overnight culture, supernatants were collected for indirect ELISA and detection of IL-17A (Figure 4A). The results demonstrated that LPS, but not CpG ODN 2007, is an inducer of IL-17A expression. We further demonstrated that TLR3 or TLR21 signaling following intramuscular inoculation of 2-week-old chicks with either poly:(IC) or CpG ODN, respectively, did not lead to IL-17A expression at 8 and 18 hrs post-inoculation compared to PBS-treated chickens (Figure 4B).

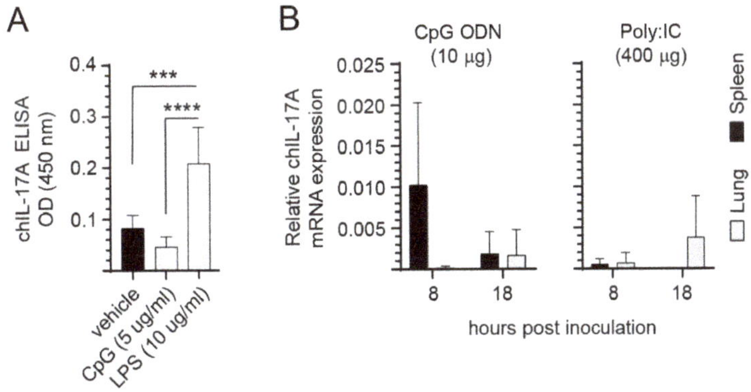

Figure 4. TLR ligands and IL-17a cytokine expression. The effect of TLR ligands on IL-17A expression both in vitro and in vivo. (**A**) Spleen mononuclear cells were isolated from 10-day-old chicks (n = 6) for in vitro stimulation with the TLR ligands CpG ODN 2007 and LPS. (**B**) Real-time PCR analysis for IL-17A expression in 2-week-old chicks (n = 36) injected intramuscularly (I/M) with 100 μL of either CpG ODN 2007 (10 μg) or Poly I:C (400 μg), respectively. Target and reference gene expression was quantified by real-time-PCR, and expression is presented relative to β-actin expression. The results are based on 12 biological replicates in each group. Wilcoxon's non-parametric test (Mann–Whitney) was used to test significance with the results shown as means ± standard deviations. *** ($p \leq 0.001$) and **** ($p \leq 0.0001$) indicates a significant difference.

3.5. Modulation of Cytokine Expression following rchIL-17A Inoculation

Three-week-old chicks were inoculated via the intramuscular route with 10 μg per chick of either the pCDNA3.1/V5-HIS-TOPO or pCDNA3.1/V5-HIS-TOPO-rchIL-17A plasmid (Figure 5). Twenty-four and 48 h later, whole spleens were collected for real-time PCR analysis. The differential gene expression analysis is presented for IL-1β (Figure 5A), TGF-β (Figure 5B), IL-2 (Figure 5C), IL-6 (Figure 5D), IL-17A (Figure 5E), IL-10 (Figure 5F), IL-12p40 (Figure 5G), and IFN-γ (Figure 5H). Pre-treatment with the rchIL-17A expressing plasmid modulated the expression of the various pro-inflammatory mediators compared to chicks treated with the control plasmid. The results demonstrated that expression of both IL-1β (Figure 5A) and TGF-β (Figure 5B) was significantly increased ($p < 0.005$) at 24, but not 48, hours post-treatment with the pCDNA3.1/rchIL-17A-V5-HIS TOPO plasmid compared to the pCDNA3.1/V5-HIS TOPO-treated chickens. No changes in IL-2 (Figure 5C), IL-6 (Figure 5D), and IL-17A (Figure 5E) mRNA transcripts were detected between pCDNA3.1/V5-HIS-TOPO-rchIL-17A- and pCDNA3.1/V5-HIS-TOPO-treated chickens. However, pre-treatment with the rchIL-17A overexpressing plasmid suppressed IL-10 (Figure 5F), IL-12p40 (Figure 5G), and IFN-γ (Figure 5H) gene expression. Specifically, the expression of IL-12p40 (Figure 5G) and IFN-γ (Figure 5H) was significantly decreased ($p < 0.005$) at both 24 and 48 hrs post-inoculation with the pCDNA3.1/rchIL-17A-V5-HIS TOPO plasmid compared to the pCDNA3.1/V5-HIS-TOPO-treated chickens. Expression of IL-10 (Figure 5F) was significantly decreased ($p < 0.05$) only at 48 h post-inoculation with the pCDNA3.1/rchIL-17A-V5-HIS TOPO plasmid compared to the pCDNA3.1/V5-HIS-TOPO-treated chickens.

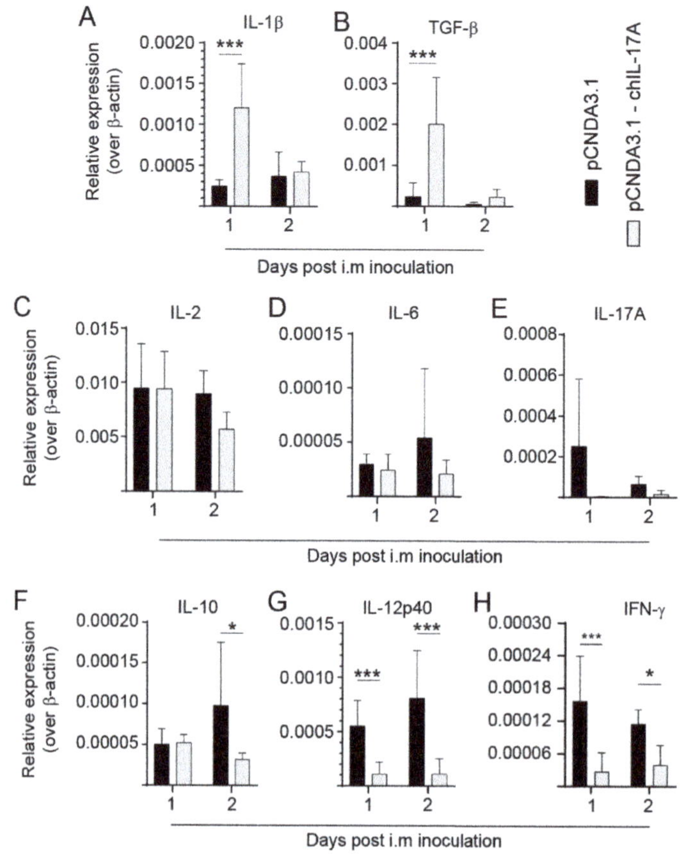

Figure 5. Differential gene expression in chickens pre-treated with the rchIL-17A overexpressing plasmid. Three-week-old chicks (n = 24) were inoculated via intramuscular route with 10 μg per chick of either the pCDNA3.1/V5-HIS-TOPO or pCDNA3.1/V5-HIS-TOPO-rchIL-17A plasmid (Figure 5). Twenty-four (n = 12) and 48 h later (n = 12), whole spleens were collected for real-time PCR analysis. The differential gene expression analysis is presented for **(A)** IL-1-β, **(B)** TGF-β, **(C)** IL-2, **(D)** IL-6, **(E)** IL-17A, **(F)** IL-10, **(G)** IL-12p40, and **(H)** IFN-γ. Target and reference gene expression was quantified by real-time PCR, and expression is presented relative to β-actin. The results are based on 6 biological replicates in each group. Wilcoxon's non-parametric test (Mann–Whitney) was used to test significance with the results shown as means ± standard deviations. * ($p \leq 0.05$) and *** ($p \leq 0.001$) indicates a significant difference.

MD Tumor Incidence in rchIL-17A-Treated and HVT-Vaccinated Chickens

Chicks were inoculated with 10 μg of the pCDNA3.1/V5-HIS-TOPO or pCDNA3.1/ rchIL-17A-V5-HIS TOPO plasmids 24 h prior to infection with an MDV-RB1B virus, and they were necropsied at 21 dpi to assess the presence of MD lymphoma in visceral organs (Figure 6A). Tumor incidence defines the number of individual chickens that presented with tumors. One hundred percent of all chickens (15/15 chickens) infected with only the vvMDV-RB1B virus developed MD tumors, while no tumors were observed in the non-infected/non-vaccinated group (Figure 6B). The incidence of MD tumors was reduced in chickens that received half the dose of HVT + pCDNA3.1/V5-HIS-TOPO-rchIL-17A (50%; 7/15) compared to the groups of chicken that were vaccinated with half the dose of HVT and subsequently challenged with pCDNA3.1/V5-HIS-TOPO/RB1B (73.3%; 10/15 chickens) or RB1B only (66.6%; 11/15 chickens) (Figure 6B). The group of chickens that received only pCDNA3.1/V5-HIS-TOPO-rchIL-17A treatment before infection with vvRB1B had a higher tumor incidence (80%; 12/15) than chickens that received half the dose of HVT and were subsequently challenged with RB1B (73.3%; 10/15 chickens) (Figure 6B), although no significant differences were observed.

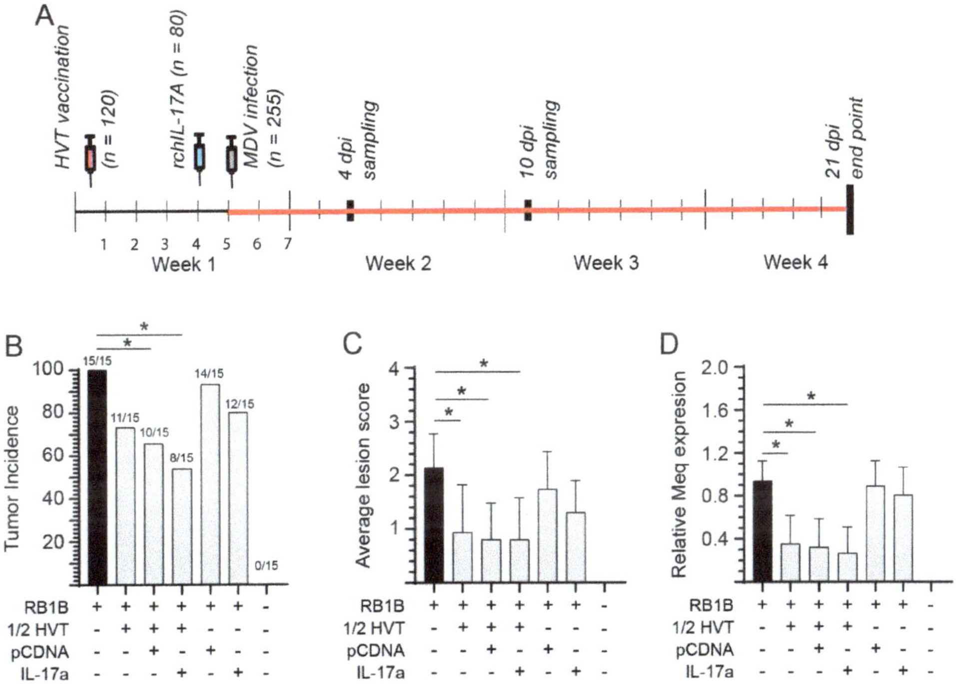

Figure 6. MDV tumor incidence at 21 dpi. Assessment of MD disease severity in vvMDV-RB1B-infected chickens based on presence of gross tumor lesions. (**A**) Schematic diagram depicting the experimental timeline for various treatments within each group. Day-old chicks were randomly allocated to each group. Sampling was performed at 4, 10, and 21 dpi. (**B**) Tumor incidence in the various groups was calculated by observing gross tumors in visceral organs of birds at 21 dpi. (**C**) Assessment of lesion score in different treatment groups based on the number of visceral and thoracic organs showing MDV-tumor lesions were counted in each bird, and the average score was calculated. (**D**) vvMDV-*Meq* gene expression at 21 dpi as quantified by real-time PCR and presented relative to β-actin expression. The results are based on at least 15 biological replicates in each group. (**B**) Fisher's exact test or (**C**,**D**) Wilcoxon's non-parametric test (Mann–Whitney) was used to test significance with the results shown as means ± standard deviations. * ($p \leq 0.05$) indicates a significant difference.

Further, at 21 dpi, the average lesion scores (the number of organs showing MD lesions per bird) within each group were evaluated (Figure 6C). The average lesion scores in chickens infected only with vvMDV-RB1B were significantly higher ($p < 0.05$) compared to chickens that were vaccinated (HVT) and subsequently infected (vvMDV-RB1B), vaccinated, (HVT)-IL-17A treated, and subsequently infected (vvMDV-RB1B) or vaccinated, (HVT)-pCDNA3.1 treated, and subsequently infected (vvMDV-RB1B). No differences were observed between chickens that were IL-17A treated and subsequently infected (vvMDV-RB1B) and those that were pCDNA3.1 treated and subsequently infected (vvMDV-RB1B) compared to vvMDV-RB1B infected chickens only (Figure 6C).

Replication of vvMDV-RB1B was assessed by real-time PCR for the expression of MDV-*Meq* in spleens at 21 dpi (Figure 6D). No MDV-*Meq* transcripts were detected in the control chickens. Chicks that were vaccinated (HVT) and infected (vvMDV-RB1B) had significantly lower ($p < 0.05$) MDV-*Meq* transcripts compared to chickens that were only vvMDV-RB1B-infected but not HVT+rchIL-17A+vvMDV-RB1B or HVT+pCDNA+vvMDV-RB1B-treated chickens at 21 dpi. No significant differences in MDV-*Meq* transcripts were observed between the HVT+vvMDV-RB1B-treated chickens and HVT+rchIL-17A+vvMDV-RB1B- or HVT+pCDNA+vvMDV-RB1B-treated chickens at 21 dpi. Furthermore, no significant differences in MDV-*Meq* transcripts were observed between the vvMDV-RB1B-infected chickens and rchIL-17A+vvMDV-RB1B or pCDNA+vvMDV-RB1B chickens.

Differential Cytokine Expression in Spleen of vvMDV-RB1B Infected Chickens Pre-Treated with the rchIL-17A

To demonstrate the potential modulatory effects of rchIL-17A in MD, real-time PCR analysis was used to assess the differential expression of genes (COX-2, IL-10, IL-17A, TGF-β, IL-12p40, and IFN-γ) in spleens at 21-dpi (Figure 7). The results demonstrated that chickens that were vaccinated (HVT) and/or treated with the rchIL-17A or empty vector and subsequently infected (vvMDV-RB1B) had significantly lower ($p < 0.01$) expression of COX-2 (Figure 7A), IL-10 (Figure 7B), and TGF-β (Figure 7C) compared to vvMDV-RB1B-infected chickens. COX-2 (Figure 7A) and TGF-β (Figure 7C; $p < 0.01$), but not IL-10 (Figure 7B; $p < 0.05$), transcripts were significantly lower in the rchIL-17A-vector-treated and subsequently infected (vvMDV-RB1B) group compared to the vvMDV-RB1B infected only chickens. Expression of IL-17A was significantly lower ($p < 0.001$) in the rchIL-17A- or empty-vector control and infected chickens (vvMDV-RB1B) compared to non-infected chickens (Figure 7D). HVT vaccination and/or pre-treatment with the rchIL-17A or empty vector followed by infection with vvMDV-RB1B resulted in an increase in IL-17A transcript levels (Figure 7D). No difference was observed between HVT and/or rchIL-17A- or empty vector-treated and infected chickens compared to the non-infected chickens (Figure 7D). IL-12p40 is essential for the induction of IFN-γ expression. While IL-12p40 expression was significantly increased compared to non-infected chickens (Figure 7E), no differences in IFN-γ transcripts were observed between vvMDV-RB1B-infected and non-infected chickens at 21 dpi (Figure 7F). Pre-treatment with the rchIL-17A or empty vector in vvMDV-RB1B-infected chickens did not result in changes in expression of either IL-12p40 (Figure 7E) or IFN-γ (Figure 7F) transcripts compared to vvMDV-RB1B-infected chickens only. The combination with HVT resulted in an increase in both IL-12p40 (Figure 7E) and IFN-γ (Figure 7F) transcripts. Specifically, chickens that were inoculated with HVT and/or that were rchIL-17A or empty vector treated and subsequently infected (vvMDV-RB1B) had significantly higher expression of both IL-12p40 (Figure 7E; $p < 0.01$) and IFN-γ (Figure 7F; $p < 0.01$) genes compared to the non-infected chickens.

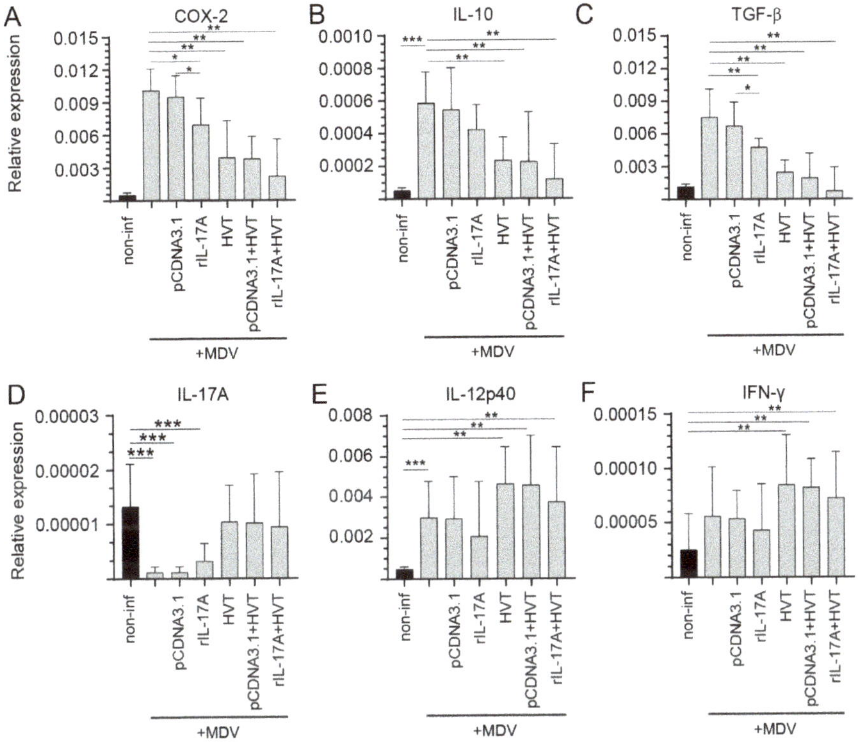

Figure 7. Expression of various cytokines in vvMDV-RB1B-infected chickens pre-treated with the rchIL-17A overexpressing plasmid at 21 dpi. Gene expression analysis at 21 dpi in vvMDV-RB1B-infected chickens that were pre-treated with the rchIL-17A overexpressing plasmid. Expression of the target genes (**A**) COX-2, (**B**) IL-10, (**C**) IL-17A, (**D**) TGF-β, (**E**) IL-12p40, and (**F**) IFN-γ was quantified by real-time PCR, and expression is presented relative to the reference gene (β-actin) in whole spleens. The results are based on at least 6 biological replicates in each group. Wilcoxon's non-parametric test (Mann–Whitney) was used to test significance with the results shown as the means ± standard deviations. * ($p \leq 0.05$) ** ($p \leq 0.01$) and *** ($p \leq 0.001$) indicates a significant difference.

4. Discussion

The IL-17A cytokine is a T cell-derived proinflammatory cytokine that has an important role in inflammation and immunity [1,7]. IL-17A is believed to be mainly produced by Th17 cells, which constitute a unique helper T-cell subset well defined in mice and humans [33]. While avian Th17 cells have yet to be characterized, IL-17A gene expression has recently been observed in mitogen stimulated avian αβ and γδ T cells and further detected via intracellular cytokine staining in various T cell subsets [5,6]. In this study, we demonstrated that: (i) infection with MDV resulted in early induction of IL-17A and IFN-γ expression in chicken splenocytes; (ii) based on FACS analysis, IL-17A+ T cells could be detected in splenocytes from vvMDV-RB1B infected chickens with the highest proportion of IL-17A+ T cells peaking at 21 dpi; and (iii) in vivo inoculation with a recombinant chIL-17A post-HVT vaccination significantly reduced tumor incidence compared to infected-only chickens compared to HVT+IL-17A or HVT+vector control. This outcome was associated with a reduction in COX-2, TGF-β, and MDV-*Meq* gene expression and an increase in IL-12p40 and IFN-γ expression.

The function and expression of IL-17A in the avian lung mucosal tissue following infection with MDV have not been previously defined. While the avian host's early response

(4 dpi) to MDV infection results in overexpression of both IL-17A and IFN-γ, their respective roles in chickens are not well understood. Due to the cell-associated nature of MDV, T cells play an essential role in immunity against this virus. We previously reported that the frequency of IFN-γ+CD8α+ γδ T cells in vvMDV-RB1B-infected chickens was significantly higher at 4 ($p < 0.05$) and 10 dpi ($p < 0.01$) than in the control group [32]. Pre-treatment with a recombinant chIFN-γ in an MDV infection model was associated with a reduction in tumor incidence [19]. However, subsequent IFN-γ knockdown based on an siRNA treatment had limited to no effects on MDV replication and tumor incidence [34]. More recently, Bertzbach et al. showed that chIFN-γ can have direct anti-viral effects against MDV and was induced by NK cells in vitro following MDV infection [35,36]. This finding indicated that other mechanisms exist that can limit MDV replication. In the present study, we demonstrated a relationship between the specific stages of MDV pathogenesis (4, 10, and 21 dpi) and percentage of IFN-γ+ and IL-17A+ T cells. Based on intracellular cytokine staining, the frequency of IFN-γ+ αβ T cells peaked at 10 dpi and was significantly lower at 4 and 21 dpi compared to 10 dpi. IL-17A+ T cells could be detected in splenocytes from vvMDV-RB1B infected chickens with the highest proportion of IL-17A+ αβ and γδ T cells peaking at 21 dpi. It is unclear whether the reduction in IFN-γ+ and increase in IL-17A+ T cells at 21 dpi relate to CD4+ T cell expansion and aggravation of lymphoma. In mice, IFN-γ is the major cytokine produced by CD27+ γδ T cells, and their CD27− counterparts preferentially secrete IL-17A [37,38]. In the present study, double positive IL-17A+IFN-γ+ T cells could not be defined due to limitations in antibody availability. Both IL-17A and IFN-γ can regulate the lineage differentiation of naïve T cells into Th1 or Th17 cells, respectively [33]. The presence of both IFN-γ+ and IL-17A+ T cells during MDV infection suggests a potential role in MDV pathogenesis.

The biological role of IL-17A has yet to be fully characterized in chickens but has been defined in mice and humans [39]. We observed a discrepancy between the abundance of IL-17A cognate protein and transcripts following MDV infection. Protein and RNA represent different steps of the multi-stepped cellular process, in which they are expressed, translated, and degraded. Specific regulatory processes can suppress mRNA transcription, thereby leading to no or limited protein expression. MDV is known to express and stimulate the expression of an array of micro RNAs that impact the host cellular machinery [40]. At mucosal sites, IL-17 may have protective roles by enhancing the Th1 response in host defenses against infectious diseases and promoting the induction of cytotoxic T lymphocyte (CTL) responses [33]. HVT-vaccinated chickens that were pre-treated with the rIL-17A had an increase in IL-12p40 and IFN-γ expression at 21-dpi compared to vvMDV-RB1B-infected chickens. This outcome was associated with a reduction in the expression of IL-10 and TGF-β at 21 dpi in IL-17A-pre-treated and HVT-vaccinated chickens. To gain some insight into the immune mechanisms triggered by treatment with IL-17A in chickens, we analyzed the cytokine expression profiles in the spleen. Both negative and positive feedback loops regulate the local and system effects of IL-17A.

Compared to negative control groups, vaccination with HVT does not result in a significant increase in IL-17A expression. Similarly, Hao et al. demonstrated that vaccination with the CVI988-RISPENS vaccine does not induce IL-17A expression [41]. This finding suggests that HVT vaccination elicits a cytokine-specific pattern of expression that works in synergy with pCDNA:IL-17A to modulate the IL-17A-mediated immunopathology observed post-MDV infection. In chickens, infection with a vvMDV-RB1B virus led to a significant increase in IL-17A expression at 4 dpi with no changes at 10 dpi and a significant reduction at 21 dpi compared to control chickens. It is possible that the increased production of proinflammatory cytokines in the absence of IL-17A during vaccination is a compensatory mechanism implemented to help protect the avian host against MDV infection. This theory suggests that IL-17A could be involved in modulating the lung mucosal environment. Viruses can utilize IL-17A to modulate the host immune system to support productive infection. *Gammaherpesviruses*, such as murine gammaherpesvirus 68 (MHV68), preferentially modulate the mouse immune system following primary infection, leading to IL-17A expression [42]. As such, during early infection, IL-17A can antagonize Th1 activity while promoting productive MHV68 viral infection. IL-17A directly suppresses the expression of IFN-γ, T-bet, and eomesodermin in T cells [43]. In MDV infection, we observed an increase in both IL-17A and IFN-γ during the early stage of infection. While antigen-specific T cell responses based on IFN-γ could be detected in MDV-infected chickens, it is not clear whether overexpression of IL-17A could limit antigen-specific T cell responses [18]. However, following MDV infection, IL-17A expression was induced during the early phase (4 dpi). More work is required to define whether IL-17A could be involved in the pathogenesis of MD.

In mammals, T cells of the γδ subset immediately produce IL-17A and induce inflammation upon pathogenic infections [44]. It is understood that immunity against herpes simplex virus type (HSV) 1 is partly imparted by the functional activity of IL-17A+ γδ T cells [45]. γδ T cells exert their immune-modulatory activity by secreting various innate factors, as well as Th1 and Th17 cytokines, which promote direct cytotoxic activity against infected cells [46]. During the early phase of MDV pathogenesis, high counts of both macrophages and T cells can be seen in the lungs of MDV-infected chickens [47]. Our FACS results indicated that γδ T cells are a major source of IL-17A during the later stages of MDV infection. It cannot be ruled out that the differentiation of IL-17A-producing γδ T cells occurs within the tumor microenvironment, but this possibility requires further experimental confirmation. IL-17A has been implicated in protective γδ T cell responses in other cancer models. Mechanistically, tumor-infiltrating IL-1β activated IL-17A+ γδ T cells, enhancing the priming and recruitment of IFN-γ-producing CD8+ T cells that exert antitumor effects [48]. We have previously reported that avian γδ T cells are activated during MDV infection and retain effector functions, based on expression of IFN-γ and cytotoxic (CD107a+) activity [32]. It is not known whether the increase in IL-17A at 4 dpi relates to an increase in T cell infection and modulation of antigen-presenting cell functions, such as macrophages [33]. However, IL-17 can promote the recruitment of innate cells to the site of inflammation [49]. We identified an increase in the percentage of IL-17A+ T cells at 21 dpi in the spleens of MDV-infected chickens. Chronic IL-17A production, at the protein level, has been shown to support tumor formation and progression [50]. This effect is derived from IL-17A-mediated signaling, which results in recruitment of innate immune cells to the site of injury [51]. IL-17-induced inflammatory mediators, such as IL-1β, IL-6, and TGF-β, promote a proangiogenic and immune suppressive tumor environment that enhances tumor growth. It is likely that the host response to MDV infection, which results in IL-17A progression, contributes to tumor formation. MDV is an oncogenic virus causing CD4+ T cell lymphoma in chickens. Transformation of CD4+ T cells has been linked to MDV-*Meq* function [52]. MDV-*Meq* has been shown to modulate the PI3K/AKT pathway [53]. In mice, the PI3K/AKT pathway regulates IL-17A expression and the induction of Th17 cells [54]. However, the mechanism for induction of IL-17A in MDV-infected chickens is unknown. Furthermore, it has yet to be shown whether IL-17A-mediated signaling during MDV results in tumor progression. Taken together, induction of IL-17A during MDV infection can have multifactorial effects.

5. Conclusions

In summary, our data revealed the effects of IL-17A during MDV infection and improved our understanding of adaptive immune responses against MDV. Preventive strategies that aim to induce strong Th1 cell immunity should be able to effectively control MDV infection. Along with Th1 responses, MDV infection also induces a Th17 immune response. More work is required to demonstrate this response in avian Th17 cells and the modulatory activity of IL-17A in avian innate and adaptive immune cell types.

Author Contributions: N.B. and S.S. designed the experiment. N.B. performed the experiment, and N.B., F.F. and M.S.-D. collected the samples. N.B., S.R., M.S.-D. and A.M.-K. processed all samples. N.B., A.M.-K. and S.R. analyzed the data, prepared the figures, and wrote the first draft of the manuscript. vvMDV-RB1B preparation was performed by F.F. The manuscript was critically reviewed by N.B., S.R., F.F., M.S.-D., A.M.-K. and S.S. All authors have read and agreed to the published version of the manuscript.

Funding: The support for the current work was provided by the Ontario Ministry of Agriculture, Food and Rural Affairs, and Natural Sciences and Engineering Research Council of Canada (NSERC).

Institutional Review Board Statement: All experiments and procedures were approved by the Animal Care Committee at the University of Guelph and complied with specifications of the Canadian Council on Animal Care. (Approval code: 4328; Approval date: 12 July 2020).

Data Availability Statement: Not applicable.

Acknowledgments: We acknowledge the isolation facility team at Ontario Veterinary College, University of Guelph, for providing constant support with the regular checks and managerial practices for the experimental chickens.

Conflicts of Interest: The authors declare no conflict of interest.

References

1. Moseley, T.; Haudenschild, D.; Rose, L.; Reddi, A. Interleukin-17 family and IL-17 receptors. *Cytokine Growth Factor Rev.* **2003**, *14*, 155–174. [CrossRef]
2. Veldhoen, M.; Hocking, R.J.; Atkins, C.J.; Locksley, R.M.; Stockinger, B. TGFβ in the context of an inflammatory cytokine milieu supports de novo differentiation of IL-17-producing T cells. *Immunity* **2006**, *24*, 179–189. [CrossRef]
3. Kaiser, P.; Poh, T.Y.; Rothwell, L.; Avery, S.; Balu, S.; Pathania, U.S.; Hughes, S.; Goodchild, M.; Morrell, S.; Watson, M.; et al. A Genomic Analysis of Chicken Cytokines and Chemokines. *J. Interf. Cytokine Res.* **2005**, *25*, 467–484. [CrossRef]
4. Veldhoen, M. Interleukin 17 is a chief orchestrator of immunity. *Nat. Immunol.* **2017**, *18*, 612–621. [CrossRef]
5. Walliser, I.; Göbel, T.W. Generation of glycosylphosphatidylinositol linked chicken IL-17 to generate specific monoclonal antibodies applicable for intracellular cytokine staining. *Dev. Comp. Immunol.* **2017**, *73*, 27–35. [CrossRef] [PubMed]
6. Walliser, I.; Göbel, T.W. Chicken IL-17A is expressed in αβ and γδ T cell subsets and binds to a receptor present on macrophages, and T cells. *Dev. Comp. Immunol.* **2018**, *81*, 44–53. [CrossRef]
7. Cua, D.J.; Tato, C.M. Innate IL-17-producing cells: The sentinels of the immune system. *Nat. Rev. Immunol.* **2010**, *10*, 479–489. [CrossRef] [PubMed]
8. Park, H.; Li, Z.; Yang, X.O.; Chang, S.H.; Nurieva, R.; Wang, Y.-H.; Wang, Y.; Hood, L.; Zhu, Z.; Tian, Q.; et al. A distinct lineage of CD4 T cells regulates tissue inflammation by producing interleukin 17. *Nat. Immunol.* **2005**, *6*, 1133–1141. [CrossRef] [PubMed]
9. Harrington, L.E.; Hatton, R.D.; Mangan, P.R.; Turner, H.; Murphy, T.L.; Murphy, K.M.; Weaver, C.T. Interleukin 17–producing CD4+ effector T cells develop via a lineage distinct from the T helper type 1 and 2 lineages. *Nat. Immunol.* **2005**, *6*, 1123–1132. [CrossRef]
10. Patera, A.C.; Pesnicak, L.; Bertin, J.; Cohen, J.I. Interleukin 17 Modulates the Immune Response to Vaccinia Virus Infection. *Virology* **2002**, *299*, 56–63. [CrossRef]
11. Peng, T.; Chanthaphavong, R.S.; Sun, S.; Trigilio, J.A.; Phasouk, K.; Jin, L.; Layton, E.D.; Li, A.Z.; Correnti, C.E.; De van der Schueren, W.; et al. Keratinocytes produce IL-17c to protect peripheral nervous systems during human HSV-2 reactivation. *J. Exp. Med.* **2017**, *214*, 2315–2329. [CrossRef] [PubMed]
12. Bagri, P.; Anipindi, V.C.; Nguyen, P.V.; Vitali, D.; Stampfli, M.; Kaushic, C. Novel Role for Interleukin-17 in Enhancing Type 1 Helper T Cell Immunity in the Female Genital Tract following Mucosal Herpes Simplex Virus 2 Vaccination. *J. Virol.* **2017**, *91*, e01234-17. [CrossRef] [PubMed]
13. Lahiri, A.; Sharif, S.; Mallick, A.I. Intragastric delivery of recombinant Lactococcus lactis displaying ectodomain of influenza matrix protein 2 (M2e) and neuraminidase (NA) induced focused mucosal and systemic immune responses in chickens. *Mol. Immunol.* **2019**, *114*, 497–512. [CrossRef]
14. Boodhoo, N.; Gurung, A.; Sharif, S.; Behboudi, S. Marek's disease in chickens: A review with focus on immunology. *Veter. Res.* **2016**, *47*, 119. [CrossRef]
15. Calnek, B.W. Pathogenesis of Marek's Disease Virus Infection. In *Marek's Disease*; Current Topics in Microbiology and Immunology; Hirai, K., Ed.; Springer: Berlin/Heidelberg, Germany, 2001; Volume 255, pp. 25–55. [CrossRef]
16. Boodhoo, N.; Kamble, N.; Kaufer, B.B.; Behboudi, S. Replication of Marek's Disease Virus Is Dependent on Synthesis of De Novo Fatty Acid and Prostaglandin E2. *J. Virol.* **2019**, *93*, e00352-19. [CrossRef]
17. Schat, K.; Calnek, B.; Fabricant, J. Characterisation of two highly oncogenic strains of Marek's disease virus. *Avian Pathol.* **1982**, *11*, 593–605. [CrossRef]

18. Boodhoo, N.; Behboudi, S. Differential Virus-Specific IFN-Gamma Producing T Cell Responses to Marek's Disease Virus in Chickens with B19 and B21 MHC Haplotypes. *Front. Immunol.* **2022**, *12*, 784359. [CrossRef]
19. Haq, K.; Elawadli, I.; Parvizi, P.; Mallick, A.I.; Behboudi, S.; Sharif, S. Interferon-γ influences immunity elicited by vaccines against very virulent Marek's disease virus. *Antivir. Res.* **2011**, *90*, 218–226. [CrossRef]
20. Lee, Y.K.; Turner, H.; Maynard, C.L.; Oliver, J.R.; Chen, D.; Elson, C.O.; Weaver, C.T. Late Developmental Plasticity in the T Helper 17 Lineage. *Immunity* **2009**, *30*, 92–107. [CrossRef]
21. Gurung, A.; Kamble, N.; Kaufer, B.B.; Pathan, A.; Behboudi, S. Association of Marek's Disease induced immunosuppression with activation of a novel regulatory T cells in chickens. *PLoS Pathog.* **2017**, *13*, e1006745. [CrossRef]
22. Kamble, N.; Gurung, A.; Kaufer, B.B.; Pathan, A.A.; Behboudi, S. Marek's Disease Virus Modulates T Cell Proliferation via Activation of Cyclooxygenase 2-Dependent Prostaglandin E2. *Front. Immunol.* **2021**, *12*, 801781. [CrossRef]
23. Abdul-Careem, M.F.; Read, L.R.; Parvizi, P.; Thanthrige-Don, N.; Sharif, S. Marek's disease virus-induced expression of cytokine genes in feathers of genetically defined chickens. *Dev. Comp. Immunol.* **2009**, *33*, 618–623. [CrossRef]
24. Boodhoo, N.; Sharif, S.; Behboudi, S. 1α,25(OH)$_2$ Vitamin D3 Modulates Avian T Lymphocyte Functions without Inducing CTL Unresponsiveness. *PLoS ONE* **2016**, *11*, e0150134. [CrossRef]
25. Bavananthasivam, J.; Alizadeh, M.; Astill, J.; Alqazlan, N.; Matsuyama-Kato, A.; Shojadoost, B.; Taha-Abdelaziz, K.; Sharif, S. Effects of administration of probiotic lactobacilli on immunity conferred by the herpesvirus of turkeys vaccine against challenge with a very virulent Marek's disease virus in chickens. *Vaccine* **2021**, *39*, 2424–2433. [CrossRef]
26. Brisbin, J.T.; Gong, J.; Parvizi, P.; Sharif, S. Effects of Lactobacilli on Cytokine Expression by Chicken Spleen and Cecal Tonsil Cells. *Clin. Vaccine Immunol.* **2010**, *17*, 1337–1343. [CrossRef]
27. Yitbarek, A.; Rodriguez-Lecompte, J.C.; Echeverry, H.M.; Munyaka, P.; Barjesteh, N.; Sharif, S.; Camelo-Jaimes, G. Performance, histomorphology, and Toll-like receptor, chemokine, and cytokine profile locally and systemically in broiler chickens fed diets supplemented with yeast-derived macromolecules. *Poult. Sci.* **2013**, *92*, 2299–2310. [CrossRef]
28. Paul, M.S.; Paolucci, S.; Barjesteh, N.; Wood, R.D.; Schat, K.A.; Sharif, S. Characterization of Chicken Thrombocyte Responses to Toll-Like Receptor Ligands. *PLoS ONE* **2012**, *7*, e43381. [CrossRef]
29. Taha-Abdelaziz, K.; Alkie, T.N.; Hodgins, D.C.; Yitbarek, A.; Shojadoost, B.; Sharif, S. Gene expression profiling of chicken cecal tonsils and ileum following oral exposure to soluble and PLGA-encapsulated CpG ODN, and lysate of *Campylobacter jejuni*. *Vet. Microbiol.* **2017**, *212*, 67–74. [CrossRef]
30. Crhanova, M.; Hradecka, H.; Faldynova, M.; Matulova, M.; Havlickova, H.; Sisak, F.; Rychlik, I. Immune Response of Chicken Gut to Natural Colonization by Gut Microflora and to Salmonella enterica Serovar Enteritidis Infection. *Infect. Immun.* **2011**, *79*, 2755–2763. [CrossRef]
31. Brisbin, J.T.; Zhou, H.; Gong, J.; Sabour, P.; Akbari, M.R.; Haghighi, H.R.; Yu, H.; Clarke, A.; Sarson, A.J.; Sharif, S. Gene expression profiling of chicken lymphoid cells after treatment with Lactobacillus acidophilus cellular components. *Dev. Comp. Immunol.* **2008**, *32*, 563–574. [CrossRef] [PubMed]
32. Matsuyama-Kato, A.; Iseki, H.; Boodhoo, N.; Bavananthasivam, J.; Alqazlan, N.; Abdul-Careem, M.F.; Plattner, B.L.; Behboudi, S.; Sharif, S. Phenotypic characterization of gamma delta (γδ) T cells in chickens infected with or vaccinated against Marek's disease virus. *Virology* **2022**, *568*, 115–125. [CrossRef]
33. Xu, S.; Cao, X. Interleukin-17 and its expanding biological functions. *Cell. Mol. Immunol.* **2010**, *7*, 164–174. [CrossRef]
34. Haq, K.; Wootton, S.K.; Barjesteh, N.; Golovan, S.; Bendall, A.; Sharif, S. Effects of interferon-γ knockdown on vaccine-induced immunity against Marek's disease in chickens. *Can. J. Vet. Res.* **2015**, *79*, 7.
35. Bertzbach, L.D.; Harlin, O.; Härtle, S.; Fehler, F.; Vychodil, T.; Kaufer, B.B.; Kaspers, B. IFNα and IFNγ Impede Marek's Disease Progression. *Viruses* **2019**, *11*, 1103. [CrossRef]
36. Bertzbach, L.D.; van Haarlem, D.A.; Härtle, S.; Kaufer, B.B.; Jansen, C.A. Marek's Disease Virus Infection of Natural Killer Cells. *Microorganisms* **2019**, *7*, 588. [CrossRef]
37. Haas, J.D.; Ravens, S.; Düber, S.; Sandrock, I.; Oberdörfer, L.; Kashani, E.; Chennupati, V.; Föhse, L.; Naumann, R.; Weiss, S.; et al. Development of Interleukin-17-Producing γδ T Cells Is Restricted to a Functional Embryonic Wave. *Immunity* **2012**, *37*, 48–59. [CrossRef]
38. Ribot, J.C.; Debarros, A.; Pang, D.J.; Neves, J.F.; Peperzak, V.; Roberts, S.J.; Girardi, M.; Borst, J.; Hayday, A.C.; Pennington, D.J.; et al. CD27 is a thymic determinant of the balance between interferon-γ- and interleukin 17–producing γδ T cell subsets. *Nat. Immunol.* **2009**, *10*, 427–436. [CrossRef]
39. Kao, C.-Y.; Huang, F.; Chen, Y.; Thai, P.; Wachi, S.; Kim, C.; Tam, L.; Wu, R. Up-Regulation of CC Chemokine Ligand 20 Expression in Human Airway Epithelium by IL-17 through a JAK-Independent but MEK/NF-κB-Dependent Signaling Pathway. *J. Immunol.* **2005**, *175*, 6676–6685. [CrossRef]
40. Teng, M.; Zhu, Z.-J.; Yao, Y.; Nair, V.; Zhang, G.-P.; Luo, J. Critical roles of non-coding RNAs in lifecycle and biology of Marek's disease herpesvirus. *Sci. China Life Sci.* **2023**, *66*, 251–268. [CrossRef]
41. Hao, X.; Li, S.; Li, J.; Yang, Y.; Qin, A.; Shang, S. An Anti-Tumor Vaccine Against Marek's Disease Virus Induces Differential Activation and Memory Response of γδ T Cells and CD8 T Cells in Chickens. *Front. Immunol.* **2021**, *12*, 645426. [CrossRef]
42. Jondle, C.N.; Johnson, K.E.; Aurubin, C.; Sylvester, P.; Xin, G.; Cui, W.; Huppler, A.R.; Tarakanova, V.L. Gammaherpesvirus Usurps Host IL-17 Signaling to Support the Establishment of Chronic Infection. *mBio* **2021**, *12*, e00566-21. [CrossRef]

43. Yuan, J.; Yu, M.; Lin, Q.-W.; Cao, A.-L.; Yu, X.; Dong, J.-H.; Wang, J.-P.; Zhang, J.-H.; Wang, M.; Guo, H.-P.; et al. Th17 Cells Contribute to Viral Replication in Coxsackievirus B3-Induced Acute Viral Myocarditis. *J. Immunol.* **2010**, *185*, 4004–4010. [CrossRef]
44. Papotto, P.H.; Ribot, J.C.; Silva-Santos, B. IL-17 + Γδ T Cells as Kick-Starters of Inflammation. *Nat. Immunol.* **2017**, *18*, 604–611.
45. Sciammas, R.; Kodukula, P.; Tang, Q.; Hendricks, R.; Bluestone, J.; Persson, K.; Mörgelin, M.; Lindbom, L.; Alm, P.; Björck, L.; et al. T Cell Receptor–γ/δ Cells Protect Mice from Herpes Simplex Virus Type 1–induced Lethal Encephalitis. *J. Exp. Med.* **1997**, *185*, 1969–1975. [CrossRef]
46. Shiromizu, C.M.; Jancic, C.C. γδ T Lymphocytes: An Effector Cell in Autoimmunity and Infection. *Front. Immunol.* **2018**, *9*, 2389.
47. Fletcher, O.J.; Tan, X.; Cortes, L.; Gimeno, I. Cost effective and time efficient measurement of CD4, CD8, major histocompatibility complex Class II, and macrophage antigen expression in the lungs of chickens. *Vet. Immunol. Immunopathol.* **2012**, *146*, 225–236. [CrossRef]
48. Ma, Y.; Aymeric, L.; Locher, C.; Mattarollo, S.R.; Delahaye, N.F.; Pereira, P.; Boucontet, L.; Apetoh, L.; Ghiringhelli, F.; Casares, N.; et al. Contribution of IL-17–producing γδ T cells to the efficacy of anticancer chemotherapy. *J. Exp. Med.* **2011**, *208*, 491–503. [CrossRef]
49. Yang, X.O.; Chang, S.H.; Park, H.; Nurieva, R.; Shah, B.; Acero, L.; Wang, Y.-H.; Schluns, K.S.; Broaddus, R.R.; Zhu, Z.; et al. Regulation of inflammatory responses by IL-17F. *J. Exp. Med.* **2008**, *205*, 1063–1075. [CrossRef]
50. Zhao, J.; Chen, X.; Herjan, T.; Li, X. The role of interleukin-17 in tumor development and progression. *J. Exp. Med.* **2019**, *217*, e20190297. [CrossRef]
51. Wang, L.; Yi, T.; Kortylewski, M.; Pardoll, D.M.; Zeng, D.; Yu, H. IL-17 can promote tumor growth through an IL-6–Stat3 signaling pathway. *J. Exp. Med.* **2009**, *206*, 1457–1464. [CrossRef]
52. Lupiani, B.; Lee, L.F.; Cui, X.; Gimeno, I.; Anderson, A.; Morgan, R.W.; Silva, R.F.; Witter, R.L.; Kung, H.-J.; Reddy, S.M. Marek's disease virus-encoded Meq gene is involved in transformation of lymphocytes but is dispensable for replication. *Proc. Natl. Acad. Sci. USA* **2004**, *101*, 11815–11820. [CrossRef]
53. Li, H.; Zhu, J.; He, M.; Luo, Q.; Liu, F.; Chen, R. Marek's Disease Virus Activates the PI3K/Akt Pathway through Interaction of Its Protein Meq with the P85 Subunit of PI3K to Promote Viral Replication. *Front. Microbiol.* **2018**, *9*, 2547. [CrossRef]
54. Kurebayashi, Y.; Nagai, S.; Ikejiri, A.; Ohtani, M.; Ichiyama, K.; Baba, Y.; Yamada, T.; Egami, S.; Hoshii, T.; Hirao, A.; et al. PI3K-Akt-mTORC1-S6K1/2 Axis Controls Th17 Differentiation by Regulating Gfi1 Expression and Nuclear Translocation of RORγ. *Cell Rep.* **2012**, *1*, 360–373. [CrossRef]

viruses

MDPI

Article

ICP8-vhs- HSV-2 Vaccine Expressing B7 Costimulation Molecules Optimizes Safety and Efficacy against HSV-2 Infection in Mice

Maria Korom, Hong Wang, Kaelin M. Bernier, Brian J. Geiss and Lynda A. Morrison *

Department of Molecular Microbiology and Immunology, Saint Louis University School of Medicine, 1100 South Grand Blvd., St. Louis, MO 63104, USA; koromm@gwu.edu (M.K.); hwang360@wustl.edu (H.W.); kaelin.bernier@bannerhealth.com (K.M.B.); brian.geiss@colostate.edu (B.J.G.)
* Correspondence: lynda.morrison@health.slu.edu

Abstract: Herpes simplex virus 2 (HSV-2) causes most sexually transmitted genital ulcerative disease. No effective prophylactic vaccine is currently available. Replication-defective (ICP8-) HSV stimulates immune responses in animals without producing progeny virus, making it potentially useful as a safe form of a live vaccine against HSV. We previously demonstrated that mice generate a stronger response to ICP8- virus encoding B7-2 costimulation molecules than to the parental replication-defective virus. We have also demonstrated enhanced immunogenicity of an ICP8-, virion host shutoff (vhs)- virus which can no longer destabilize viral and host mRNAs. Here, we constructed a triple mutant, ICP8-vhs-B7-2+ strain, and compared it to both double mutant viruses. Immunization of mice with a single dose of ICP8-B7-2+ or ICP8-vhs-B7-2+ virus decreased challenge virus replication in the vaginal mucosa, genital disease, and mortality more effectively than immunization with the ICP8-vhs- virus. Immunization with ICP8-B7-2+ or ICP8-vhs-B7-2+ virus also effectively suppressed subsequent HSV-2 infection of the nervous system compared to immunization with the ICP8-vhs- virus. ICP8-B7-2+ and ICP8-vhs-B7-2+ strains induced more IFN gamma-producing CD8 T cells and memory CD8 T cells than did ICP8-vhs- virus, potentially explaining the enhanced protective effects. Thus, B7 costimulation molecules expressed from a replication-defective vaccine can enhance vaccine efficacy, even in an immunocompetent host.

Keywords: herpes simplex virus; HSV-2; vaccine; costimulation; genital; antibodies; T cells

Citation: Korom, M.; Wang, H.; Bernier, K.M.; Geiss, B.J.; Morrison, L.A. ICP8-vhs- HSV-2 Vaccine Expressing B7 Costimulation Molecules Optimizes Safety and Efficacy against HSV-2 Infection in Mice. *Viruses* **2023**, *15*, 1570. https://doi.org/10.3390/v15071570

Academic Editors: Pietro Hiram Guzzi, Marianna Milano and Jayanta Kumar Das

Received: 3 June 2023
Revised: 11 July 2023
Accepted: 12 July 2023
Published: 18 July 2023

1. Introduction

Sexually transmitted infections with herpes simplex virus 2 (HSV-2) are the leading cause of genital ulcerative disease. The global burden of HSV-2 infection is staggering, with over half a billion persons affected worldwide [1,2]. HSV-2 infections result in a significant amount of morbidity in the United States; nearly one in five adults have been exposed to HSV-2 [3], and more than 770,000 new infections occur each year [4]. Indeed, the proportion of infected individuals can approach 70% in some demographics [2]. HSV-2 causes ulcerative lesions in anogenital skin and mucosa and is frequently shed in the absence of symptoms. Primary or recurrent infections late in pregnancy pose a significant perinatal risk to babies born to infected mothers. The infected newborn frequently suffers widespread infection with the potential for permanent neurological sequalae and even death. Ulcerative disease associated with HSV-2 also increases the propensity for the acquisition of HIV in exposed individuals. In addition, the psychosocial impacts of recurrent genital infections can be traumatic and isolating. A vaccine to mitigate these infections and prevent transmission is an important, unmet medical need.

Vaccine development against HSV-2 has focused for decades on the development of the potential product with the greatest safety, primarily subunit vaccines composed of viral cell attachment proteins [5]. However, despite showing preclinical promise, a phase III trial of HSV-2 gD in adjuvant modestly reduced oral and genital HSV-1 infections but had no efficacy against HSV-2 [6]. Immune responses to HSV-2 in naturally acquired infections show a great deal of antigenic breadth, encompassing not only surface components of virions important in antibody-mediated neutralization and antibody-dependent cellular cytotoxicity, but also internal structural components and non-structural proteins that are favored in T cell recognition [7–9]. In light of decades of experience with glycoprotein vaccine, a new approach that increases the number and type of protein targets and presents them in ways that stimulate the immune system may be essential. Because HSV-2, as with any self-perpetuating species, carries its own mechanisms of defense against immune recognition, the inactivation of critical defenses for maximum immune stimulation may help optimize antiviral immune responses evoked through vaccination.

Live-attenuated viruses as vaccines are a next-generation approach that shows promise. They are more immunogenic than subunit vaccines [10–13]. Nonetheless, they must be rendered as safe as possible without unduly compromising immunogenicity to prevent risk to vaccinees, particularly persons with potential underlying immune deficits. Some live-attenuated viruses such as ICP0- or gE- HSV-2 [11,14] are effective in protecting animal models, but replication and establishment of latency may be insufficiently attenuated. Viruses lacking a glycoprotein essential for cellular entry are grown in cells that produce a protein that complements the genetic defect. These "single-cycle" viruses undergo one round of replication in the host but cannot initiate a second round. The single-cycle gH- virus is immunogenic and reduces viral recurrences in guinea pigs [15] but proved disappointing in a clinical trial [16]. Strong cellular and humoral immune responses have been achieved using a single-cycle mutant virus ΔgD-2 which protects against large doses of HSV-2 in a variety of models while improving safety over live attenuated viruses [17]. Another form of potentially safer live virus vaccine is deleted in a gene essential for virus replication. It expresses numerous HSV proteins but is replication-defective. One example, *dl*5-29 (UL5-ICP8-) [18], effectively protects against HSV-2 challenge in animal models [12,13,19]. However, this virus grows slowly in culture and requires two large doses to fully protect even mice [12,20]. In addition, *dl*5-29 proved insufficiently immunogenic in phase I trials, especially when administered to seropositive women [21,22]. Newer versions, *dl*5-29-41L or *dl*5-29-41.1 (UL5-ICP8-vhs-) [20,23,24], seek to increase immunogenicity by reducing immune evasion promoted by the virion host shutoff (vhs) protein [25,26]. Although the inactivation of *vhs* in HSV-1 strongly increases immune responses and protection [27], the impact curiously is weaker in the context of the HSV-2 replication-defective vaccine [20].

Independently, the immunogenicity of a live, replication-defective HSV-2 vaccine has been improved by engineering the virus to encode host B7-2 costimulation molecules (ICP8-B7-2+), a critical signal in T cell activation which has been demonstrated to boost T cell responses to vaccination in mice [28]. In addition, B7-2-expressing virus shows strong protective efficacy compared with ICP8- virus [28]. With the goal of optimizing immunogenicity and protective capacity, we sought to determine how these forms of replication-defective viruses compare, attempting to identify which demonstrates the best efficacy while maintaining safety.

2. Materials and Methods

2.1. Cells and Virus Growth

S2 cells, a Vero cell line stably expressing ICP8 upon infection [29], were used to propagate ICP8-deficient 5BlacZ, 5BΔlacZ and its derivatives. Vero cells were used to generate stocks of HSV-2 strain G-6 [30], a plaque-purified derivative of strain G. For immunizations, the supernatant of infected cell monolayers was collected and subjected to high-speed centrifugation to generate virus free of cell debris as previously described [31]. Virus titers were determined on S2 or Vero cells by standard plaque assay [32]. In experiments requiring HSV-2 strain *dl*5-29 and its derivatives, V529 cells were used because they express ICP8 and UL5 [19,20]. S2 and V529 cells and *dl*5-29 and its derivatives were obtained from David Knipe, Harvard Medical School.

2.2. Construction and Isolation of Mutant Viruses

Certain previously studied, replication-defective HSV-2 vaccine strains also contain a deletion of vhs or encode B7 costimulation molecules [20,23,28]. To facilitate direct comparison of their efficacy we recreated these strains in a homogenous genetic background. The replication-defective mutant, 5BlacZ, does not produce the essential viral gene product ICP8 due to the insertion of the *E. coli lacZ* gene into the UL29 open reading frame [33]. Because lacZ is potentially immunogenic, we mutated 5BlacZ to remove the *lacZ* gene by cotransfection into S2 cells of full-length 5BlacZ DNA along with plasmid p8BSΔXhoI which contains the UL29 open reading frame with an XhoI-XhoI deletion (Figure 1). Plaques under X-gal overlay were screened for white plaques, indicating possible loss of *lacZ* sequences. The identity of candidate recombinant virus 5BΔlacZ, plaque purified to homogeneity, was confirmed by PCR; 5BΔlacZ was used as the basis for all recombinant virus vaccine strains generated in this study. Next, we sought to disrupt *UL41* and/or insert murine CD86 encoding B7-2 costimulation molecules. To engineer a mutation in the *UL41* (*vhs*) ORF of 5BΔlacZ, plasmid pDL41SB5.B containing the *vhs* locus with an XcmI-XcmI deletion was cotransfected with full-length 5BΔlacZ DNA and isolated plaques were screened by PCR for the presence of the deletion in *vhs*. The resulting virus was named Δ29Δ41. To engineer 5BΔlacZ to encode B7-2 costimulation molecules, the UL37/38 intergenic region (IGR) of HSV-2 strain 186 was amplified by PCR and cloned into pBS-KS+ to create pBS-IGR29. A cassette containing the murine B7-2 (CD86) open reading frame driven by the human cytomegalovirus immediate-early enhancer/promoter (IEp) was excised from p101086.7 by BglII digestion and ligated into plasmid pBS-IGR29 which had been modified by insertion of a BglII linker at the BsmI site. The new plasmid, pBS-IGR29-B7-2, was cotransfected with full-length 5BΔlacZ DNA into S2 cells. Cells infected with potential recombinant virus expressing B7-2 costimulation molecules were enriched by panning on Petri dishes coated with anti-B7-2 monoclonal antibody and plaque isolates derived from them were screened by flow cytometry (see below). The identity of a plaque-purified, B7-2-expressing isolate was confirmed by PCR. The resultant virus, Δ29B7-2+, is similar to 5B86 [34], except that 5B86 contains the CD86 cassette inserted into a KpnI-KpnI deletion in the *UL23* thymidine kinase (tk) gene rather than the UL37/38 IGR. The resulting recombinant virus, Δ29Δ41, resembles *dl*5-29-41L [20] except that it contains no *E. coli lacZ* gene and no deletion in the UL5 open reading frame. To engineer a virus containing both the deletion in *vhs* and the CD86 insertion, full-length Δ29Δ41 DNA was cotransfected with plasmid pBS-IGR29-B7-2. Cells infected with the potential recombinant virus were enriched by panning and plaque isolation was performed. Plaque isolates were screened by flow cytometry and a B7-2+ isolate was confirmed by PCR and named Δ29Δ41B7-2+. All recombinant viruses were plaque-purified to homogeneity and the region of their insertion or deletion was verified by sequencing. For ease of labeling figures, the Δ29 designation has been dropped, leaving Δ41, B7-2+, and Δ41B7-2+.

2.3. Panning and Flow Cytometry

Petri dishes were coated with anti-mouse B7-2 antibody (BD Biosciences PharMingen, San Diego, CA, USA) (1 μg/mL in 50 mM Tris, pH 9.5) for 1 h at room temperature, and then incubated overnight at 4 °C. Plates were washed extensively with PBS and blocked with 2% newborn calf serum in PBS before use. S2 cells infected with the progeny of a cotransfection (above) were collected by gentle scraping 24 h post-infection and added to Petri dishes at a concentration of 1.2×10^6 cells/plate. After incubation at 37 °C for 1 h, plates were swirled and unbound cells were removed by pipetting. Plates were washed gently with PBS, and then bound cells were scraped into DME + 10% FCS. Collected cells were pelleted and sonicated, and the mixture was diluted for plaque isolation. Plaque isolates were sub-cultured in 24-well plates and collected when CPE reached 100%. A portion of the infected cells from each plaque was pooled with 4 others. The pools were incubated with anti-CD86-PE and analyzed for B7-2 expression by flow cytometry

(Figure S1). Each member of a positive pool was then analyzed individually to identify recombinant viruses. Plaque isolates were iteratively purified to homogeneity.

2.4. RNA Isolation and Quantitative RT-PCR

Monolayer cultures of 1.5×10^6 to 1.8×10^6 S2 cells were mock infected or infected at a multiplicity of infection (moi) 10 in the presence of 10 μg/mL of actinomycin D (Act D). At 6 h post infection, cytoplasmic RNAs were harvested using an RNeasy Mini kit (Qiagen, Germantown, MD, USA), including the on-column DNase digestion step. RNA Nano Labchips (Agilent, Santa Clara, CA, USA) was used to assess RNA integrity and purity. Five hundred ng of each RNA sample were reverse transcribed using anchored oligo$(dT)_{18}$ primers and a Transcriptor First Strand cDNA Synthesis Kit (Roche, Indianapolis, IN, USA) in 20 μL volume according to the manufacturer's instructions. Real-time PCR reactions detecting GAPDH mRNA and 18S rRNA were performed on 0.1 μL of cDNA using FastStart SYBR Green Master Mix (Roche) and an ABI 7500 FAST Real-time PCR system (Applied Biosystems, Foster City, CA, USA). Reactions were performed in duplicate in 25 μL volume. For GAPDH, the primers used were 5′-GAACGGGAAGCTTGTCATCAATGG-3′ and 5′-TGTGGTCATGAGTCCTTCCACGAT-3′, which amplify a 343 bp product. For 18S rRNA the primers used were 5′-GTAACCCGTTGAACCCCATT-3′ and 5′-CCATCCAATCGGTAG TAGCG-3′ [35], which amplify a 151 bp product. PCR parameters consisted of 10 min FastStart Taq activation at 95 °C, followed by 40 cycles of 95 °C for 20 s, and 60 °C for 1 min. Specificity was verified by melting curve analysis. GAPDH signal was normalized to the 18S rRNA signal using the $2(-\Delta\Delta Ct)$ method [36,37]. The GAPDH mRNA level in mock-infected S2 cells was set at 100% and the GAPDH mRNA level remaining in virus-infected samples was calculated as a percentage of mock.

2.5. Plaque Size Measurement

Infected monolayers incubated for 48 h in the presence of a medium containing human serum were fixed and stained with Giemsa. Plaques were photographed using a Leica DM IRB microscope. The area of 50 randomly selected plaques was determined using Leica Application Suite V4 by tracing around the circumference of each plaque and converting pixels to mm^2.

2.6. Immunizations

Female BALB/c mice were purchased from the National Cancer Institute and were rested for one week before use. BALB.B mice were purchased from the Jackson Laboratory and bred in the Department of Comparative Medicine at Saint Louis University School of Medicine. All mice were housed at Saint Louis University School of Medicine Department of Comparative Medicine under specific-pathogen-free conditions in strict accordance with good animal practice as defined by Institutional and Public Health Service guidelines and with work approved by the Institutional Animal Care and Use Committee.

For immunizations, hind flanks of the mice at 6 weeks of age were injected subcutaneously (s.c.) with a single low (2×10^4 PFU), medium (1×10^5 PFU), or high (5×10^5 PFU) dose of virus suspended in a 40 μL total volume of normal saline. Some mice received an equivalent amount of supernatant concentrated from uninfected cell cultures (control supernatant) as a negative control for immunization. A 30 g needle was used for immunizations to minimize discomfort.

2.7. ELISpot Assay

For assessment of T cell responses to immunization, inguinal and paraaortic lymph nodes were harvested from BALB.B mice 6 d after immunization with control supernatant or 1×10^5 PFU of vaccine virus. Single-cell suspensions were prepared, and functional HSV-specific T cells were enumerated by IFN-γ enzyme-linked ImmunoSpot (ELISpot) assay. Lymph node cells (5×10^5 and 2×10^5 cells per well for control supernatant and Δ29Δ41, 4×10^5, and 1.5×10^5 cells per well for Δ29B7-2+ and Δ29Δ41B7-2+) were added to multiscreen-HA plates (Millipore) coated with anti-IFN-γ capture antibody. Cells were cultured in the presence of peptide gB498-505 [38] at 0.2 μM as a CD8 T cell stimulus for 20 h. After the incubation, the membranes were washed, and spots were visualized with an anti-IFN-γ detection antibody, followed by streptavidin-alkaline phosphatase and BCIP (5-bromo-4-chloro-3-indolylphosphate)-nitroblue tetrazolium substrate. An ImmunoSpot plate reader (version 5.0; Cellular Technology, Ltd., Shaker Heights, OH, USA) was used to quantify spots.

IFN-γ ELISpot assay was also employed to assess memory and HSV-specific T cell responses. One month after s.c. immunization, splenocytes were isolated and cultured at the concentrations indicated above with 0.2 μM gB peptide for 20 h on multiscreen-HA plates coated with anti-IFN-γ capture antibody. Spots were developed as described above.

2.8. Quantification of Serum Antibodies

To determine the concentration of HSV-specific serum antibodies induced by vaccination, groups of mice were immunized with the vaccine strains or control supernatant or were left unimmunized. Blood was collected from the tail vein of mice 22 d after immunization. The serum remaining after clot retraction was analyzed by enzyme-linked immunosorbent assay (ELISA), as previously described [39]. The secondary antibody used was anti-mouse immunoglobulin G (IgG) biotin (R & D Systems, Minneapolis, MN, USA) which was detected using streptavidin–horseradish peroxide followed by *O*-phenylenediamine dihydrochloride substrate (Sigma-Aldrich, Burlington, MA, USA). Plates were read at 490 nm on a Bio-Rad 680 plate reader. Antibody titers were determined by comparison to standard curves generated with serum containing known concentrations of IgG captured on plates coated with goat anti kappa light chain antibody (Caltag, Burlingame, CA, USA).

To determine the neutralizing activity of antibodies in serum, 2-fold serial dilutions of serum in microtiter plates were mixed with an equal volume containing approximately 50 PFU of HSV-2 G-6 and guinea pig complement (Cedarlane, Burlington, NC, USA; final concentration 1:12) for 2 h at 37 °C. Contents of the wells were then transferred to Vero cell monolayers in 24-well plates and incubated for 1 h at 37 °C. Wells were washed once with PBS and overlaid for standard plaque assay. The neutralizing antibody titer was recorded as the highest serum dilution which reduced plaque number by >50% compared with the control diluent.

2.9. In Vivo Challenge

Mice were challenged 4 weeks after immunization. At 7 d and 1 d prior to the challenge, mice were injected s.c. in the neck ruff with 3 mg Depo-Provera (Pfizer, New York, NY, USA) suspended in a 100 μL volume of normal saline. Prior to the challenge, mice were anesthetized by intraperitoneal injection of ketamine/xylazine. Infection occurred by intravaginal (i.vag.) inoculation of 5×10^5 PFU G-6 in a 5 μL volume. To quantify the virus, shed from the genital epithelium, vaginal vaults were swabbed twice with calcium alginate swabs at 9 h and 1 to 5 d post infection. Duplicate swabs for each time point were placed together in 1 mL phosphate-buffered saline and stored frozen until use. The virus was quantified on Vero cell monolayers by standard plaque assay. Body weight, signs of disease, and survival were monitored daily post challenge. Mice were weighed individually, and the mean daily change from initial body weight was calculated for each group. Disease scores were assigned by a masked observer based on the following scale: 0, no apparent signs of disease; 1, slight erythema and edema of the external genitals; 2, prominent erythema and edema of the genitals; and 3, severe erythema and edema with lesions on the genitals. The mean daily disease score was calculated for each group. Hind-limb paralysis was also assessed. Mice were euthanized if they were discovered to have lost more than 20% of their body weight or had become paralyzed. To analyze virus replication in neural tissues, the brains, brainstems, and spinal cords were dissected from a cohort of mice 5 d after the challenge. Tissues were stored frozen until use. The tissues were subsequently thawed and disrupted using a mini-bead beater (BioSpec, Bartlesville, OK, USA), and then diluted for standard plaque assay.

2.10. Statistics

The significance of the difference in antibody concentrations and virus titers on individual days was determined by ANOVA with Bonferroni correction, as was the difference in the number of IFN-γ-producing T cells. Proportions of mice with hind-limb paralysis were compared using the Fisher exact method. The Mann–Whitney U nonparametric test was used to assess the significance of the difference in disease scores on individual days post challenge.

3. Results

3.1. Construction, Isolation, and Characterization of Recombinant Viruses

Previous replication-defective HSV-2 vaccine strains varied by several parameters: 5B86 [34] contains the *E. coli* lacZ gene in the ICP8 locus and is tk- due to the insertion of CD86 (encoding B7-2). *dl*5-29-41L [20] contains *lacZ* in the *vhs* locus and a deletion in UL5. To avoid the potentially misleading effects of these differences and allow for legitimate direct comparison, we built a new set of replication-defective vaccine strains. First, we replaced the *lacZ*-disrupted ICP8 locus in 5BlacZ (Figure 1A) with an ICP8 ORF containing a deletion known to interfere with the ICP8 function (Figure 1B). This virus, 5BΔlacZ, contains no bacterial genes and formed the basis for all our ICP8-, replication-defective vaccine strains. To create a *dl*5-29-41L counterpart, we replaced the *vhs* open reading frame in 5BΔlacZ with a version that contains a deletion (Figure 1C) known to inactivate vhs [40]. This virus was named Δ29Δ41. To create a 5B86 counterpart, we ligated a CD86 (B7-2) expression cassette [34] into the UL37/UL38 IGR (Figure 1D), which is neutral with respect to HSV-2 replication and virulence (Korom and Morrison, unpublished result). This virus was named Δ29B7-2+. Finally, to combine the alterations to 5BΔlacZ, we replaced the *vhs* ORF in Δ29B7-2+ with *vhs* containing the deletion, thus creating virus Δ29Δ41B7-2+ (Figure 1D). All viruses were plaque-purified and their modifications were verified by sequencing the relevant region(s).

Vero cells infected with Δ29B7-2+ and Δ29Δ41B7-2+ expressed B7-2 costimulation molecules on their surface as detected by flow cytometry (Figure 2A). To analyze vhs activity in the mutant viruses we conducted quantitative reverse transcriptase real-time PCR using primers for GAPDH and cDNA template prepared from mRNA of cells 6 h post-infection. Cells infected with Δ29Δ41and Δ29Δ41B7-2+ contained more GAPDH mRNA than cells infected with Δ29B7-2+ which has an intact *vhs* gene (Figure 2B), verifying that the XcmI-XcmI deletion reduced vhs RNase activity (Figure 2B).

Interestingly, the *vhs* deletion resulted in smaller plaque size for both Δ29Δ41 and Δ29Δ41B7-2+ compared with Δ29B7-2+ (Figure 3A,B) and resulted in lower titers on a complementing cell line (Figure 3C).

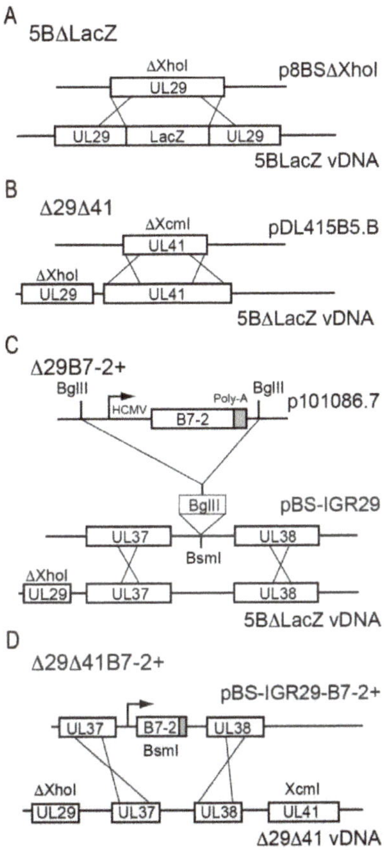

Figure 1. Construction and characterization of recombinant viruses. (**A**) Recombination of XhoI-deleted *UL29* into the DNA of virus 5BlacZ, removing the *lacZ* sequence and generating virus 5BΔLacZ. (**B**) Recombination of XcmI-deleted *UL41* into the DNA of virus 5BΔLacZ, generating virus Δ29Δ41. (**C**) Line 1 shows the insertion cassette containing the HCMV IEp fused to the B7-2 ORF. Line 2 shows insertion of this cassette into the UL38/39 IGR, and line 3 represents recombination of this cassette with 5BΔlacZ viral DNA to generate virus Δ29B7-2+. (**D**) Recombination of the cassette containing the B7-2 ORF (in the UL38/39 IGR) into Δ29Δ41 viral DNA to generate the Δ29Δ41B7-2+ virus.

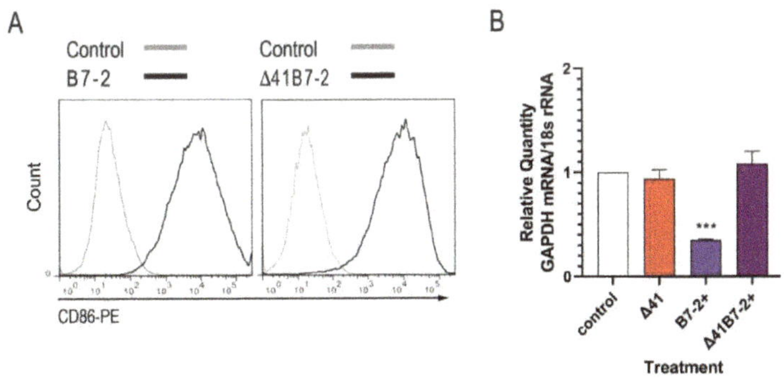

Figure 2. B7-2 expression and vhs assay. (**A**) S2 cells were mock-infected or infected with Δ29B7-2+ or Δ29Δ41B7-2+ at moi of 5 for 18 h, then collected, stained with anti-CD86-PE, and analyzed by flow

cytometry. (**B**) S2 cells were mock-infected or infected with the indicated viruses at moi 10. Six hours post infection, mRNA was isolated, reverse transcribed, and GAPDH expression was analyzed by real-time PCR. The GAPDH signal was normalized to the 18S rRNA signal, the GAPDH mRNA level in mock-infected Vero cells was set at 100%, and the GAPDH mRNA level in virus-infected samples was calculated as a percentage of mock. Results are from one of 2 independent experiments performed. ***, $p = 0.0021$.

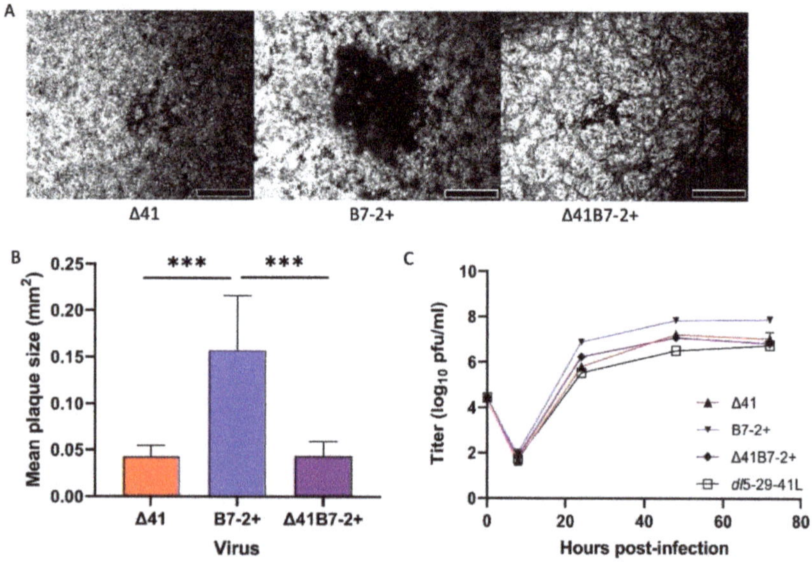

Figure 3. Plaque size comparison. S2 cells were infected with the indicated viruses and incubated for 48 h before fixation and staining. Plaques were photographed at 50× magnification. (**A**) Representative plaques are shown. Scale bar = 200 μm. (**B**) Plaque sizes were calculated using ImageJ software. Data represent the mean ± SD ($n = 50$ for each virus). ***, $p < 0.001$; Δ41 v. Δ41B7-2+, not significant. (**C**) V529 monolayers were infected at moi 0.3 in duplicate, then scraped and collected at incremental times post infection. Virus titers were determined on fresh V529 cell monolayers.

3.2. *Effect of Immunization with Recombinant Viruses on Protection from HSV-2 Challenge*

We had previously demonstrated effective protection of mice from HSV-2 challenge after a single low-dose immunization with 5B86, a virus similar to Δ29B7-2+ but expressing β-galactosidase. *dl*5-29-41L is functionally equivalent to Δ29Δ41, being replication-defective and having *vhs* disrupted. We sought to compare the efficacy of vhs- to B7-expressing virus and to determine whether the provision of B7-2 in the Δ29Δ41 background would increase protective capacity. Thus, we immunized groups of BALB/c mice once with a low, medium, or high dose of Δ29Δ41, Δ29B7-2+, or Δ29Δ41B7-2+ virus, or control supernatant of uninfected cells. Mice were then challenged i.vag. one month later with a heterologous wild-type strain of HSV-2. Over the first 4 d post challenge, Δ29B7-2+ and Δ29Δ41B7-2+ immunizations reduced challenge virus replication in the genital mucosa to a greater extent than Δ29Δ41 (Figure 4).

The differences in replication were significant on days 2 and 3 in mice immunized with the low dose (Figure 4A) and on days 1 through 4 in the high-dose immunization group (Figure 4C). In mice immunized with the medium dose (Figure 4B), only Δ29B7-2+ reduced challenge virus replication on the first day after the challenge, but both Δ29B7-2+ and Δ29Δ41B7-2+ viruses reduced challenge virus replication in the mucosa significantly better than Δ29Δ41 on days 2 through 4. By 5 d post challenge, replication of the challenge virus was still robust in the genital mucosa of control mice, but all vaccine strains had limited replication to barely detectable levels (Figure 4).

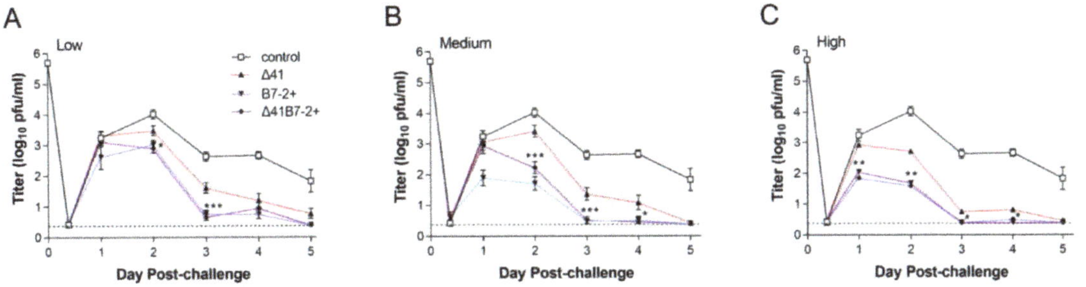

Figure 4. Titers of challenge virus shed from the genital mucosa. Groups of 10 BALB/c mice in each of two independent experiments were immunized once s.c. with a single (**A**) low (2×10^4 PFU), (**B**) medium (1×10^5 PFU), or (**C**) high (5×10^5 PFU) dose of supernatant-derived virus or an amount of control supernatant (control) equivalent to the high dose of virus. All mice were challenged 1 month after immunization by i.vag. infection with HSV-2 strain G-6. Titers of virus collected by vaginal swab of 6 mice per group ($n = 12$ total) at the indicated times post infection were determined by standard plaque assay. Data represent the geometric mean ± SEM for the compiled samples. *, $p = 0.049$–0.01; **, $p = 0.009$–0.001; ***, $p = 0.0009$–<0.0001 for Δ41 compared with Δ41B7-2+.

In control mice, HSV-2 induced severe inflammation of the genital mucosa and lesions, whereas prior immunization with any of the vaccine strains limited genital inflammation. Δ29B7-2+ and Δ29Δ41B7-2+ viruses protected mice better than Δ29Δ41 at all three immunizing doses tested (Figure 5).

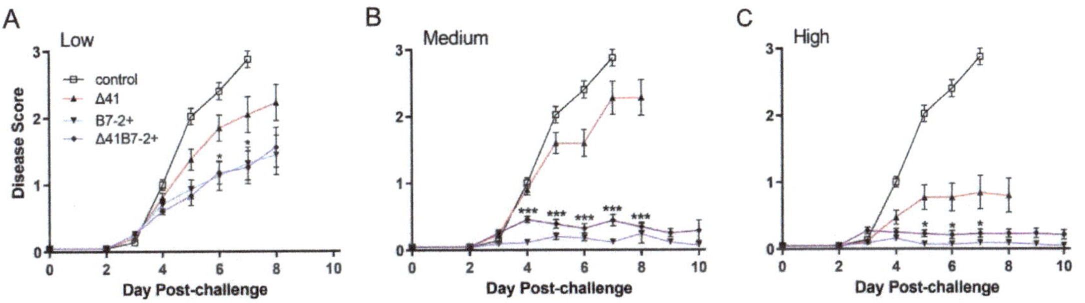

Figure 5. Genital inflammation and disease after i.vag. challenge of immunized mice. Mice as described in Figure 4 were immunized with a (**A**) low, (**B**) medium, or (**C**) high dose of virus or an amount of control supernatant corresponding to the high dose of virus. After the challenge, mice were observed daily for signs of inflammation and lesions on their external genitalia. Data represent the arithmetic mean ± SEM for all samples compiled from two independent experiments ($n = 20$). *, $p = 0.0389$–0.0178; ***, $p < 0.0001$ for Δ41 compared with Δ41B7-2+.

The difference between viruses expressing B7-2 and the Δ29Δ41 group was most significant after medium dose immunization, with protection from genital inflammation significantly enhanced beginning 4 d post challenge (Figure 5B); however, significantly enhanced protection was also observed at the low and high vaccine doses (Figure 5A,C). Using maintenance of body weight as an indicator of general health, we observed that low-dose immunization with any of the vaccine strains did not protect mice from weight loss after the challenge (Figure 6A).

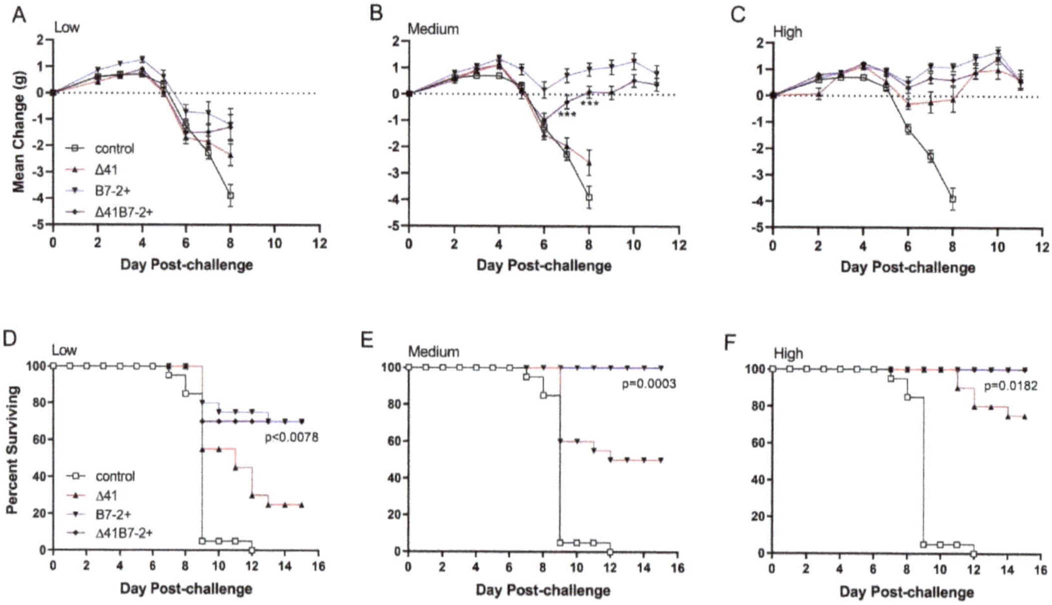

Figure 6. Body weight and survival after i.vag. challenge of immunized mice. Mice as described in Figure 4 were immunized with the low, medium, or high dose of virus or an amount of control supernatant corresponding to the high dose of virus. Body weight (**A–C**) and survival (**D–F**) were monitored over time after the challenge. Weight data represent the arithmetic mean ± SEM for all mice compiled from two independent experiments (n = 20). ***, p = 0.0002–<0.0001 for Δ41 compared with Δ41B7-2+.

In mice receiving medium dose vaccine, differences in body weight over time post-challenge reflected differences in genital inflammation and disease in that those mice immunized with Δ29B7-2+ or Δ29Δ41B7-2+ showed only a transient decrease in body weight, whereas those immunized with Δ29Δ41 lost weight at a rate similar to control mice (Figure 6B). All vaccine strains, when given at the high dose, significantly protected mice from weight loss after the challenge compared with the control group (Figure 6C). Consistent with the weight loss profile, most mice receiving the low dose of Δ29B7-2+ or Δ29Δ41B7-2+ vaccine survived challenge virus infection, but most mice immunized with low-dose Δ29Δ41 succumbed (Figure 6D). Deaths occurred in Δ29Δ41-immunized mice at the medium and high doses as well, but all mice immunized with Δ29B7-2+ or Δ29Δ41B7-2+ viruses survived the challenge (Figure 6E,F). Mortality resulting from challenge virus infection was associated with neurological debilitation (Table 1). The medium and high doses of Δ29B7-2+ and Δ29Δ41B7-2+ vaccines protected mice completely from hind-limb paralysis, but a portion of mice immunized with Δ29Δ41 became paralyzed from challenge virus infection.

Table 1. Percentage of mice with hind-limb paralysis.

		Immunization Group		
Vaccine Dose	**Control**	**Δ41**	**B7-2+**	**Δ41B7-2+**
Low	nd [a]	50	30	35
Medium	nd	50	0 [b]	0
High	75	20	0	0

[a] nd, not done. [b] p = 0.0004 compared with Δ29Δ41.

Hind-limb paralysis and death strongly suggested entry of challenge virus into the nervous system from the genital mucosa. To address this possibility, additional groups of mice were immunized with the medium dose of vaccine viruses, which was the dose demonstrating the greatest differences in genital and neurological signs of disease. Five days after i.vag. challenge, the titer of wild-type HSV-2 in regions of the nervous system was determined (Figure 7).

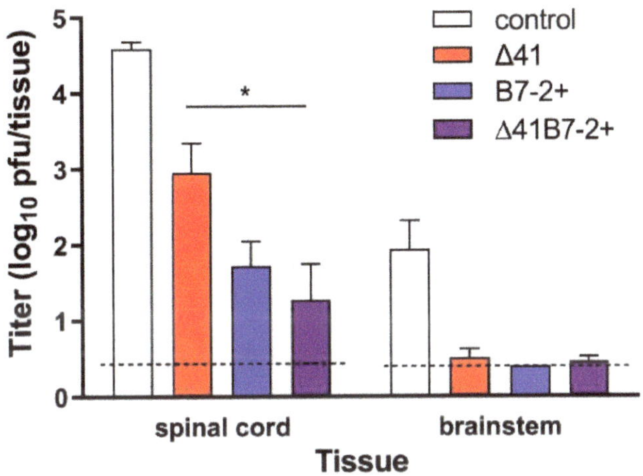

Figure 7. Levels of challenge virus in the nervous system. Mice were immunized with the medium dose of the indicated virus or control supernatant and challenged i.vag. one month later with HSV-2 G-6. After 5 d the mice were euthanized, the indicated regions of the CNS were dissected and homogenized, and the virus titer in them was determined by standard plaque assay. Data represent the geometric mean ± SEM for all samples per group, compiled from 3 independent experiments (n = 10 to 12). *, $p < 0.05$.

Compared with mice receiving control supernatant, prior immunization with Δ29Δ41 resulted in less challenge virus in the spinal cords ($p = 0.014$). Immunization with viruses expressing B7-2 further reduced the amount of challenge virus detectable in the spinal cord. Almost no challenge virus was detected in the brainstem of mice immunized with any of the vaccine strains at 5 d post challenge, though the challenge virus had already reached the brainstem in control mice (p = 0.0009–0.0002). Thus, all the vaccine strains provided significant protection against acute infection of the nervous system, with the best protection afforded by viruses expressing B7-2.

3.3. *Immune Correlates of Protection*

To define immune responses that correlated with protection mediated by the Δ29B7-2+ and Δ29Δ41B7-2+ viruses, we analyzed HSV-2-specific antibodies and T cell responses stimulated by vaccination. HSV-2-specific serum IgG was evoked by vaccination in a dose-dependent manner (Figure 8A).

All vaccine strains induced equivalent virus-specific IgG responses in serum at the low dose, but the medium dose of B7-expressing viruses stimulated a slightly more robust response (Figure 8A). The same sera were used to test the capacity to neutralize virus infectivity. Interestingly, neutralizing antibodies developed to the greatest extent after immunization with the Δ29B7-2+ virus (Figure 8B). The capacity of the vaccine viruses to stimulate HSV-specific T cell responses was analyzed 6 d after immunization by IFN-γ ELISpot. BALB.B mice were immunized to take advantage of an immunodominant, CD8 T cell epitope recognized by H-2^b-haplotype mice [38]. Immunization with the Δ29B7-2+ and particularly the Δ29Δ41B7-2+ virus elicited more IFN-γ-producing, CD8 T cells than Δ29Δ41 in the draining lymph nodes 6 d later (Figure 9A), with a significant difference observed when the total number of IFN-γ-producing, CD8 T cells was considered (Figure 9B).

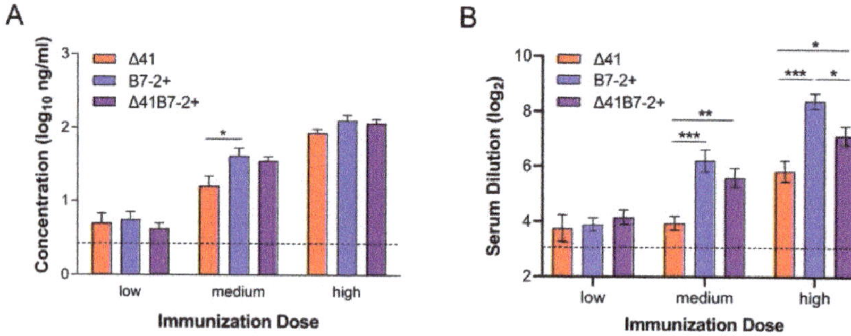

Figure 8. HSV-specific antibody in the sera of immunized mice. Serum samples were collected 22 d after immunization of mice receiving low, medium, or high doses of virus. (**A**) Concentration of HSV-specific IgG was determined by ELISA. (**B**) Neutralizing antibody titer was determined by plaque inhibition assay. Data represent the geometric mean ± SEM of 20 samples compiled from two independent experiments (n = 20 total). *, p = 0.027–0.0157; **, p = 0.0023; ***, $p < 0.0001$.

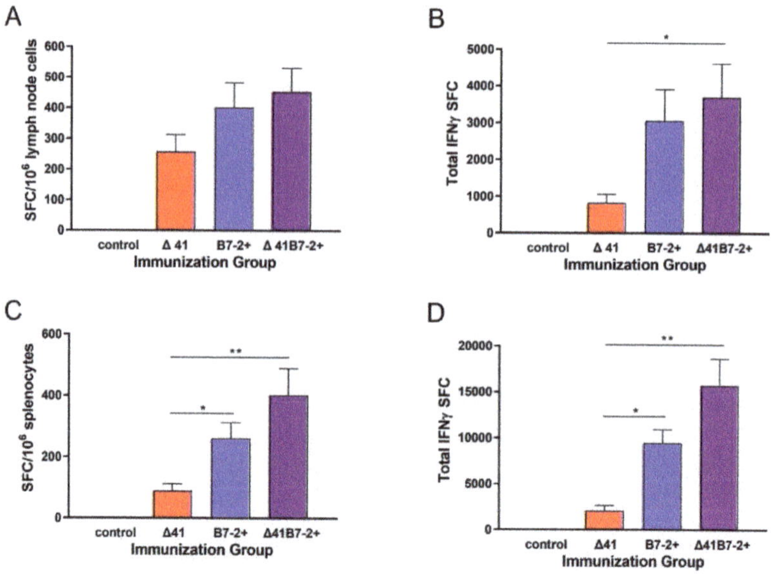

Figure 9. IFN-γ-producing T cells induced by immunization. Groups of BALB.B mice were immunized with 1×10^5 PFU of the indicated replication-defective virus or control supernatant. Cells from the draining lymph nodes were isolated 6 d after immunization, stimulated in vitro with 0.2 mM of peptide representing the CD8 epitope gB498–505, and analyzed in an IFN-γ ELISpot assay. Data represent the arithmetic mean ± SEM of (**A**) spot-forming cells (SFC) per million lymph node cells or (**B**) the absolute number of IFN-γ-producing SFC in the draining lymph nodes per mouse. Data were compiled from 3 independent experiments (n = 9 for the control group and n = 11 for each vaccine group). Alternatively, splenocytes were isolated 1 mo after immunization and stimulated as above, then analyzed in the IFN-γ ELISpot assay to determine (**C**) SFC per million splenocytes or (**D**) the absolute number of SFC per mouse. Data were compiled from 3 independent experiments (n = 7 for the control group and n = 8 to 9 for each vaccine group). *, p = 0.0475–0.0212; **, p = 0.0016; ***, $p < 0.0001$.

Memory CD8 T cells, however, were markedly more prevalent in the spleens one month after immunization of mice with Δ29B7-2+ or Δ29Δ41B7-2+ than with Δ29Δ41, whether measured as epitope-specific cells per 10^6 cells (Figure 9C), or the total number of epitope-specific cells per spleen (Figure 9D). Thus, more HSV-specific, memory CD8 T cells were available at the time of the challenge in mice immunized with Δ29B7-2+ and Δ29Δ41B7-2+.

4. Discussion

An effective vaccine against HSV-2 must be safe while simultaneously stimulating strong and effective immune responses. Various alterations of replication-defective viral genomes such as disruption of vhs activity or inclusion of B7 costimulation molecules have resulted in more immunogenic and effective vaccine strains than their predecessor strains [20,27,28,34,41]. Direct comparison of the efficacy of these manipulations to the basic replication-defective HSV-2 vaccine paradigm, however, demands that extraneous differences be minimized. Therefore, we constructed all vaccine strains in the same replication-defective (ICP8-deleted) background to evaluate the effectiveness of the virus with vhs deleted to those expressing B7-2. Furthermore, these viruses lacked any bacterial gene (*lacZ*) that could itself be immunogenic and potentially influence anti-viral immune induction. The central finding of our comparison is that replication-defective viruses expressing B7 costimulation molecules are superior in protective efficacy to replication-defective, vhs- viruses.

As was previously observed [24], viruses bearing a deletion in HSV-2 *UL41* had a smaller plaque phenotype on complementing cells than viruses capable of expressing vhs. This correlated with a reduction of maximal titer that could be achieved on complementing cells, an observation that bears on the ability to generate a sufficient quantity of a live virus vaccine for manufacture. Nonetheless, the deletion of *UL41* has benefits for vaccine design. Compromise of vhs activity enhances recognition of infected fibroblasts [42] and relieves the block to activation in infected dendritic cells, potentially a critical feature of a whole virus vaccine [26]. Interestingly, loss of vhs affects replication-defective HSV-2 more severely than it does HSV-1 strains, which may reflect the stronger, faster activity of HSV-2 vhs previously described [43–46]. Indeed, Reszka et al. [24] showed that the substitution of HSV-1 vhs into *dl*5-29 resulted in higher virus yield than *dl*5-29Δ41 but did not affect its immunogenicity. The vhs protein is also a target of HSV-immune T cells [7]. Potentially the Δ29Δ41B7-2+ vaccine strain could be further optimized by creating a smaller deletion in *UL41* which preserves most of the protein as an immunologic target while still compromising vhs activity, or by doing so in a substituted HSV-1 *UL41* gene.

Comparison of these viruses as vaccines in mice yielded several interesting observations. First, in a previous study, Hoshino et al. had shown that replication-defective, vhs- HSV-2 vaccine only transiently extended the survival time of mice compared to mice immunized with replication-defective virus alone [23]. Here, we confirmed that deletion of vhs did not confer any consistent, additional protective advantage for Δ29Δ41B7-2+ compared with Δ29B7-2+, though it did increase safety by creating a major insult at a second locus. Second, expression of B7-2, whether in the context of ICP8- virus or ICP8-vhs- virus, increased protective efficacy as measured by HSV-2 shed from the genital mucosa of mice, signs of genital inflammation and disease, infection of the nervous system, and maintenance of body weight and survival. Third, the medium dose of vaccine most clearly distinguished the vaccine strains. While all showed a pronounced capacity to protect at the high immunization dose compared with control vaccination, at the medium dose the Δ29Δ41 virus more closely resembled control immunization, particularly in terms of disease progression. Because of this ability to distinguish between the vaccine strains, we chose the medium dose to investigate the vaccine's capacity to protect against acute infection of the nervous system. Once again, the B7-expressing viruses could be distinguished from Δ41Δ29 in heightened protection against infection of the spinal cord by 5 d post challenge. Lastly, a single dose of B7-2-expressing vaccine provided significant protection against some aspects of challenge virus infection in mice compared with control supernatant or vhs- virus, even when administered at the lowest dose. Whether these vaccine strains can protect mice from the establishment of latency by challenge virus is a worthy future direction. The establishment of latency by the vaccine strains themselves was not investigated here because DNA of analogous replication-defective strain *dl*5-29 was not detected in sensory ganglia after intramuscular or intradermal immunization, and only low copy numbers were observed after intranasal immunization of mice [19]. Nonetheless, this point must be investigated with B7-expressing strains in subsequent trials to firmly establish their safety profile.

Consistent with the role of B7 costimulation molecules in potentiating differentiation of naïve T cells in response to antigen, we observed a substantial increase in the CD8 T cell response to viruses that express B7-2. Interestingly, this enhancement was most prominent in the memory T cell response rather than in the acute phase. The difference between acute and memory phases could be a result of the particular day after immunization that we looked at and/or to a cytokine milieu that drives initial T cell expansion in response to Δ29Δ41 but does not support efficient conversion to or expansion of memory cells. A similar enhancement in nascent CD8 T cell responses to an ICP8-vhs-B7+ HSV-1 strain compared with the ICP8-vhs- virus was previously observed [41]. Further investigation into the mechanism by which virus-expressed B7 molecules enhance responses to the vaccine is warranted, particularly the cell type(s) in which B7 is expressed, and those that can act as antigen-presenting cells to potentiate the immune response.

Supplementary Materials: The following supporting information can be downloaded at: https://www.mdpi.com/article/10.3390/v15071570/s1, Figure S1: iterative isolation of B7-expressing virus by panning of infected cells and flow cytometry.

Author Contributions: Conceptualization, L.A.M.; methodology, M.K. and L.A.M.; formal analysis, M.K. and L.A.M.; investigation, M.K., H.W., K.M.B., B.J.G. and L.A.M.; data curation, M.K., K.M.B. and L.A.M.; writing—original draft preparation, M.K. and L.A.M.; writing—review and editing, L.A.M.; visualization, M.K., K.M.B. and L.A.M.; supervision, M.K. and L.A.M.; project administration, L.A.M.; funding acquisition, L.A.M. All authors have read and agreed to the published version of the manuscript.

Funding: This research was funded by the National Institutes of Health award R21 AI090223 to L.A.M.

Institutional Review Board Statement: The animal study protocol was approved by the Institutional Animal Care and Use Committee (protocol code 2093 and last date of approval 14 October 2021).

Informed Consent Statement: Not applicable.

Data Availability Statement: Underlying original data are available upon request.

Acknowledgments: The authors are grateful to David Knipe for providing virus *dl*5-29, *dl*5-29-41L, and complementing V529 cells. We thank Sherri Koehm and Joy Eslick for assistance with flow cytometry, and Sindhu Sangabathula and Srikanth Chiravuri for technical assistance.

Conflicts of Interest: The authors declare no conflict of interest.

References

1. James, C.; Harfouche, M.; Welton, N.J.; Turner, K.M.; Abu-Raddad, L.J.; Gottlieb, S.L.; Looker, K.J. Herpes simplex virus: Global infection prevalence and incidence estimates, 2016. *Bull. World Health Organ.* **2020**, *98*, 315–329. [CrossRef]
2. Looker, K.J.; Magaret, A.S.; Turner, K.M.; Vickerman, P.; Gottlieb, S.L.; Newman, L.M. Global estimates of prevalent and incident herpes simplex virus type 2 infections in 2012. *PLoS ONE* **2015**, *10*, e114989. [CrossRef]
3. Bradley, H.; Markowitz, L.E.; Gibson, T.; McQuillan, G.M. Seroprevalence of herpes simplex virus types 1 and 2—United States, 1999–2010. *J. Infect. Dis.* **2014**, *209*, 325–333. [CrossRef]
4. Satterwhite, C.L.; Torrone, E.; Meites, E.; Dunne, E.F.; Mahajan, R.; Ocfemia, M.C.; Su, J.; Xu, F.; Weinstock, H. Sexually transmitted infections among US women and men: Prevalence and incidence estimates, 2008. *Sex. Transm. Dis.* **2013**, *40*, 187–193. [CrossRef] [PubMed]
5. Aschner, C.B.; Herold, B.C. Alphaherpesvirus Vaccines. *Curr. Issues Mol. Biol.* **2021**, *41*, 469–508. [CrossRef]

6. Belshe, R.B.; Leone, P.A.; Bernstein, D.I.; Wald, A.; Levin, M.J.; Stapleton, J.T.; Gorfinkel, I.; Morrow, R.L.; Ewell, M.G.; Stokes-Riner, A.; et al. Efficacy results of a trial of a herpes simplex vaccine. *N. Engl. J. Med.* **2012**, *366*, 34–43. [CrossRef]
7. Jing, L.; Haas, J.; Chong, T.M.; Bruckner, J.J.; Dann, G.C.; Dong, L.; Marshak, J.O.; McClurkan, C.L.; Yamamoto, T.N.; Bailer, S.M.; et al. Cross-presentation and genome-wide screening reveal candidate T cells antigens for a herpes simplex virus type 1 vaccine. *J. Clin. Investig.* **2012**, *122*, 654–673. [CrossRef] [PubMed]
8. Laing, K.J.; Dong, L.; Sidney, J.; Sette, A.; Koelle, D.M. Immunology in the Clinic Review Series; focus on host responses: T cell responses to herpes simplex viruses. *Clin. Exp. Immunol.* **2012**, *167*, 47–58. [CrossRef]
9. Laing, K.J.; Magaret, A.S.; Mueller, D.E.; Zhao, L.; Johnston, C.; De Rosa, S.C.; Koelle, D.M.; Wald, A.; Corey, L. Diversity in CD8(+) T cell function and epitope breadth among persons with genital herpes. *J. Clin. Immunol.* **2010**, *30*, 703–722. [CrossRef]
10. Bernstein, D.I.; Pullum, D.A.; Cardin, R.D.; Bravo, F.J.; Dixon, D.A.; Kousoulas, K.G. The HSV-1 live attenuated VC2 vaccine provides protection against HSV-2 genital infection in the guinea pig model of genital herpes. *Vaccine* **2019**, *37*, 61–68. [CrossRef]
11. Halford, W.P.; Püschel, R.; Gershburg, E.; Wilber, A.; Gershburg, S.; Rakowski, B. A live-attenuated HSV-2 ICP0 virus elicits 10 to 100 times greater protection against genital herpes than a glycoprotein D subunit vaccine. *PLoS ONE* **2011**, *6*, e17748. [CrossRef] [PubMed]
12. Hoshino, Y.; Dalai, S.K.; Wang, K.; Pesnicak, L.; Lau, T.Y.; Knipe, D.M.; Cohen, J.I.; Straus, S.E. Comparative efficacy and immunogenicity of replication-defective, recombinant glycoprotein, and DNA vaccines for herpes simplex virus 2 infections in mice and guinea pigs. *J. Virol.* **2005**, *79*, 410–418. [CrossRef] [PubMed]
13. Hoshino, Y.; Pesnicak, L.; Dowdell, K.C.; Burbelo, P.D.; Knipe, D.M.; Straus, S.E.; Cohen, J.I. Protection from herpes simplex virus (HSV)-2 infection with replication-defective HSV-2 or glycoprotein D2 vaccines in HSV-1-seropositive and HSV-1-seronegative guinea pigs. *J. Infect. Dis.* **2009**, *200*, 1088–1095. [CrossRef]
14. Brittle, E.E.; Wang, F.; Lubinski, J.M.; Bunte, R.M.; Friedman, H.M. A replication-competent, neuronal spread-defective, live attenuated herpes simplex virus type 1 vaccine. *J. Virol.* **2008**, *82*, 8431–8441. [CrossRef] [PubMed]
15. McLean, C.S.; Erturk, M.; Jennings, R.; Challanain, D.N.; Minson, A.C.; Duncan, I.; Boursnell, M.E.; Inglis, S.C. Protective vaccination against primary and recurrent disease caused by herpes simplex virus (HSV) type 2 using a genetically disabled HSV-1. *J. Infect. Dis.* **1994**, *170*, 1100–1109. [CrossRef]
16. de Bruyn, G.; Vargas-Cortez, M.; Warren, T.; Tyring, S.K.; Fife, K.H.; Lalezari, J.; Brady, R.C.; Shahmanesh, M.; Kinghorn, G.; Beutner, K.R.; et al. A randomized controlled trial of a replication defective (gH deletion) herpes simplex virus vaccine for the treatment of recurrent genital herpes among immunocompetent subjects. *Vaccine* **2006**, *24*, 914–920. [CrossRef]
17. Petro, C.; González, P.A.; Cheshenko, N.; Jandl, T.; Khajoueinejad, N.; Bénard, A.; Sengupta, M.; Herold, B.C.; Jacobs, W.R. Herpes simplex type 2 virus deleted in glycoprotein D protects against vaginal, skin and neural disease. *Elife* **2015**, *4*, e06054. [CrossRef]
18. Da Costa, X.; Kramer, M.F.; Zhu, J.; Brockman, M.A.; Knipe, D.M. Construction, phenotypic analysis, and immunogenicity of a UL5/UL29 double deletion mutant of herpes simplex virus 2. *J. Virol.* **2000**, *74*, 7963–7971. [CrossRef]
19. Da Costa, X.J.; Jones, C.A.; Knipe, D.M. Immunization against genital herpes with a vaccine virus that has defects in productive and latent infection. *Proc. Natl. Acad. Sci. USA* **1999**, *96*, 6994–6998. [CrossRef]
20. Dudek, T.; Mathews, L.C.; Knipe, D.M. Disruption of the U(L)41 gene in the herpes simplex virus 2 dl5-29 mutant increases its immunogenicity and protective capacity in a murine model of genital herpes. *Virology* **2008**, *372*, 165–175. [CrossRef]
21. Dropulic, L.; Wang, K.; Oestreich, M.; Pietz, H.; Garabedian, D.; Jegaskanda, S.; Dowdell, K.; Nguyen, H.; Laing, K.; Koelle, D.; et al. A Replication-Defective Herpes Simplex Virus (HSV)-2 Vaccine, HSV529, is Safe and Well-Tolerated in Adults with or without HSV Infection and Induces Significant HSV-2-Specific Antibody Responses in HSV Seronegative Individuals. *Open Forum Infect. Dis.* **2017**, *4*, S415–S416. [CrossRef]
22. Dropulic, L.K.; Oestreich, M.C.; Pietz, H.L.; Laing, K.J.; Hunsberger, S.; Lumbard, K.; Garabedian, D.; Turk, S.P.; Chen, A.; Hornung, R.L.; et al. A Randomized, Double-Blinded, Placebo-Controlled, Phase 1 Study of a Replication-Defective Herpes Simplex Virus (HSV) Type 2 Vaccine, HSV529, in Adults With or Without HSV Infection. *J. Infect. Dis.* **2019**, *220*, 990–1000. [CrossRef]
23. Hoshino, Y.; Pesnicak, L.; Dowdell, K.C.; Lacayo, J.; Dudek, T.; Knipe, D.M.; Straus, S.E.; Cohen, J.I. Comparison of immunogenicity and protective efficacy of genital herpes vaccine candidates herpes simplex virus 2 dl5-29 and dl5-29-41L in mice and guinea pigs. *Vaccine* **2008**, *26*, 4034–4040. [CrossRef]
24. Reszka, N.J.; Dudek, T.; Knipe, D.M. Construction and properties of a herpes simplex virus 2 dl5-29 vaccine candidate strain encoding an HSV-1 virion host shutoff protein. *Vaccine* **2010**, *28*, 2754–2762. [CrossRef] [PubMed]
25. Murphy, J.A.; Duerst, R.J.; Smith, T.J.; Morrison, L.A. Herpes simplex virus type 2 virion host shutoff protein regulates alpha/beta interferon but not adaptive immune responses during primary infection in vivo. *J. Virol.* **2003**, *77*, 9337–9345. [CrossRef]
26. Samady, L.; Costigliola, E.; MacCormac, L.; McGrath, Y.; Cleverley, S.; Lilley, C.E.; Smith, J.; Latchman, D.S.; Chain, B.; Coffin, R.S. Deletion of the virion host shutoff protein (vhs) from herpes simplex virus (HSV) relieves the viral block to dendritic cell activation: Potential of vhs- HSV vectors for dendritic cell-mediated immunotherapy. *J. Virol.* **2003**, *77*, 3768–3776. [CrossRef] [PubMed]
27. Geiss, B.J.; Smith, T.J.; Leib, D.A.; Morrison, L.A. Disruption of virion host shutoff activity improves the immunogenicity and protective capacity of a replication-incompetent herpes simplex virus type 1 vaccine strain. *J. Virol.* **2000**, *74*, 11137–11144. [CrossRef]

28. Vagvala, S.P.; Thebeau, L.G.; Wilson, S.R.; Morrison, L.A. Virus-encoded b7-2 costimulation molecules enhance the protective capacity of a replication-defective herpes simplex virus type 2 vaccine in immunocompetent mice. *J. Virol.* **2009**, *83*, 953–960. [CrossRef]
29. Gao, M.; Knipe, D.M. Genetic evidence for multiple nuclear functions of the herpes simplex virus ICP8 DNA-binding protein. *J. Virol.* **1989**, *63*, 5258–5267. [CrossRef] [PubMed]
30. Morrison, L.A.; Zhu, L.; Thebeau, L.G. Vaccine-induced serum immunoglobin contributes to protection from herpes simplex virus type 2 genital infection in the presence of immune T cells. *J. Virol.* **2001**, *75*, 1195–1204. [CrossRef] [PubMed]
31. Morrison, L.A.; Knipe, D.M. Mechanisms of immunization with a replication-defective mutant of herpes simplex virus 1. *Virology* **1996**, *220*, 402–413. [CrossRef] [PubMed]
32. Knipe, D.M.; Spang, A.E. Definition of a series of stages in the association of two herpesviral proteins with the cell nucleus. *J. Virol.* **1982**, *43*, 314–324. [CrossRef] [PubMed]
33. Da Costa, X.J.; Bourne, N.; Stanberry, L.R.; Knipe, D.M. Construction and characterization of a replication-defective herpes simplex virus 2 ICP8 mutant strain and its use in immunization studies in a guinea pig model of genital disease. *Virology* **1997**, *232*, 1–12. [CrossRef]
34. Thebeau, L.G.; Vagvala, S.P.; Wong, Y.M.; Morrison, L.A. B7 costimulation molecules expressed from the herpes simplex virus 2 genome rescue immune induction in B7-deficient mice. *J. Virol.* **2007**, *81*, 12200–12209. [CrossRef]
35. Wylie, K.M.; Schrimpf, J.E.; Morrison, L.A. Increased eIF2alpha phosphorylation attenuates replication of herpes simplex virus 2 vhs mutants in mouse embryonic fibroblasts and correlates with reduced accumulation of the PKR antagonist ICP34.5. *J. Virol.* **2009**, *83*, 9151–9162. [CrossRef]
36. Livak, K.J.; Schmittgen, T.D. Analysis of relative gene expression data using real-time quantitative PCR and the 2(-Delta Delta C(T)) Method. *Methods* **2001**, *25*, 402–408. [CrossRef]
37. Pfaffl, M.W. A new mathematical model for relative quantification in real-time RT-PCR. *Nucleic Acids Res.* **2001**, *29*, e45. [CrossRef]
38. Wallace, M.E.; Keating, R.; Heath, W.R.; Carbone, F.R. The cytotoxic T-cell response to herpes simplex virus type 1 infection of C57BL/6 mice is almost entirely directed against a single immunodominant determinant. *J. Virol.* **1999**, *73*, 7619–7626. [CrossRef]
39. Morrison, L.A.; Da Costa, X.J.; Knipe, D.M. Influence of mucosal and parenteral immunization with a replication-defective mutant of HSV-2 on immune responses and protection from genital challenge. *Virology* **1998**, *243*, 178–187. [CrossRef]
40. Smith, T.J.; Morrison, L.A.; Leib, D.A. Pathogenesis of herpes simplex virus type 2 virion host shutoff (vhs) mutants. *J. Virol.* **2002**, *76*, 2054–2061. [CrossRef] [PubMed]
41. Schrimpf, J.E.; Tu, E.M.; Wang, H.; Wong, Y.M.; Morrison, L.A. B7 costimulation molecules encoded by replication-defective, vhs-deficient HSV-1 improve vaccine-induced protection against corneal disease. *PLoS ONE* **2011**, *6*, e22772. [CrossRef]
42. Tigges, M.A.; Leng, S.; Johnson, D.C.; Burke, R.L. Human herpes simplex virus (HSV)-specific CD8+ CTL clones recognize HSV-2-infected fibroblasts after treatment with IFN-gamma or when virion host shutoff functions are disabled. *J. Immunol.* **1996**, *156*, 3901–3910. [CrossRef] [PubMed]
43. Everly, D.N., Jr.; Read, G.S. Mutational analysis of the virion host shutoff gene (UL41) of herpes simplex virus (HSV): Characterization of HSV type 1 (HSV-1)/HSV-2 chimeras. *J. Virol.* **1997**, *71*, 7157–7166. [CrossRef] [PubMed]
44. Everly, D.N., Jr.; Read, G.S. Site-directed mutagenesis of the virion host shutoff gene (UL41) of herpes simplex virus (HSV): Analysis of functional differences between HSV type 1 (HSV-1) and HSV-2 alleles. *J. Virol.* **1999**, *73*, 9117–9129. [CrossRef]
45. Hill, T.M.; Sadler, J.R.; Betz, J.L. Virion component of herpes simplex virus type 1 KOS interferes with early shutoff of host protein synthesis induced by herpes simplex virus type 2 186. *J. Virol.* **1985**, *56*, 312–316. [CrossRef] [PubMed]
46. Hill, T.M.; Sinden, R.R.; Sadler, J.R. Herpes simplex virus types 1 and 2 induce shutoff of host protein synthesis by different mechanisms in Friend erythroleukemia cells. *J. Virol.* **1983**, *45*, 241–250. [CrossRef]

viruses

Article

Prediction of Antigenic Distance in Influenza A Using Attribute Network Embedding

Fujun Peng [1], Yuanling Xia [2] and Weihua Li [1,*]

[1] School of Information Science and Engineering, Yunnan University, Kunming 650500, China; john7@mail.ynu.edu.cn
[2] State Key Laboratory for Conservation and Utilization of Bio-Resources in Yunnan, Yunnan University, Kunming 650500, China; xiayl@ynu.edu.cn
* Correspondence: liweihua@ynu.edu.cn

Abstract: Owing to the rapid changes in the antigenicity of influenza viruses, it is difficult for humans to obtain lasting immunity through antiviral therapy. Hence, tracking the dynamic changes in the antigenicity of influenza viruses can provide a basis for vaccines and drug treatments to cope with the spread of influenza viruses. In this paper, we developed a novel quantitative prediction method to predict the antigenic distance between virus strains using attribute network embedding techniques. An antigenic network is built to model and combine the genetic and antigenic characteristics of the influenza A virus H3N2, using the continuous distributed representation of the virus strain protein sequence (ProtVec) as a node attribute and the antigenic distance between virus strains as an edge weight. The results show a strong positive correlation between supplementing genetic features and antigenic distance prediction accuracy. Further analysis indicates that our prediction model can comprehensively and accurately track the differences in antigenic distances between vaccines and influenza virus strains, and it outperforms existing methods in predicting antigenic distances between strains.

Keywords: influenza; H3N2; antigenic distance; hemagglutinin; attribute network embedding

Citation: Peng, F.; Xia, Y.; Li, W. Prediction of Antigenic Distance in Influenza A Using Attribute Network Embedding. *Viruses* **2023**, *15*, 1478. https://doi.org/10.3390/v15071478

Academic Editor: Pietro Hiram Guzzi, Marianna Milano and Jayanta Kumar Das

Received: 28 April 2023
Revised: 23 June 2023
Accepted: 28 June 2023
Published: 29 June 2023

1. Introduction

Influenza A virus is a negative-sense single-stranded RNA virus, and its viral membrane is mainly composed of two surface glycoproteins: hemagglutinin (HA) and neuraminidase (NA). Existing influenza A viruses are classified into 18 HA subtypes and 11 NA subtypes according to different combinations of HA and NA. Each subtype has distinct pathogenic characteristics and antigenicity. Currently, the H3N2 and H1N1 subtypes are the main seasonal influenza viruses circulating in humans. Among them, the H3N2 virus was first discovered during the 1968 Hong Kong influenza pandemic and has since been continuously circulating globally [1,2]. Seasonal influenza is expected to cause 290,000–640,000 respiratory-related fatalities worldwide yearly [3].

HA is the primary surface protein of influenza A virus and is crucial for the virus's entry into host cells. The receptor-binding site on the HA protein can bind to sialic acid receptors on host cells, thereby mediating the entry of viral particles into cells and causing infection in humans [4]. Consequently, many vaccines have been developed targeting the receptor-binding site of the HA protein to protect us from future seasonal influenza virus infections [2,5]. The quaternary structure of the HA protein is a homotrimer, with each monomer composed of two subunits, HA1 and HA2. The HA1 subunit contains antigenic determinant regions and critical binding sites for the virus to attach to host cells. Regrettably, mutations in the HA1 subunit can influence the virus's antigenicity and receptor binding affinity, and the mutation rate is typically higher than in other regions [6]. Antigenic drift caused by accumulated sequence mutations greatly impedes the progress in the development of drug treatments and vaccines against potential influenza viruses.

Therefore, the rapid detection of antigenic variation and accurate quantification of antigenic variation are crucial for designing and screening effective vaccines.

Numerous research studies have been conducted using HA protein sequence or structure to generate theoretical models and infer antigenic similarity based on sequence similarity [7–18]. Liao et al. [8] used the difference in each non-conserved residue between each pair of HA1 protein sequences as a feature to predict the antigenic distance between HA sequences based on a multiple linear regression model. Ref. [18] used a fitting model to infer two fitness components of the strains that were prevalent in a given year. Then, based on the fitness and frequency of each strain, they predicted the frequency of their descendant strains in the following year. Additionally, the study by Christopher et al. [16] first showed a good correlation between the results of the experimental antigenicity measurements and the antigenic distance prediction based on sequences. Other studies [19–21] further predicted the antigenic variations between influenza vaccines and circulating strains by exploring amino acid sequence mutations that identify epitope regions and their association with seasonal influenza. In order to enhance the performance of mutation prediction for particular residue sites of the influenza A virus, Yin et al. [19] built a time series sample by simulating the evolutionary path of HA sequences. Ref. [22] encoded protein sequences into numerical matrices, and subsequently input these matrices into downstream machine learning models, which was shown to improve the accuracy of predicting influenza antigenicity.

The immune efficacy of influenza vaccines mainly depends on how closely the vaccine and circulating virus strains match one another antigenically, so antigenic difference analysis is crucial for selecting vaccine strains [10]. The hemagglutination inhibition (HI) assay has currently established itself as a standard technique for determining the antigenic distance between current circulating influenza virus strains and reference vaccines [23]. The titer acquired from the HI assay is used to generate antigenic cartography that visualizes the antigenic characteristics of different virus strains on a two-dimensional plane. In antigenic maps, we can intuitively observe the antigenic distances and similarities between different viral strains, and then analyze the virus transmission patterns and vaccine strategies accordingly. Smith et al. [24], based on Lapedes and Farber's [25] metric multidimensional scaling method (MDS), plot antigenic characteristics on the antigenic cartography. The Euclidean distance in antigenic cartography directly describes the antigenic distance, thus providing a reliable quantitative interpretation of antigenic differences.

In order to improve conventional cross-reactivity experiments, Cai et al. [26] believe that modeling to reduce temporal bias in the distribution of HI data is important in antigenic mapping. They used a low-rank matrix completion method to complete the HI titer matrix and then applied the improved MDS method (MC-MDS) to generate the antigenic map. Neher et al. [20] introduced virus potency to explain the systematic changes in HI titers of a virus in multiple sera and used it as the basis for implementing and validating the standardized log-transformed titers relative to the homologous titer. Based on this observation, Lee [27] and Qiu [28] suggested improved antigenic distance calculation methods to reduce the impact of variations in experimental conditions.

Some efforts have been devoted to exploring the relationship between the genetic and antigenic characterization of influenza viruses. According to Koel et al. [29], genetic changes in H3N2 viruses are relatively persistent, while changes in antigenic features recognizable by the human immune system occur intermittently. This suggests that influenza viruses evolve in a way that evades recognition by the human immune system, making the development of effective vaccines more challenging. To conduct a cross-study of genetic and antigenic characterization, ref. [30] mapped the antigenic distance from HI titer data onto the HA lineage. They used available antigenic and genomic sequence data to explain the antigenic novelty and virus transmission rates across the population, and then determined the antigenic changes between clades of high growth. Bedford et al. [31] combined antigenic maps with genetic information on the four human influenza virus subtypes and found that the H3N2 subtype's antigenic phenotype evolves faster than the other three subtypes. Moreover, there is a strong correlation between the antigenic drift of each influenza strain and the number of new influenza cases each year.

Previous works have tended to analyze the genetic and antigenic patterns of viruses in isolation. This work aims to use a unified space composed of antigenic and genetic features to model and analyze the evolutionary dynamics of influenza viruses, with the main challenges being how to integrate genetic information represented by amino acid sequences into the space and how to predict antigenic distance quantitatively. To this end, we propose an effective framework to jointly model antigenic and genetic features through antigenic network representation learning with the ProtVec of HA1 sequences as node attributes and antigenic distance converted by HI titer as edge weights. Then, our model learned effective virus representations by introducing node attribute affinity to predict antigenic distances. We applied our proposed model to the H3N2 dataset, performed antigenic distance prediction tasks using the workflow shown in Figure 1, and studied the antigenic evolutionary dynamics of the H3N2 virus. The model takes two inputs: the attribute matrix capturing node attributes, embedded using ProtVec [32], and the link weight matrix representing antigen distances. By calculating the similarity between attribute matrices, it reveals relationships and patterns among node attributes. The link weight matrix quantifies distances between strains using an antigenic distance formula. As the output, the model provides embedding vector representations of strain nodes. These vectors encode the essential characteristics and relationships of strains in a low-dimensional latent space.

Compared with previous methods, we significantly reduced the root-mean-square error of the prediction results and the classification index of antigenic differences. Through vector analysis based on representation learning, we found that the Pearson correlation between genetic distance and antigenic distance (between antigenic clusters) was 0.8694 to 0.9573, while the correlation within antigenic clusters was only 0.7119 to 0.8556. This suggests high global correspondence and some local differences between influenza genetic and antigenic evolution. Eventually, we found that a historical genetic variation of 0.05 ± 0.004813 led to antigenic drift events of H3N2 influenza virus.

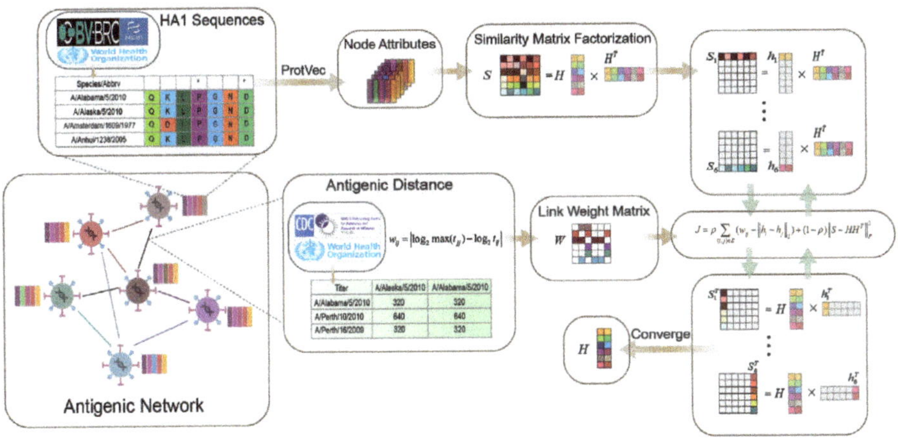

Figure 1. Antigenic network representation learning frame based on AANE.

2. Materials and Methods

2.1. Datasets

We used Trevor Bedford's benchmark dataset [31], which contains 402 H3N2 virus isolates dated from 1968 to 2011.

2.1.1. HI Titers

HI titer data were obtained from official documents published by the World Health Organization. Under the framework of the World Health Organization's global influenza surveillance network, influenza centers in various countries collect viral isolates of seasonal influenza and use the HI assay to calculate titers for antigenic characterization. The range of HI titers typically spans from 10 to 10,240, as lower dilutions may be subject to potential non-specific inhibition, while higher titers are usually not used. Due to the lack of sensitivity of HI assays beyond a certain threshold, accurate data for experimental results on H3N2 strains mainly exist between strains separated by no more than 14 years. At the same time, some assay data only retain the threshold value, such as "<40". Specifically, we arrange the antigen and antibody of the HI assay experiments in a matrix according to their chronological order. This results in three types of data: (I) conventional accurate HI titer data showing a band-shaped distribution close to the diagonal of the matrix; (II) data that are lower or higher than a certain threshold, which is typically slightly deviated from the diagonal; and (III) entries lacking experimental data, which are more likely to occur in positions that are significantly off the diagonal.

The HI titer dataset used in our experiments includes 402 viral isolates and 114 antiserum samples. Of these isolates, 187 are from Europe and 215 are from other parts of the world. We obtained a total of 8599 individual HI titer values in the dataset, of which 1110 (12.9%) are type (II) data, i.e., values lower than a certain threshold indicated by "$< t$", where $t \in \{5, 10, 20, 40, 80, 160\}$. Considering experimental variations in each study, we removed the top 10% of experimental values for each isolate pair $|t_{ij} - \overline{t_{ij}}|$ arranged in descending order and calculated the average of the remaining values as the mean titer between isolate i and reference isolate j.

In actuality, there are three difficulties in interpreting the results of hemagglutination inhibition (HI) assays. Firstly, in order to obtain optimal and reliable results, measurements must be performed under specific experimental conditions such as incubation time, red blood cell concentration, and red blood cell type, which can lead to differences in the results obtained under different experimental conditions. Secondly, HI titers are influenced by the affinity of hemagglutinin for red blood cells, and, to some extent, reflect differences in affinity [33]. In addition, HI assays depend heavily on the antibodies that bind near the receptor-binding domain, and therefore tend to measure responses to specific epitopes. Finally, due to the impossibility of performing HI assays for all pairs of antigens and antisera reactions, the combination of multiple datasets often leads to an incomplete HI titer matrix.

Smith [24] proposed a way of measuring the antigenic distance between two viruses based on HI titer data, where the antigenic distance between strain i and j is defined as follows:

$$w_{ij} = b_j - \log_2(t_{ij}) \tag{1}$$

where b_j represents the maximum titer value of the serum j, i.e., $b_j = \log_2 \max(t_j)$. When the HI titer value is of the (I) type, t represents the maximum dilution of serum j that is necessary to inhibit the cell agglutination caused by the virus strain i. When the HI titer data are of the "$< t$" type, $t_{ij} = t$.

To enhance the traditional cross-reactivity assay method and reduce the impact of differences in receptor binding affinities among different virus strains, we were inspired by Neher's method [20] of calculating relative titers and used the following formula to calculate antigenic distance:

$$w_{ij} = \left|\log_2 \max(t_{jj}) - \log_2(t_{ij})\right| \tag{2}$$

According to the above formula, the antigenic distance between strain i and antiserum j is converted into the deviation value between the relative titer experimental value of antiserum j itself and the titer experimental value of strain i relative to antiserum j, where t_{jj} and t_{ij} denotes the titer data between strains. This method, as a supplement, can effectively enhance the accuracy and reproducibility of cross-reactivity experiments and has better application effects in analyzing and comparing the receptor binding affinity between different virus strains. For two virus strains with an antigenic distance greater than 4, their antigenicity is considered different, leading to immune escape. Otherwise, they are regarded as antigenically similar. After processing and calculating according to the above method, we finally identified 1543 antigenic difference pairs and 2744 antigenic similarity pairs among 402 strains.

The logarithmic linear correlation concentration ratio defined by the shape space theory of Lapedes and Farber [25] is currently the most widely used method for calculating antigenic distance.

$$w_{ij} = \log \sqrt{\frac{t_{ii}t_{jj}}{t_{ij}t_{ji}}} \tag{3}$$

There are only 1403 entries for antigenic distance, as calculated by Formula (3). However, trying to make comparable predictions from the sparse and coarser-grained data in these 1403 entries is more difficult. We validated our proposed method's significant consistency with the popular method by performing a correlation analysis between the standardized logarithmic transformed antigenic distance calculations and the calculation method described in Formula (3) (measured with the Pearson correlation coefficient (PCC), R = 0.9629, 95% confidence interval: 94.86% to 97.68%, see Figure 2a). Similarly, we also calculated a correlation of 0.9815 between our proposed method and the method used by Smith. Based on these findings, it can be concluded that the standardized logarithmic transformation method for antigenic distance calculations can expand the utilization of influenza monitoring data, and we can better predict major antigenic changes.

A number of studies [20,27,34] have made efforts to explore the relationship between point mutations and influenza outbreaks based on a small number of amino acid characteristics. However, these models only measure the contribution of the selected amino acids as individuals. On the one hand, the lack of background of composite effects is due to the fact that amino acid changes in HA form a three-dimensional structure spatially. A previous study [35] analyzed the contributions of individual mutations and their related combined effects through CNN models. In addition, different amino acid residues have a significant impact on the antigenicity of H3N2 virus, and how to measure the contribution of different residues is one of the key issues in sequence processing.

2.1.2. HA Sequences

The HA sequences of 402 virus strains were collected from databases such as NCBI and GISAID. The lengths of HA1 sequences were 329 for influenza A/H3N2. Sequences containing missing or abnormal amino acids (i.e., "-") were manually and automatically removed, and aligned using Mega. From 1968 to 2011, there were 0 to 30 amino acid changes between these strains. We defined the genetic distance between two HA sequences as the sum of pairwise distances between their 329 amino acids. To construct the antigenic network and represent the attribute information of its nodes, we embedded the HA sequences into a node attribute information matrix based on ProtVec. ProtVec [32] applies Skip–Gram to learn the distributed embedding vector representations of influenza virus amino acid sequences, representing three contiguous amino acid sequences as 100-dimensional vectors. By training only on protein sequences, the ProtVec feature extraction method is able to capture various meaningful physical and chemical properties, and can serve as an informative and dense representation of biological sequences in protein family classification. Specifically, each HA sequence is represented as a list of 327 contiguous 3 g, and a 327 × 100 matrix is generated as the node attribute information in Figure 2b. To represent 3 g sequences containing "-" in our study, we utilize the "<unk>" vector from ProtVec.

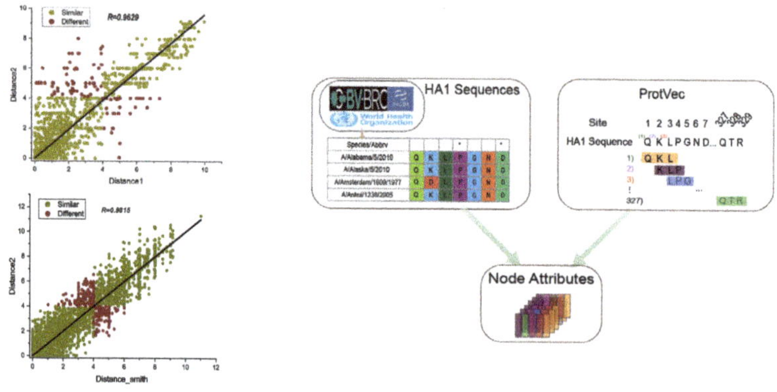

(a) (b)

Figure 2. (**a**) The graph illustrates the relationship between the antigenic distance determined by Formula (3) (x-axis) and the proposed calculation method in this paper (y-axis). The green circles represent consistent antigenicity comparisons (i.e., similar or dissimilar) between the two calculation methods, while the red circles represent inconsistent results. The black solid line is the best linear fit with zero intercept, and the correlation between the two antigenic distance calculation methods is 0.9629 with a 95% confidence interval of 94.86% to 97.68%. There is a correlation of 0.9815 between our proposed method and the method used by Smith. (**b**) Using H3N2 HA1 sequence data for ProtVec continuously distributed the representation as node attribute information. To construct the antigenic diversity network and represent the node attribute information, we embedded the HA1 amino acid sequences of 329 in length into a node attribute information matrix using ProtVec. ProtVec applies Skip–Gram to learn the distributed embedding vector representation of influenza virus amino acid sequences, representing continuous triplets of amino acid sequences as 100-dimensional vectors. Through ProtVec representation, each node's attribute information matrix can be obtained.

2.2. *Antigenic Network Representation Learning*

In recent years, some interpretable attributed network embedding algorithms have emerged. For example, accelerated attributed network embedding (AANE) [36] decomposes the heterogeneous information similarity matrix and penalizes embedding differences between adjacent nodes to preserve the similarity of nodes in the original network space on the new continuous vector representation. Inspired by this observation, this paper discusses the possibility of integrating the node attribute information similarity matrix represented by the HA sequence of influenza virus H3N2 and the topological structure represented by antigenic distance into the network embedding representation learning, and whether this method can help to better learn node vectors in the antigenic network. In addition, the rapid mutation and seasonal epidemics require the scalability of antigenic network embedding representation. AANE embeds noisy network topology and node attributes to improve the model's time complexity and convergence speed.

The graph-based attribute network embedding method demonstrates strong robustness in handling missing data. In cases where certain antigenic distances are missing, the embedding method can utilize the attribute information of the nodes to impute the missing values, thereby partially compensating for the loss of information. By using the ProtVec representation of the genetic sequences as node attributes, we were able to exploit the informative features of each strain. This not only enriches the available information within the network but also facilitates the extraction of meaningful patterns and relationships between strains. Meanwhile, graph networks inherently possess the ability to model non-linear interactions and dependencies, which may be crucial for capturing the complexity of virus strain evolution. This flexibility allows our model to better capture the intricate patterns and dynamics present in the antigenic distance data.

We define the embedding learning of the antigenic network as follows for an antigenic network $G = (V, E, W)$, where $V = \{v_1, v_2, ..., v_n\}$ represents the set of n strain nodes. The attributes of each node are represented by the matrix $p_i \in \mathbb{R}^{(k-2)\times m}$, which is obtained by embedding the amino acid sequence of length k using ProtVec. There is also a set of edges between strain nodes in the network $E = \{e_{ij}\}_{i,j=1}^{n}$, and each edge $(i, j) \in E$ is associated with a non-negative antigenic distance $w_{ij} \in W$. e_{ij} is an unobserved edge when $w_{ij} = -1$. This study assumes that the antigenic distances between strains are non-negative and symmetric, used to measure the edge weights between antigenic network nodes. We use AANE for embedding learning to obtain a d-dimensional vector $h_i \in H$ for nodes v_i. This embedding method produces a representation of every node in the network space that combines genetic and antigenic properties, allowing H to maintain the adjacency of both the topological structure and node attributes simultaneously.

The strain vectors learned through network embedding representation learning should have the following four advantages: (1) low dimensionality, i.e., the embedding dimension d should be smaller than the length of the original sequence, in order to improve the efficiency of downstream tasks; (2) preserving the antigenic features of the original antigenic network structure, i.e., nodes with structural similarity should still be similar in the new space; (3) preserving the genetic features of the original antigenic network, i.e., the similarity of original HA1 sequence should be well captured to complement rather than degrade the expression of the network structure; and (4) the pairwise similarity of node embedding vectors should reflect the pairwise similarity of nodes in the original antigenic network. Compared with pure network embedding, network embedding representation H, which concurrently maintains the topological structure and node attribute information, performs better on the link weight prediction challenge.

2.2.1. Network Topological Structure Modeling

To maintain the proximity of nodes in the network while improving the performance of antigenic distance prediction, we first assume that strains with similar topological structures or connected by smaller weighted edges are more likely to have similar embedding vector representations. To accomplish this, we suggest the following loss function to reduce the differences in embedding between linked strain nodes:

$$L_G = \sum_{(i,j)\in E} \left(w_{ij} - \left\|h_i - h_j\right\|_2\right) \tag{4}$$

where h_i and h_j are the vector representations of node v_i and node v_j, and w_{ij} is the edge weight between them. The key idea of this loss function is to minimize $w_{ij} - \left\|h_i - h_j\right\|_2$. For the antigenic network, a smaller antigenic distance w_{ij} needs to force the difference between the vector representations h_i and h_j of two strains to be smaller. By using the 2-norm $\left\|h_i - h_j\right\|_2$ as the difference metric between vectors, it can not only characterize the distance between vectors but also alleviate the negative effects of outliers and missing data.

2.2.2. Node Attribute Proximity Modeling

According to social science theories such as homophily and social influence [37,38], node attribute information is closely related to network topology. Thus, the similarity of nodes in the network space should be consistent with the similarity of nodes in the attribute space. Inspired by symmetric matrix factorization, the product of H and H^T approximates the node attribute similarity matrix S. The basic idea is to force the dot product of the embedding representation of h_i and h_j to be similar to the corresponding attribute similarity matrix. Therefore, to maintain node attribute proximity, we define the loss function as follows:

$$L_S = \left\|S - HH^T\right\|_F^2 = \sum_{i=1}^{n}\sum_{j=1}^{n}\left(s_{ij} - h_i h_j^T\right)^2 \tag{5}$$

where the matrix $S \in \mathbb{R}^{n\times n}$ represents the 2-norm between the node attribute matrices, capturing the attribute similarity and the differences between different strains in the joint space. Specifically, given the node attribute information $a, b \in \mathbb{R}^{(k-2)\times m}$ based on ProtVec representation, the formula for calculating node attribute similarity is as follows:

$$s_{ij} = \sqrt{\sum_i^{k-2} \sum_j^m (a_{ij} - b_{ij})^2} \tag{6}$$

2.2.3. Antigenic Network Embedding Representation Learning

According to Equations (4) and (5), we have implemented two loss functions, L_G and L_S, to fit the similarity between network topology and node attributes. To make them complement each other and form a unified network representation space, our optimization objective is the following function:

$$J = \rho \sum_{(i,j)\in E} (w_{ij} - \|h_i - h_j\|_2) + \left\|S - HH^T\right\|_F^2 \tag{7}$$

ρ functions as both a regularization parameter that balances the number of clusters and a global parameter that defines the contribution of network structure and attribute information to the node representation. The intuitive explanation is that when it approaches 0, the network topology cannot affect the final node representation H, and each strain will reflect the similarity of node attributes. When ρ is sufficiently large, the vector representation of all nodes will tend to reflect the network's topology fully. We need to use the Euclidean distance in the new network space to calculate the predicted antigenic distance between any two strains v_i and v_j. To ensure the accuracy and reliability of our predictions, we will follow widely adopted methods for validating network embedding quality.

3. Results

We will address the following three questions through experiments: (1) Does supplementing genetic information represented by HA1 sequences better predict antigenic distances based on antigenic network learning compared to using only antigenic features? (2) Can the AANE method used in antigenic network learning achieve more accurate antigenic distance prediction than other attribute network embedding learning methods? We will also further discuss the sensitivity of model parameters, including the regularization parameter ρ of the loss function and the dimension d of the node representation vector. (3) By modeling genetic and antigenic features in a combined space, can we gain new insights into the evolution of H3N2 viruses?

3.1. Baseline

To answer the aforementioned questions, we will use the following models as baselines and describe them in detail:

- Node2vec [39] is a Skip–Gram-based algorithm that generates node sequences through biased random walks. Hyperparameters p and q are used to control the random walks, and we adjust them according to the original paper. When p and q are both 1, node2vec is equivalent to DeepWalk. Node2Vec only uses antigenic distance for an edge-weighted biased random walk.
- LINE [40] constructs the network using only antigenic distance and generates context nodes through breadth-first search, where a node's neighboring nodes are limited to those that are at most two hops away. LINE models both first-order and second-order similarities for each node and concatenates the two learned embedding vectors according to the actual scenario.
- LINE1 constructs the network using only antigenic distance and models first-order similarity to learn node representation, which mainly constrains directly connected nodes.

- LINE2 also constructs the network using only antigenic distances, but focuses on the neighborhood similarity of nodes and learns node representation by preserving second-order similarity.
- Attri2vec [41] learns node representation by performing linear or non-linear mapping on node attributes. To preserve structural similarity, it uses the DeepWalk learning mechanism so that nodes with similar random walk contexts have similar dense representations in the subspace. This is achieved by maximizing the probability of the appearance of context nodes conditioned on the target representation.
- GCN [42] is a graph-specific model that applies convolution on graph nodes to generate representations for each node.
- By employing the masked self-attention layer, GAT [43] overcomes certain limitations present in existing methods. The key aspect of GAT lies in stacking multiple layers, where each layer can implicitly assign varying weights to neighboring nodes without the need for costly matrix operations or prior knowledge of the graph structure.
- GraphSAGE [44] utilizes local neighborhood sampling to aggregate features and generate embeddings. Subsequently, a minibatch forward propagation algorithm is employed to train the data.
- GALA [45] proposes a symmetric graph convolutional autoencoder for generating low-dimensional latent representations of graphs. Compared to existing graph autoencoders, our model features a newly designed symmetric decoder that effectively utilizes the graph structure for reconstructing node features.
- TADW [46] not only considers the structural information of nodes but also utilizes the text information of nodes. It implements the DeepWalk idea through matrix factorization and introduces node text information to improve the expression of embedding vectors.

3.2. Evaluation of Antigenic Distance Prediction Performance

The performance of antigenic distance prediction is evaluated using two metrics: the root mean square error (RMSE) and the Pearson correlation coefficient (PCC) between the predicted and actual antigenic distances. These metrics reflect different aspects of prediction performance: RMSE amplifies the differences between larger errors and quantifies the degree of proximity between the prediction and the average true value, while PCC measures the relative trend between the two. Given n true antigenic distances w_{ij} and predicted antigenic distances d_{ij}, the corresponding metrics are defined as follows:

$$RMSE = \sqrt{\frac{1}{n} \sum_{(i,j)\in E} (w_{ij} - d_{ij})^2} \tag{8}$$

$$PCC = \frac{\sum_{(i,j)\in E} (w_{ij} - \overline{w_{ij}})(d_{ij} - \overline{d_{ij}})}{\sqrt{\sum_{(i,j)\in E} (w_{ij} - \overline{w_{ij}})^2} \sqrt{\sum_{(i,j)\in E} (d_{ij} - \overline{d_{ij}})^2}} \tag{9}$$

Furthermore, we will evaluate the model's ability to detect antigenic drift based on the prediction results. We will assess this capability using four evaluation metrics: accuracy, precision, recall, and F1 score.

With a default embedding dimension, d, of 50, we set the baseline method's settings as recommended in the original paper. All experimental results are arithmetic averages of 10 tests. We divide the dataset into a training set and a test set (75% and 25%), and perform multiple rounds of training and evaluation through 3-fold cross-validation by dividing the training set into three subsets and using these subsets alternately as validation sets. Based on the preliminary findings from Figure 3, we found that the methods based on network embedding representation learning achieved significantly better performance than LINE1 in predicting antigenic distances on the H3N2 dataset. We attribute this phenomenon to the following two reasons: (1) the LINE model was applied to millions of data points, which is vastly different from the sample size used in this experiment; and (2) in contrast, LINE1 only models first-order proximity, which cannot capture enough information for link weight prediction tasks.

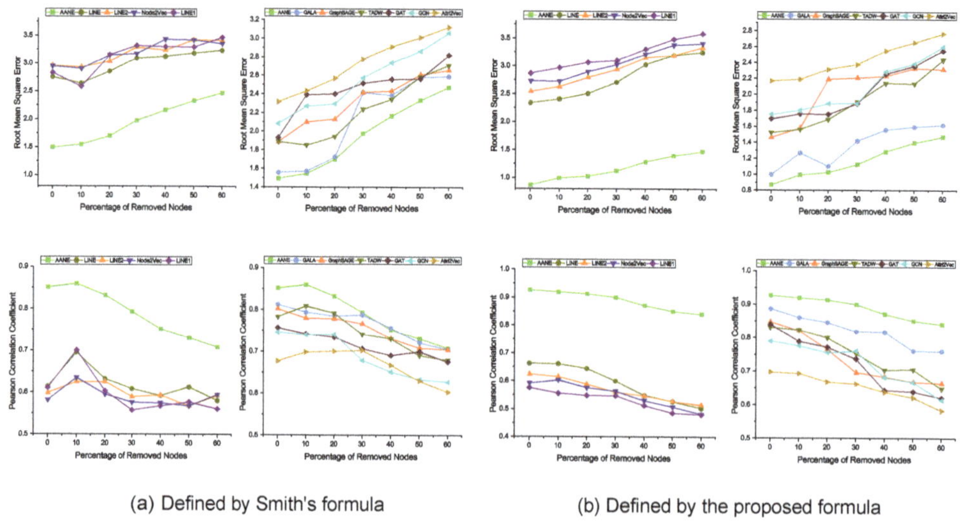

Figure 3. Figure depicts the performance of the antigenic network embedding learning model based on AANE in terms of the RMSE (top) and PCC (bottom) metrics for antigenic distance prediction tasks, using the H3N2 antigenic network dataset (1968–2011) with $d = 50$. This model outperforms all other models. The x-axis represents the percentage of randomly removed nodes from the network (from 0% to 60%), and the y-axis represents the corresponding evaluation metrics. In (**a**) and (**b**), the predicted results of antigenic distance calculation are shown using the formula defined by Smith and the proposed normalized logarithmic transformation formula in this paper, respectively. In the left subplots of (**a**) and (**b**), the models that utilize antigenic distance as the only link weight for antigenic distance prediction are compared with AANE (green line) in terms of RMSE and PCC results. In the right subplots of (**a**) and (**b**), the models that utilize antigenic distance as the link weight and the ProtVec matrix encoding HA as the node attribute for network embedding learning are compared with AANE in terms of RMSE and PCC results (green line).

Next, we then evaluate the impact of merging node attribute information. In order to make a fair comparison with models that utilize node attributes, we use Node2Vec, LINE1, LINE2, and LINE as controls, which only consider antigenic distance as the link weight for node vector learning. These methods learn node vectors only through network structure and then predict antigenic distance. As shown in Figure 3a,b, on the dataset of all 402 nodes, the model based on attribute network representation learning outperforms in terms of RMSE and PCC. This confirms our hypothesis that the combination of genetic and antigenic features proposed by us contributes to antigenic distance prediction tasks. Horizontally, Figure 3a,b suggests that utilizing both network structure and node attribute information is beneficial for downstream link prediction tasks.

To further evaluate how much it improves the performance of antigenic distance prediction in the antigenic network, as well as to verify the robustness of the AANE model, we further reduced the number of nodes in the network by 10%, 20%, 30%, 40%, 50%, and 60%. Each experiment was repeated 5 times and the average was taken to test the model's robustness in predicting missing antigenic distances. For example, as the number of deleted nodes increased, LINE (grayish green line in Figure 3b) and Node2Vec (bluish violet line in Figure 3b) showed an overall decreasing trend in RMSE, with a final decrease in the predictive performance of 24.6% and 24.9%, respectively.

According to the experimental results in Figure 3a,b, AANE consistently outperformed Attri2Vec, GCN, GAT, GraphSAGE, GALA, and TADW. All of these approaches execute node embedding learning and represent the network using node attributes and link weights. This illustrates how effective AANE is. For example, on the dataset consisting of all nodes, AANE achieved a 38.1% and 13.1% improvement over TADW and GALA, respectively. Although TADW is effective in learning information-based node embeddings using rich node text features, its mechanism is not as straightforward, i.e., a clear objective is not provided for how the network structure and node attributes interact with each other.

We compared the predicted antigenic distances of all models with the actual antigenic distances by linear fitting (Figure 4). The results show that the attribute network embedding method had a greater advantage in reducing FN data and increasing TN data. Meanwhile, AANE's predicted values had good robustness and a more uniform distribution when linearly fitted with the actual distances. The reason for Node2vec's performance exceeding our expectations may be explained here. We believe that Node2vec does not reflect structural information well. Due to the limited sample size and walk length, it is almost impossible to include two structurally similar nodes in the same sequence through biased walks when the distance between them is very large. This is also related to hyperparameter selection. We set p as 1 and q as 2. The larger q is, the more the embedding tends to express homogeneity. As the number of nodes decreases, nodes with experimental data are often those that are close to their own antigenic distances. Therefore, when expressing network structure, it tends to embed peripheral and central nodes as similar vectors (TN- and FN-predicted values in Figure 4b,f occupy a considerable part). In fact, these predicted values, which fit the original data distribution more closely, can obtain better predictive ability. We followed the author's recommendation and did not stack higher layers of GCN. Perhaps there will be better performance with more than three layers of GCN, but this is not within the scope of this paper. GCN has the same drawback as Attri2Vec—the over-smoothing problem (that is, after multiple layers of stacking, the node's representation vectors tend to be consistent, and the nodes are difficult to distinguish). Due to the low-pass filter effect of GCN, the aggregated features continuously merge the node features, which tend to be the same after multiple iterations. GAT has more parameters than GCN and is trained in a full-batch manner. It only considers 1-hop neighbors and does not utilize higher-order neighbors. When higher-order neighbors are used, excessive smoothing is prone to occur. Attri2Vec has a strong bias toward node attributes, so it is not practical to maximize the attribute information difference among strains with no more than 30 different amino acids even after ProtVec representation.

We tested the ability of all models to successfully predict antigenic escape (as described in Section 2.1, where the antigenic distance between two strains is higher than the antigenic escape threshold (d_{ij} = 4)). The main evaluation metrics for qualitative results are accuracy, precision, recall, and F1 score, as shown in Figure 5. Our results show that the AANE model accurately predicted the antigenic distance between two strains with an accuracy of 91.25%, and other metrics also showed significant advantages.

As shown in Figures 3 and 5, we will present a comprehensive analysis including quantitative evaluation and statistical measures to validate the accuracy and reliability of our defined distances. We compared the predictive performance of the antigenic distances obtained with the computational methods defined by Smith on all benchmark models. As for Equation (3), we chose to discard this set of comparisons because of the small number of entries obtained. We could not obtain more substantial results with these network embedding methods. We found erratic metric fluctuations in the sequential reduction in the number of nodes on the antigenic dataset defined by Smith, a phenomenon that occurs in most of the benchmark methods. As we feared, the formula defined by Smith tended to obtain certain fixed values during the calculations performed on the titer data we collected (also found in Figures 2a), even though it allowed us to obtain more entries of antigenic data. This comparative analysis will help to confirm the validity of our proposed method and to obtain a clearer picture of the differences between the "actual" antigenic distances in our study and the established criteria.

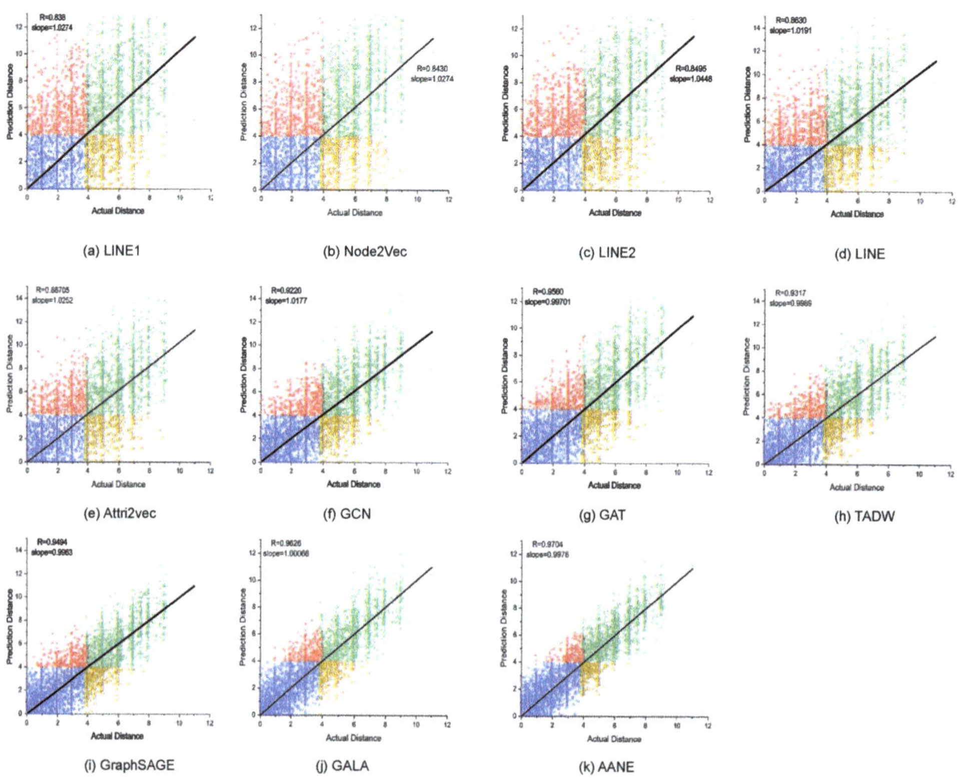

Figure 4. Linear regression analysis of predicted values (y-axis) versus actual values (x-axis) for different models (solid black line). Green dots represent true-positive (TP) predictions; blue dots represent true-negative (TN) predictions; red dots represent false-positive (FP) predictions and yellow dots represent false-negative (FN) predictions.

3.3. Parameter Sensitivity Study

In this section, we looked into the effects of two significant parameters, ρ and d. As described in Section 2.2.3, ρ in AANE balances the contribution of network structure and node attributes. To study the effect of ρ, we changed it from 10^{-3} to 10^{6}. As there are up to 35 or more different combinations of hyperparameters (a, b), we only give the optimal value of another parameter b under a specific parameter a. Table 1 shows the RMSE and PCC results of antigenic distance prediction under different ρ values. Setting $\rho = 10^{-3}$ almost ignores the influence of network structure information, and nodes tend to have the same embedding vector representation. As ρ increases, AANE predicts the antigenic distance based on the topological structure, and the performance gradually improves. As shown in the figure, when ρ is close to 10^{5}, the performance of antigenic distance prediction peaks. When ρ continues to increase, the performance will decrease, as larger ρ values tend to make all nodes too dependent on sparse structural information. However, we cannot directly infer from this so-called optimal value that genetic information only contributes 0.001% to the distance prediction task because the two dimensions are intuitively very different. Nevertheless, from the overall trend of change, it indicates to some extent that genetic information contributes to the advantage of using attribute network embedding for antigenic distance prediction, as shown by the improvement in the RMSE value.

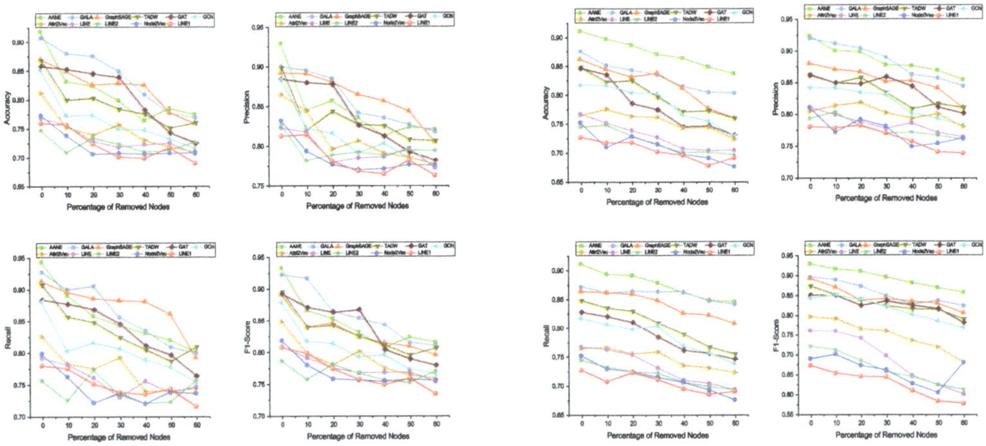

(a)Defined by Smith's formula

(b) Defined by the proposed formula

Figure 5. The antigenic distances predicted by the model were converted to antigenic differences (using $D_{(i,i+1)} = 4$ as the threshold for binary classification) and measured on the H3N2 dataset with different classification metrics: accuracy, precision, recall, and F1 score. (**a**) and (**b**) represent the results of predictions using the antigenic distance data calculated using the formula defined by Smith and the antigenic distance data calculated using Equation (3), respectively.

Table 1. Different combinations of the regularization parameter ρ and embedding dimension d affect the prediction results of antigenic distance.

Parameters	RMSE	PCC
$(\rho = 10^{-3}, d = 150)$ *	2.1994	0.6956
$(\rho = 10^{-2}, d = 150)$	2.0167	0.7354
$(\rho = 10^{-1}, d = 150)$	1.9503	0.7462
$(\rho = 1, d = 150)$	1.8559	0.7662
$(\rho = 10, d = 120)$	1.7025	0.7897
$(\rho = 10^2, d = 120)$	1.6603	0.8018
$(\rho = 10^3, d = 120)$	1.2973	0.8660
$(\rho = 10^4, d = 100)$	0.8899	0.9336
$(\rho = 10^5, d = 50)$	0.8678	0.9373
$(\rho = 10^6, d = 20)$	1.5160	0.8311
$(d = 20, \rho = 10^6)$	1.6102	0.8086
$(d = 32, \rho = 10^6)$	0.9746	0.9197
$(d = 50, \rho = 10^5)$	0.8730	0.9362
$(d = 64, \rho = 10^5)$	0.9120	0.9303
$(d = 80, \rho = 10^3)$	1.1248	0.8965
$(d = 100, \rho = 10^4)$	0.8950	0.9326
$(d = 120, \rho = 10^3)$	1.3108	0.8652
$(d = 150, \rho = 10^2)$	1.7906	0.7753

* The first parameter represents the determined parameter, and the second parameter represents the optimal value under the condition of the first parameter.

Following the rules we established when building the network representation model (Section 2.2), dimension d should be less than 329. Specifically, we changed the embedding dimension from 20 to 150, and Table 1 shows the prediction performance on the dataset. From the results, we found that by increasing d, the performance of the method first increases and then remains stable. This indicates that low-dimensional representations perform well in capturing most of the meaningful information. In reality, determining an appropriate dimension is not easy, especially for antigenic distance prediction. Although a lower dimension has lower time and space complexity, it will undoubtedly lose a lot of information originally present in the network. Higher dimensions may improve reconstruction accuracy to some extent, but at the same time, the Euclidean distance between vectors of higher dimensions will likely become larger. Based on our experimental results, we can conclude that model performance is relatively stable within a small range of node embedding dimensions, and performance declines when the node embedding dimension is too small or too large.

3.4. Antigenic Evolution Dynamic Analysis

The effectiveness of the model has enabled us to explore the dynamics of influenza antigenic evolution based on vector representation in the joint space. As shown in Figure 6, we first performed preliminary clustering by year and calculated the average antigenic distance $D_{(i,i+1)}$ between all pairs of strains from year i and $i+1$ for the 44-year period, resulting in 41 data points (except for 1978 in the H3N2 dataset). Then, we merged adjacent clusters that had the smallest antigenic distance ($D_{(i,i+1)} < 4$) without antigenic variation, and each new cluster was named after the earliest year in the cluster. For example, 1982 and 1983 strains had the smallest antigenic distance (0.045804) and were merged into a new cluster, followed by recalculating the average antigenic distance to 1981 and 1984 strains. All strains were finally grouped into seven significant antigenic drift episodes, as seen in Figure 6b. We calculated the antigenic distance between each pair of strains in each event. In an antigenic drift event E including n strains, the antigenic variation level of strain i was defined as

$$C_i = \frac{\sum\limits_{j \in E} d_{ij}}{n-1} \tag{10}$$

where d_{ij} represents the antigenic distance between strain i and all other strains j within the same event E. The strain with the smallest antigenic variation value within the current cluster is chosen as the dominant strain (which has the smallest average antigenic distance to all other strains in the cluster) and is used to name the event. Between 1968 and 2011, we discovered seven significant antigenic drift events: BI68, BI73, LY79, VI87, MA93, FU00, and ST09.

Based on the clustering results, we quantified the relationship between antigenic and genetic distances among strains. The genetic distance was calculated from the amino acid sequence differences between strains. In our study, the differences in the number of identified antigenic drift events compared to the research conducted by Smith et al. [24] could be attributed to various factors, including the datasets used, the definition of antigenic distance, and the specific criteria employed to define antigenic drift events. Furthermore, we looked at the roughly evolutionary relationship between H3N2's genetic and antigenic characteristics. First, for the global time scale, the Pearson correlation coefficient between genetic and antigenic distances was 0.8559, indicating a roughly linear relationship between genetic and antigenic differences during the inter-epochal evolution of influenza, which is consistent with the relationship observed by Smith et al. [24]. We randomly selected some strains within and between each cluster event to calculate their average genetic and antigenic distances. Surprisingly, the genetic–antigenic evolutionary relationship between clusters showed a stronger linear pattern than that within clusters (Figure 7). Moreover, the genetic and antigenic evolution between the seven adjacent antigenic drift events showed a linear correlation of 0.8694 to 0.9573, while the evolution within clusters was characterized by discontinuous development (see Figure 8). Furthermore, we calculated that an average of 0.05 ± 0.004813 units of genetic variation led to the occurrence of an antigenic drift event. However, the distribution of genetic distances between different antigenic drift events varied greatly, and the distribution within clusters was more concentrated, even though they were usually very small. This aligns with earlier studies [47,48], which suggested that strong selection and neutral antigenic evolution alternated during antigenic drift events. As a result, the new vector space learned by the antigenic network representation learning method can explain the short-term and long-term patterns of the relationship between genetic and antigenic distances. Moreover, this method greatly improves the resolution and accuracy of antigenic differences.

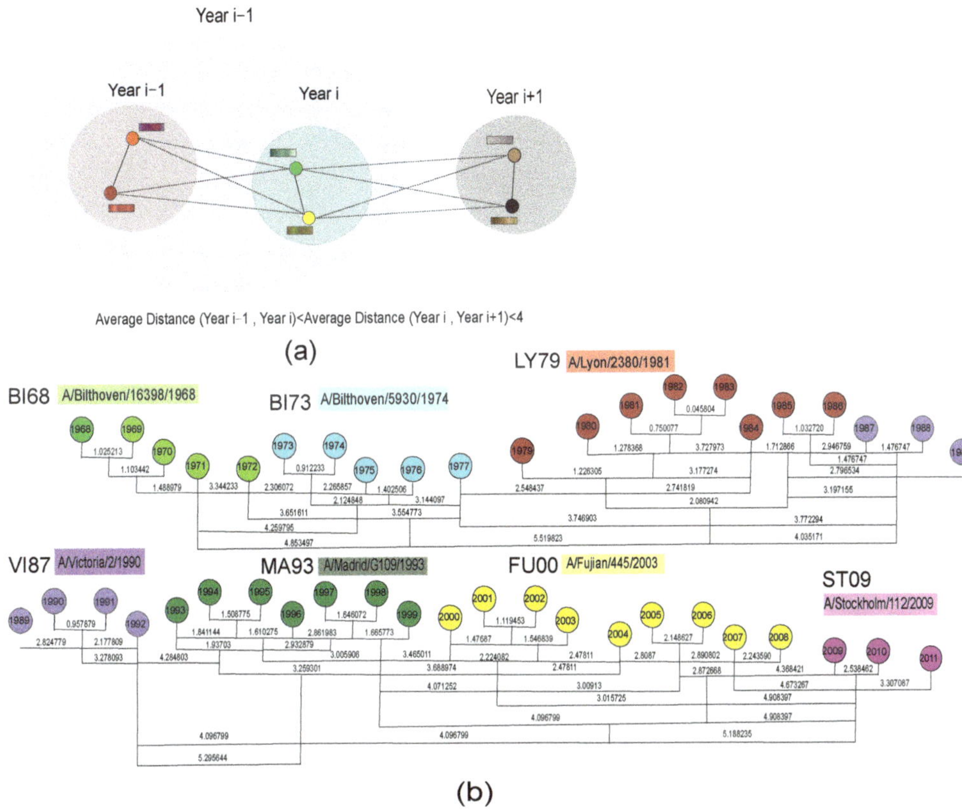

Figure 6. Antigenic clustering over the past four decades (1968–2011). (**a**) During the entire clustering process, adjacent clusters with the smallest antigenic distance in the current collection of all clusters are selected successively and merged into a new cluster without antigenic variation ($D_{(i,i+1)} < 4$). (**b**) Each circle represents all strains in a given year, and the numerical values between two circles represent the average antigenic distance between clusters during the updating process. Adjacent clusters with similar antigenicity are merged into new clusters, and the strain with the smallest antigenic variation in each cluster is used to name the final cluster.

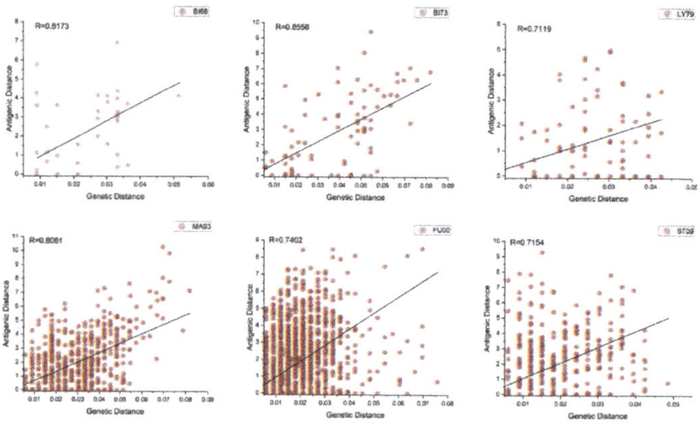

Figure 7. Comparison of the relationship between genetic distance and antigenic distance in the same cluster.

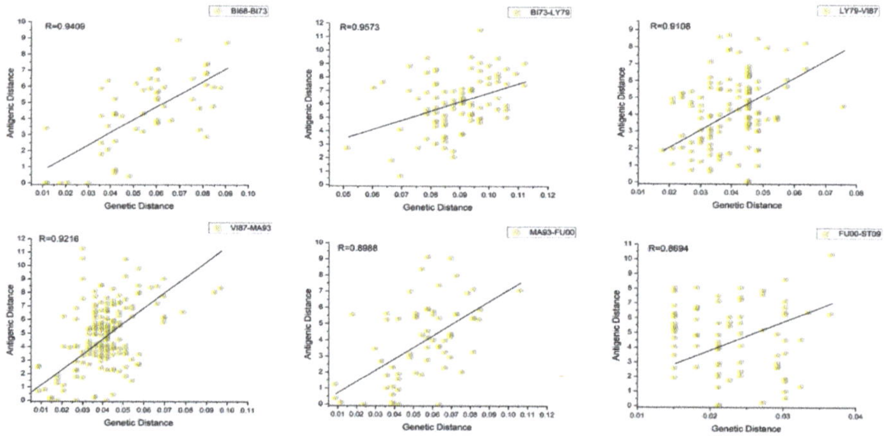

Figure 8. Comparison of the relationship between genetic distance and antigenic distance in adjacent clusters.

4. Discussion

Certainly, there have been some notable achievements in the field. For instance, ref. [49] proposed a novel approximation method for antigenic distance, which is based on deep learning in the feature space induced by hemagglutinin protein sequences and convolutional neural networks (CNNs). On the other hand, ref. [50] evaluated the predictive capability of their method by conducting laboratory measurements. We recognize the importance of a thorough comparison and evaluation with other relevant methods. However, the experimental results indicate that our study is an initial exploration of the proposed method and provides a foundation for future investigations and advances.

Historical experience shows that antigenic data cannot significantly reflect genetic data. This article proposes a method based on the AANE model for network representation learning to integrate HA sequence data into the antigenic network structure. This method can quantitatively predict the antigenic differences between strains well. Since the HA sequence and titer data provide different sources of information, it is crucial to capture their key features to learn the comprehensive representation of strains in the antigenic network.

For instance, specific epitope positions on the HA protein and the amino acids that make up the epitope may differ due to host species and genotype. The affinity matrix can map specific amino acid mutations between different nodes to the perspective of the entire sequence, which is another reason why we think it should be introduced. Its biological characterization distinguishes the differences in the three continuous amino acid sequences.

In this paper, the antigenic distance is inferred using both genetic and antigenic data, which differs from the positions inferred solely from the HI data. If the HI data are abundant, we expect smaller differences in prediction when genetic data are included (possibly only for H3N2). In contrast, if the HI data are limited, we expect genetic data to play a greater role in determining antigenic positions. As described in Section 3.3, genetic and antigenic features are not complementary, but rather genetic features complement antigenic features to explain quantitative differences in antigenicity. This is in line with our original intention of incorporating genetic information into the network structure to improve antigenic distance prediction performance. Furthermore, our validation of the contribution of hyperparameters of the loss function to the prediction task also supports the same viewpoint.

Little is known about the relationship between influenza virus antigenic phenotype changes and genetic sequence changes. A helpful framework for investigating influenza is provided by the work of Bedford et al. [31], which may be used to identify which alterations to virus genes lead to changes in antigenicity. Our model for antigenic network learning based on AANE optimizes the observed different patterns (such as homogeneity and structure) in the network, proposes a robust objective function, and makes some assumptions about the relationship between the underlying network structure and the prediction task. We can determine its effectiveness and scalability by observing how the learned vectors reflect the relationship between genetic evolution and antigenic evolution. Information about the virus's "antigenic dynamics" can be reflected in the evolutionary perspective captured between antigenic drift events, especially in the long-term and short-term patterns of antigenic evolution.

5. Conclusions

The main contributions of this paper can be summarized as follows. We propose an effective attribute network embedding framework to learn low-dimensional representations of strains from both the node attribute affinity matrix and topological structure information. By expanding the utilization of influenza surveillance data and the representation of sequence biology significance, we validated the basis of joint spatial modeling, which supports the combination of genetic and antigenic features on real datasets. Through the learned low-dimensional representations, we can better predict the antigenic distance between any strains in the network and explore the new dynamics of antigenic evolution.

Author Contributions: Conceptualization, W.L. and Y.X.; methodology, W.L. and F.P.; software, F.P.; validation, F.P.; formal analysis, F.P. and W.L.; investigation, W.L.,Y.X. and F.P.; resources, W.L.; data curation,W.L.; writing—original draft preparation, F.P.; writing—review and editing, W.L. , F.P. and Y.X.; visualization, F.P.; supervision, W.L.; project administration, W.L.; funding acquisition, W.L. All authors have read and agreed to the published version of the manuscript.

Funding: This work is supported by the National Natural Science Foundation of China (Grant No. 32060151), and the Yunnan Provincial Foundation for Leaders of Disciplines in Science and Technology, China (Grant No. 202305AC160014).

Institutional Review Board Statement: Not applicable.

Informed Consent Statement: Not applicable.

Data Availability Statement: The antigenic dataset of HI titers was compiled by Bedford et al. and can be obtained at https://datadryad.org/stash/dataset/doi:10.5061/dryad.rc515 (accessed on 28 April 2023); the code implementation and documents related to this paper can be obtained at https://github.com/john-darwin/Antigenic-Distance-Prediction (accessed on 29 June 2023).

Conflicts of Interest: The authors declare no conflict of interest.

References

1. Agor, J.K.; Özaltın, O.Y. Models for predicting the evolution of influenza to inform vaccine strain selection. *Hum. Vaccines Immunother.* **2018**, *14*, 678–683. [CrossRef]
2. Allen, J.D. Ross, T.M. H3N2 influenza viruses in humans: Viral mechanisms, evolution, and evaluation. *Hum. Vaccines Immunother.* **2018**, *14*, 1840–1847. [CrossRef]
3. Iuliano, A.D.; Roguski, K.M.; Chang, H.H.; Muscatello, D.J.; Palekar, R.; Tempia, S.; Cohen, C.; Gran, J.M.; Schanzer, D.; Cowling, B.J.; et al. Estimates of global seasonal influenza-associated respiratory mortality: A modelling study. *Lancet* **2018**, *391*, 1285–1300. [CrossRef] [PubMed]
4. Kumlin, U.; Olofsson, S.; Dimock, K.; Arnberg, N. Sialic acid tissue distribution and influenza virus tropism. *Influenza Other Respir. Viruses* **2008**, *2*, 147–154. [CrossRef] [PubMed]
5. Neu, K.E.; Dunand, C.J.H.; Wilson, P.C. Heads, stalks and everything else: How can antibodies eradicate influenza as a human disease? *Curr. Opin. Immunol.* **2016**, *42*, 48–55. [CrossRef] [PubMed]
6. Nelson, M.I.; Holmes, E.C. The evolution of epidemic influenza. *Nat. Rev. Genet.* **2007**, *8*, 196–205. [CrossRef] [PubMed]
7. Skowronski, D.M.; Sabaiduc, S.; Leir, S.; Rose, C.; Zou, M.; Murti, M.; Dickinson, J.A.; Olsha, R.; Gubbay, J.B.; Croxen, M.A.; et al. Paradoxical clade-and age-specific vaccine effectiveness during the 2018/19 influenza A (H3N2) epidemic in Canada: Potential imprint-regulated effect of vaccine (I-REV). *Eurosurveillance* **2019**, *24*, 1900585. [CrossRef] [PubMed]
8. Liao, Y.C.; Lee, M.S.; Ko, C.Y.; Hsiung, C.A. Bioinformatics models for predicting antigenic variants of influenza A/H3N2 virus. *Bioinformatics* **2008**, *24*, 505–512. [CrossRef]
9. Qiu, J.; Qiu, T.; Yang, Y.; Wu, D.; Cao, Z. Incorporating structure context of HA protein to improve antigenicity calculation for influenza virus A/H3N2. *Sci. Rep.* **2016**, *6*, 31156. [CrossRef]
10. Qiu, T.; Yang, Y.; Qiu, J.; Huang, Y.; Xu, T.; Xiao, H.; Wu, D.; Zhang, Q.; Zhou, C.; Zhang, X.; et al. CE-BLAST makes it possible to compute antigenic similarity for newly emerging pathogens. *Nat. Commun.* **2018**, *9*, 1772. [CrossRef]
11. Gupta, V.; Earl, D.J.; Deem, M.W. Quantifying influenza vaccine efficacy and antigenic distance. *Vaccine* **2006**, *24*, 3881–3888. [CrossRef]
12. Sun, H.; Yang, J.; Zhang, T.; Long, L.P.; Jia, K.; Yang, G.; Webby, R.J.; Wan, X.F. Using sequence data to infer the antigenicity of influenza virus. *MBio* **2013**, *4*, e00230-13. [CrossRef]
13. Daly, J.M.; Elton, D. Potential of a sequence-based antigenic distance measure to indicate equine influenza vaccine strain efficacy. *Vaccine* **2013**, *31*, 6043–6045. [CrossRef] [PubMed]
14. Anderson, C.S.; DeDiego, M.L.; Thakar, J.l; Topham, D.J. Novel sequence-based mapping of recently emerging H5NX influenza viruses reveals pandemic vaccine candidates. *PLoS ONE* **2016**, *11*, e0160510. [CrossRef]
15. Li, X.; Deem, M.W. Influenza evolution and H3N2 vaccine effectiveness, with application to the 2014/2015 season. *Protein Eng. Des. Sel.* **2016**, *29*, 309–315. [CrossRef]
16. Anderson, C.S.; McCall, P.R.; Stern, H.A.; Yang, H.; Topham, D.J. Antigenic cartography of H1N1 influenza viruses using sequence-based antigenic distance calculation. *BMC Bioinform.* **2018**, *19*, 51. [CrossRef] [PubMed]
17. Zhou, X.; Yin, R.; Kwoh, C.K.; Zheng, J. A context-free encoding scheme of protein sequences for predicting antigenicity of diverse influenza A viruses. *BMC Genom.* **2018**, *19*, 145–154. [CrossRef] [PubMed]
18. Łuksza, M.; Lässig, M. A predictive fitness model for influenza. *Nature* **2014**, *507*, 57–61. [CrossRef]
19. Yin, R.; Luusua, E.; Dabrowski, J.; Zhang, Y.; Kwoh, C.K. Tempel: Time-series mutation prediction of influenza A viruses via attention-based recurrent neural networks. *Bioinformatics* **2020**, *36*, 2697–2704. [CrossRef]
20. Neher, R.A.; Bedford, T.; Daniels, R.S.; Russell, C.A.; Shraiman, B.I. Prediction, dynamics, and visualization of antigenic phenotypes of seasonal influenza viruses. *Proc. Natl. Acad. Sci. USA* **2016**, *113*, E1701–E1709. [CrossRef]
21. Neher, R.A.; Russell, C.A.; Shraiman, B.I. Predicting evolution from the shape of genealogical trees. *Elife* **2014**, *3*, e03568. [CrossRef]
22. Yin, R.; Thwin, N.N.; Zhuang, P.; Lin, Z.; Kwoh, C.K. IAV-CNN: A 2D convolutional neural network model to predict antigenic variants of influenza A virus. *IEEE/ACM Trans. Comput. Biol. Bioinform.* **2021**, *19*, 3497–3506. [CrossRef]
23. Hirst, G.K. Studies of antigenic differences among strains of influenza A by means of red cell agglutination. *J. Exp. Med.* **1943**, *78*, 407–423. [CrossRef]
24. Smith, D.J.; Lapedes, A.S.; De Jong, J.C.; Bestebroer, T.M.; Rimmelzwaan, G.F.; Osterhaus, A.D.; Fouchier, R.A. Mapping the antigenic and genetic evolution of influenza virus. *Science* **2004**, *305*, 371–376. [CrossRef]
25. Lapedes, A.; Farber, R. The geometry of shape space: Application to influenza. *J. Theor. Biol.* **2001**, *212*, 57–69. [CrossRef]
26. Cai, Z.; Zhang, T.; Wan, X.F. A computational framework for influenza antigenic cartography. *PLoS Comput. Biol.* **2010**, *6*, e1000949. [CrossRef] [PubMed]
27. Lees, W.D.; Moss, D.S.; Shepherd, A.J. A computational analysis of the antigenic properties of haemagglutinin in influenza A H3N2. *Bioinformatics* **2010**, *26*, 1403–1408. [CrossRef]
28. Qiu, T.; Qiu, J.; Yang, Y.; Zhang, L.; Mao, T.; Zhang, X.; Xu, J.; Cao, Z. A benchmark dataset of protein antigens for antigenicity measurement. *Sci. Data* **2020**, *7*, 212. [CrossRef] [PubMed]

29. Koel, B.F.; Burke, D.F.; Bestebroer, T.M.; Van Der Vliet, S.; Zondag, G.C.; Vervaet, G.; Skepner, E.; Lewis, N.S.; Spronken, M.I.; Russell, C.A.; et al. Substitutions near the receptor binding site determine major antigenic change during influenza virus evolution. *Science* **2013**, *342*, 976–979. [CrossRef]
30. Steinbrück, L.; Klingen, T.R.; McHardy, A.C. Computational prediction of vaccine strains for human influenza A (H3N2) viruses. *J. Virol.* **2014**, *88*, 12123–12132. [CrossRef] [PubMed]
31. Bedford, T.; Suchard, M.A.; Lemey, P.; Dudas, G.; Gregory, V.; Hay, A.J.; McCauley, J.W.; Russell, C.A.; Smith, D.J.; Rambaut, A. Integrating influenza antigenic dynamics with molecular evolution. *elife* **2014**, *3*, e01914. [CrossRef] [PubMed]
32. Asgari, E.; Mofrad, M.R. Continuous distributed representation of biological sequences for deep proteomics and genomics. *PLoS ONE* **2015**, *10*, e0141287. [CrossRef] [PubMed]
33. Hensley, S.E.; Das, S.R.; Bailey, A.L.; Schmidt, L.M.; Hickman, H.D.; Jayaraman, A.; Viswanathan, K.; Raman, R.; Sasisekharan, R.; Bennink, J.R.; et al. Hemagglutinin receptor binding avidity drives influenza A virus antigenic drift. *Science* **2009**, *326*, 734–736. [CrossRef] [PubMed]
34. Veljkovic, V.; Paessler, S.; Glisic, S.; Prljic, J.; Perovic, V.R.; Veljkovic, N.; Scotch, M. Evolution of 2014/15 H3N2 influenza viruses circulating in US: Consequences for vaccine effectiveness and possible new pandemic. *Front. Microbiol.* **2015**, *6*, 1456. [CrossRef]
35. Lee, E.K.; Tian, H.; Nakaya, H.I. Antigenicity prediction and vaccine recommendation of human influenza virus A (H3N2) using convolutional neural networks. *Hum. Vaccines Immunother.* **2020**, *16*, 2690–2708. [CrossRef]
36. Huang, X.; Li, J.; Hu, X. Accelerated attributed network embedding. In Proceedings of the 2017 SIAM International Conference on Data Mining, Houston, TX, USA, 27–29 April 2017; pp. 633–641.
37. Pan, S.; Wu, J.; Zhu, X.; Zhang, C.; Wang, Y. Tri-party deep network representation. *Network* **2016**, *11*, 12.
38. Liao, L.; He, X.; Zhang, H.; Chua, T.S. Attributed social network embedding. *IEEE Trans. Knowl. Data Eng.* **2018**, *30*, 2257–2270. [CrossRef]
39. Grover, A.; Leskovec, J. node2vec: Scalable feature learning for networks. In Proceedings of the 22nd ACM SIGKDD International Conference on Knowledge Discovery and Data Mining, San Francisco, CA, USA, 13–17 August 2016; pp. 855–864.
40. Tang, J.; Qu, M.; Wang, M.; Zhang, M.; Yan, J.; Mei, Q. Line: Large-scale information network embedding. In Proceedings of the 24th International Conference on World Wide Web, Florence, Italy, 18–22 May 2015; pp. 1067–1077.
41. Zhang, D.; Yin, J.; Zhu, X.; Zhang, C. Attributed network embedding via subspace discovery. *Data Min. Knowl. Discov.* **2019**, *33*, 1953–19808. [CrossRef]
42. Kipf, T.N.; Welling, M. Semi-supervised classification with graph convolutional networks. *arXiv* **2016**, arXiv:1609.02907.
43. Veličković, P.; Cucurull, G.; Casanova, A.; Romero, A.; Lio, P.; Bengio, Y. Graph attention networks. *arXiv* **2017**, arXiv:1710.1090.
44. Hamilton, W.; Ying, Z.; Leskovec, J. Inductive representation learning on large graphs. *arXiv* **2017**, arXiv:1706.02216.
45. Park, J.; Lee, M.; Chang, H.J.; Lee, K.; Choi, J.Y. Symmetric graph convolutional autoencoder for unsupervised graph representation learnings. In Proceedings of the IEEE/CVF International Conference on Computer Vision, Seoul, Republic of Korea, 27 October–2 November 2019; pp. 6519–6528.
46. Yang, C.; Liu, Z.; Zhao, D.; Sun, M.; Chang, E.Y. Network representation learning with rich text information. In Proceedings of the 24th International Conference on Artificial Intelligence, Buenos Aires, Argentina, 25–31 July 2015; pp. 2111–2117.
47. McHardy, A.C.; Adams, B. The role of genomics in tracking the evolution of influenza A virus. *PLoS Pathog.* **2009**, *5*, e1000566. [CrossRef] [PubMed]
48. Wolf, Y.I.; Viboud, C.; Holmes, E.C.; Koonin, E.V.; Lipman, D.J. Long intervals of stasis punctuated by bursts of positive selection in the seasonal evolution of influenza A virus. *Biol. Direct* **2006**, *1*, 34. [CrossRef]
49. Forghani, M.; Khachay, M. Convolutional neural network based approach to in silico non-anticipating prediction of antigenic distance for influenza virus. *Viruses* **2020**, *12*, 1019. [CrossRef] [PubMed]
50. Zeller, M.A.; Gauger, P.C.; Arendsee, Z.W.; Souza, C.K.; Vincent, A.L.; Anderson, T.K. Machine learning prediction and experimental validation of antigenic drift in H3 influenza A viruses in swine. *MSphere* **2021**, *6*, e00920-20. [CrossRef]

viruses

MDPI

Article

Development of a Multi-Epitope Universal mRNA Vaccine Candidate for Monkeypox, Smallpox, and Vaccinia Viruses: Design and In Silico Analyses

Nino Rcheulishvili [1,†], Jiawei Mao [1,†], Dimitri Papukashvili [1], Shunping Feng [1], Cong Liu [1], Xidan Yang [1,2], Jihui Lin [1,2], Yunjiao He [1,*] and Peng George Wang [1,*]

1 Department of Pharmacology, School of Medicine, Southern University of Science and Technology, Shenzhen 518000, China; nino@sustech.edu.cn (N.R.); maojw@sustech.edu.cn (J.M.); dimitri@sustech.edu.cn (D.P.); 12133106@mail.sustech.edu.cn (S.F.); 11930759@mail.sustech.edu.cn (C.L.); xidanyang993@163.com (X.Y.); linjihui@swmu.edu.cn (J.L.)
2 School of Nursing, Southwest Medical University, Luzhou 646000, China
* Correspondence: heyj@sustech.edu.cn (Y.H.); wangp6@sustech.edu.cn (P.G.W.)
† These authors contributed equally to this work.

Abstract: Notwithstanding the presence of a smallpox vaccine that is effective against monkeypox (mpox), developing a universal vaccine candidate against monkeypox virus (MPXV) is highly required as the mpox multi-country outbreak has increased global concern. MPXV, along with variola virus (VARV) and vaccinia virus (VACV), belongs to the Orthopoxvirus genus. Due to the genetic similarity of antigens in this study, we have designed a potentially universal mRNA vaccine based on conserved epitopes that are specific to these three viruses. In order to design a potentially universal mRNA vaccine, antigens A29, A30, A35, B6, and M1 were selected. The conserved sequences among the three viral species—MPXV, VACV, and VARV—were detected, and B and T cell epitopes containing the conserved elements were used for the design of the multi-epitope mRNA construct. Immunoinformatics analyses demonstrated the stability of the vaccine construct and optimal binding to MHC molecules. Humoral and cellular immune responses were induced by immune simulation analyses. Eventually, based on in silico analysis, the universal mRNA multi-epitope vaccine candidate designed in this study may have a potential protection against MPXV, VARV, and VACV that will contribute to the advancement of prevention strategies for unpredictable pandemics.

Keywords: monkeypox; mpox; MPXV; universal vaccine; multi-epitope mRNA vaccine; immunoinformatics

Citation: Rcheulishvili, N.; Mao, J.; Papukashvili, D.; Feng, S.; Liu, C.; Yang, X.; Lin, J.; He, Y.; Wang, P.G. Development of a Multi-Epitope Universal mRNA Vaccine Candidate for Monkeypox, Smallpox, and Vaccinia Viruses: Design and In Silico Analyses. *Viruses* **2023**, *15*, 1120. https://doi.org/10.3390/v15051120

Academic Editors: Pietro Hiram Guzzi, Marianna Milano and Jayanta Kumar Das

Received: 14 April 2023
Revised: 3 May 2023
Accepted: 4 May 2023
Published: 7 May 2023

1. Introduction

Unforeseeable outbreaks of infectious diseases are causing a rise in the worldwide risk to public health. Monkeypox virus (MPXV) has emerged in May 2022 and affected more than 86,000 people until now [1]. Monkeypox (mpox) is a zoonotic disease that has been largely neglected although there were cases in its endemic areas—West and Central Africa. MPXV, along with variola virus (VARV), vaccinia virus (VACV), and cowpox virus, belongs to the genus *Orthopoxvirus*, family Poxviridae. MPXV is a large, 200–300 nm, brick-shaped virus that exists in two different infectious forms: extracellular enveloped virions (EVs) and intracellular mature virions (MVs). EVs have an extra envelope compared to MVs. The genome of MPXV is about 197 kb linear double-stranded DNA (dsDNA) that encodes approximately 190 proteins [2–4]. Due to their intricate structure, many viral antigens and their functions still need to be studied.

Except for Africa, there was an outbreak in the United States (US) in 2003 when imported prairie dogs from Ghana spread the virus, and, consequently, there were 47 confirmed cases [5,6]. Interestingly, the prairie dogs were housed with African rodents, so the virus could have been transmitted via the rodents [7]. In addition, single cases of mpox were identified in different countries such as the US [8,9], Israel, the United Kingdom (UK), and Singapore and all were linked to travel to Nigeria [10]. What caused the MPXV virus outbreak this time? This and many other questions are unanswered until now. However, it is assumed that eradicating smallpox and ceasing vaccination in 1980 globally led to increased cases of mpox. This theory is quite convincing as more than 70% of people are unvaccinated against the smallpox virus today, and most cases of mpox take place in younger people. Interestingly, most of the cases occur in men who have sex with men (MSM), and close physical contact plays a key role in transmission [11–13]. Indeed, the sex-related infection rate was always observed even before the current outbreak—males predominated over females, while children accounted for the majority of cases [7]. The fatality rates range from 1% to 11% [14], and the disease is more severe in young children [15,16]. There are two genetic clades of MPXV: the Central African (Democratic Republic of the Congo (DRC), previously known as Zaire) and the West African clades. Out of these two clades, the Central African clade is more virulent and deadly. Unlike the coronavirus disease 2019 (COVID-19) and influenza viruses, the MPXV is a dsDNA [17] virus and is more stable. Generally, the mutation rate of DNA viruses is much lower than RNA viruses. Therefore, the assumption of an MPXV mutation as the only reason for the current outbreak would not be rational. However, a few mutations still take place, and it should not be completely neglected. Even though a smallpox vaccine is quite effective against the MPXV virus, there is still a necessity for developing a new, universal vaccine based on conserved elements. Moreover, apparently, the smallpox vaccine cannot completely protect from MPXV [18]. Additionally, in the worst-case scenario, if certain mutations take place in MPXV and the effectiveness of the smallpox vaccine to MPXV decreases, the situation will also worsen. Although the daily confirmed cases of mpox are significantly decreasing, it is globally spread, and the re-emergence of the same clade or even the spread of the Central African clade is anticipated. Remarkably, messenger RNA (mRNA)-based vaccines have revolutionized the field of vaccinology due to their favorable safety profile, low-cost manufacturing, high potency, and rapid development among other impressive advantages [19–21]. mRNA vaccine contains the antigen sequence that is translated into the corresponding protein after the introduction into the host body. After the antigen is released, it is recognized by antigen-presenting cells (APCs) and phagocytosed. The vaccine antigen is then processed into small peptides that are presented on the cell surface via major histocompatibility complex (MHC) I and II, and the cellular and humoral immune responses are induced [20].

In this research, five antigen proteins were selected from MPXV, VACV, and VARV. The corresponding sequences were retrieved and aligned. Conserved sequences were detected in each protein. Relevant T cell epitopes that were tested through experimentation were searched in the immune epitope database (IEDB), sorted according to the optimal results, and selected, while B cell epitopes were predicted via IEDB due to the lack of B cell epitope data in the database. The multi-epitope mRNA vaccine made up of the conserved epitopes of A29, A30, A35, B6, and M1 proteins (encoded by the genes *A29L*, *A30L*, *A35R*, *B6R*, and *M1R*, respectively) was constructed. Various properties of the vaccine were assessed in silico. Based on the obtained results, the mRNA construct proposed in this study has potentially high efficiency and elicits protection from MPXV, VARV, and VACV. The proposed construct represents a favorable candidate for developing the mRNA vaccine for global purposes. The contribution of the design of this potentially universal vaccine candidate is one step forward in the advancement of vaccine development and pandemic alertness.

2. Methods

2.1. Antigen Selection

Five antigens that are common for MPXV, VARV, and VACV were selected: A29 (A30 in VARV, A27 in VACV), A30 (A31 in VARV, A28 in VACV), A35 (A36 in VARV, A33 in VACV), M1 (M1 in VARV, L1 in VACV), and B6 (B7 in VARV, B5 in VACV). These proteins are highly conserved among the orthopoxviruses and serve important functions in immune response [3,22–27]. All five antigens were selected based on their functions in immune response and their successful application in previous vaccine studies [24,27–29]. The selected antigens along with their functions are given in Table 1.

Table 1. Selected antigens among MPXV, VARV, and VACV, and their location in the virus, function, and characteristics.

Name (MPXV)	Name (VARV)	Name (VACV)	Location	Function and Characteristics
A29	A30	A27	MV	Surface membrane fusion protein; Binds to cell surface heparan; Neutralizing antibody target
A30	A31	A28	MV	Envelope protein; Virus entry into a host; Cell–cell fusion (syncytial formation); Neutralizing antibody target
A35	A36	A33	EV	Envelope glycoprotein; Formation of actin-containing microvilli and cell-to-cell spread of virion; Neutralizing antibody target; Target of complement-mediated cytolysis
B6	B7	B5	EV	Palmitylated glycoprotein; Required for efficient cell spread; Complement control
M1	M1	L1	MV	Myristylated surface membrane protein; Virus entry into a host; Neutralizing antibody target

Notes: MPXV, monkeypox virus; VARV, variola virus; VACV, vaccinia virus.

2.2. Selection of Conserved Regions and Epitopes

The amino acid sequences of A29, A30, A35, B6, and M1 of MPXV and the corresponding proteins of VARV and VACV were retrieved from the National Center for Biotechnology Information (NCBI) database in FASTA format. In total, 372 sequences were downloaded for the antigen A29: 159 A29 (MPXV), 120 A30 (VARV), and 93 A27 (VACV). This protein is 110 amino acids in length, thus, the sequences longer or shorter than 110 amino acids were removed, and the rest (176 sequences) were aligned together. A total of 433 sequences were retrieved for the antigen A30: 363 A30 (MPXV), 8 A31 (VARV), and 62 A28 (VACV). As the full length of this protein is 146 amino acids, all the shorter or longer sequences were removed, and the remaining 158 sequences were aligned together. A total of 365 sequences of the antigen A35 were downloaded: 153 A35 (MPXV), 109 A36 (VARV), and 103 A33 (VACV). As the full-length sequence of A35 is 181 amino acids, A36 is 184, and A33 is 185, all the records shorter than 181 and longer than 185 were removed, and the remaining 194 sequences were aligned. In total 365 sequences of antigen B6 were retrieved: 180 B6 (MPXV), 6 B7 (VARV), and 179 B5 (VACV). The full length of this protein is 317 amino acids, thus, all the sequences shorter or longer than 317 amino acids were removed, and the rest of the sequences (196) were aligned. A total of 391 sequence records of antigen M1 were retrieved from NCBI: 82 M1 (MPXV), 6 M1 (VARV), and 303 L1 (VACV), which were aligned together. As the full length of this protein is 250 amino acids, all the sequences shorter or longer than 250 were removed, and the remaining 346 sequences were aligned together. The conserved sequences were identified via the bioinformatics software Jalview 2.11.1.4 [30]. The filter of conservation threshold was given 10 "below threshold".

Experimentally tested T cell epitopes and MHC ligands of each antigen were found on IEDB [31] according to the species of orthopoxviruses (MPXV, VARV, and VACV) and host (human). IEDB is an excellent database supported by experimental data of humans and animals on antibody and T cell epitopes. IEDB also incorporates immunoinformatics tools for epitope prediction [31]. As there were no B cell epitopes found on IEDB, the Bepipred linear epitope prediction tool (v2.0) was used to predict B cell epitopes [32]. Ultimately, experimentally tested T cell epitopes were retrieved from IEDB; epitopes containing conserved sequences were ranked and the most studied ones were selected. On the other hand, linear B cell epitopes were predicted using the consensus sequence of each antigen.

2.3. Vaccine Design

For the construction of the multi-epitope mRNA vaccine, the epitopes were arranged orderly. B cell epitopes were joined with a flexible linker KK [33,34], while a GGGS linker was used to connect the T cell epitopes [35]. The leading sequence of tissue plasminogen activator (tPA) that augments the antigen presentation [36] was used as a signal sequence (MDAMKRGLCCVLLLCGAVFVSPS). The incorporation of the amino acid sequence of interleukin 6 (IL-6) was done to enhance the immunogenicity of the vaccine [37]. The pan-HLA DR binding epitope (PADRE sequence) was connected to the IL-6 sequence via a GGGS linker and to an epitope of the B cell via EAAAK [34,38]. A polyhistidine (6x) tag was placed on the C-terminal of the vaccine sequence for facilitating the fusion protein detection [39]. The complete open reading frame of the proposed vaccine is given in the Supplementary Materials.

2.4. Prediction of Vaccine Properties

To check whether the antigen could provoke an allergic response, the AlgPred server, a highly accurate tool, was used to predict its allergenicity [40]. As the prediction tool, a hybrid approach—Support Vector Machine (SVMc) algorithm + IgE epitope + ARPs BLAST + MAST—was selected [40,41]. This method detects the allergenicity of the protein based on the composition of amino acids and dipeptides using SVM, which is a motif-based technique using the software MAST. Ultimately, the tool specifies the antigen as an allergen if it contains a segment identical to the known IgE epitopes or similar to allergen-representative proteins [40,41]. For the prediction of antigenicity, server Vaxijen v2.0, which is based on an alignment-independent prediction of protective antigens, was used [42]. Vaxijen categorizes viral, bacterial, and tumor antigens according to their physicochemical characteristics [43].

ProtParam was used to calculate the physicochemical properties [44]. The following parameters were characterized: the molecular weight (MW) of the multi-peptide, the atomic and amino acid composition, the theoretical isoelectric point (pI), the instability, the estimated half-life, the grand average of hydropathicity (GRAVY), and the aliphatic indexes. MW was calculated by adding the average isotopic masses of amino acids and the average isotopic mass of one water molecule. In addition to the MW, the pI, which plays a crucial role in pH-dependent properties, was computed based on the pKa value of amino acids, which plays an essential role in characterizing the pH-dependency of the protein. The estimated half-life denotes the duration required from the synthesis of the protein until its decay and reduction to half of its original amount within the cell. The instability index is used to evaluate the stability of the protein in a test tube. An instability index of less than 40 is considered an indication of stability. When this measure is >40, the protein of interest is considered unstable. To compute GRAVY, the hydropathy values of each amino acid are summed up and then divided by the total residue number in the protein. A higher number denotes that the amino acids are more hydrophobic [44]. An aliphatic index is used to characterize the protein relative volume that is taken up by amino acids that have aliphatic side chains, and it is considered a positive factor in augmenting its thermostability.

2.5. Tertiary Structure Prediction and Evaluation of Its Quality; Discontinuous B Cell Epitope Prediction

RoseTTAFold was used to generate the tertiary structure of the vaccine protein [45]. By combining the sequence data of a one-dimensional protein with the two-dimensional data of the distances between the amino acids, as well as the prediction of the three-dimensional atomic structure, this tool is able to forecast the configuration and interplay of the protein [45]. After preparing the structure of the vaccine protein, it was refined with GalaxyRefine [46]. Through the use of molecular dynamics simulation, this immunoinformatics approach reconstructs and reorganizes the amino acid side chains, while also relaxing the overall protein structure [46]. Subsequently, the Ramachandran plot and ERRAT were used to verify the quality of the protein tertiary structure. The Ramachandran plot visualizes the energetically allowed regions for backbone dihedral angles ψ against φ of amino acid residues in protein structures, which allows the testing of the quality of protein structures [47]. The ERRAT score serves as a quality indicator for the non-bonded interactions, where a larger score suggests a superior quality of the protein tertiary structure. ElliPro was employed for the prediction of B cell conformational epitopes based on the 3D structure of the protein [48].

2.6. Immune Simulation

Humoral and cellular immune responses driven against the proposed multi-epitope vaccine protein were analyzed via the C-ImmSim server (https://kraken.iac.rm.cnr.it/C-IMMSIM/; accessed on 31 August 2022). This immunoinformatics approach provides the service of immune simulation and characterizes the immune system responses, both humoral and cellular, towards vaccination. For predicting immune epitopes and analyzing immune interactions, C-ImmSim applies a position-specific scoring matrix in combination with machine learning techniques. After uploading the sequence of vaccine antigen in FASTA format, the server predicted the immune responses [49]. The "Allele Frequency Net Database" identified the most widespread HLA-A, HLA-B, and DRB alleles globally. The outcome indicated HLA-A*02:01, HLA-A*01:01, HLA-B*07:02, HLA-B*08:01, DRB1*07:01, and DRB1*15:01. The vaccine was administered thrice with an interval of four weeks between each injection. The simulation volume was adjusted at 10, and the simulation progressed through 270 steps. The vaccine used during the simulation did not comprise LPS. The random seed was 12345, and time periods were set at 1, 85, and 169 [50].

2.7. In Silico Validation of Vaccine Protein Binding to the Host Receptors

In order to assess the binding capacity of vaccine protein to antigen recognition receptors, molecular docking was carried out with a ClusPro server. ClusPro calculates the docking interaction between two protein structures and provides a list of potential complexes in order of priority. It is performed based on a predicted conformation of ligand, orientation, and position as well as binding affinity analysis. Ultimately, the complexes that exhibit good electrostatic and desolvation energies are chosen [51]. A molecular docking of vaccine construct with MHC-I (HLA-A*02:01) (6TDS) and MHC-II (HLA-DRB1*01:01) (1AQD) host receptors was carried out. The following epitopes of the vaccine construct were docked with the MHC molecules: TLFPGDDDL (A29/A30/A27), NTLSERISSK (M1/M1/L1)—MHC-I ligands; FFIVVATAAVCLLFI (A30/A31/A28), LSMITMSAFLIVRLN (A35/A36/A33), ASYISCTANSWNVIP (B6/B7/B5), and KIQNVIIDECY (M1/M1/L1)—MHC-II ligands. The tertiary structure of these epitopes of the vaccine construct was predicted with AlphaFold2 [52,53]. The PDB files 6TDS and 1AQD were edited and cleaned to remove heteroatoms, bound peptides, and water molecules.

3. Results

3.1. Selecting Conserved Epitopes

The selected antigens in the virus are illustrated in Figure 1. All the conserved sequences and epitopes of each antigen are given in the Supplementary Materials (Tables S1–S10). The strategy for designing a potentially universal MPXV mRNA vaccine based on the conserved epitopes is given in Figure 2. The ultimately selected epitopes that were used for the vaccine design along with the MPXV multi-epitope universal mRNA vaccine construct and the plasmid vector are given in Figure 3.

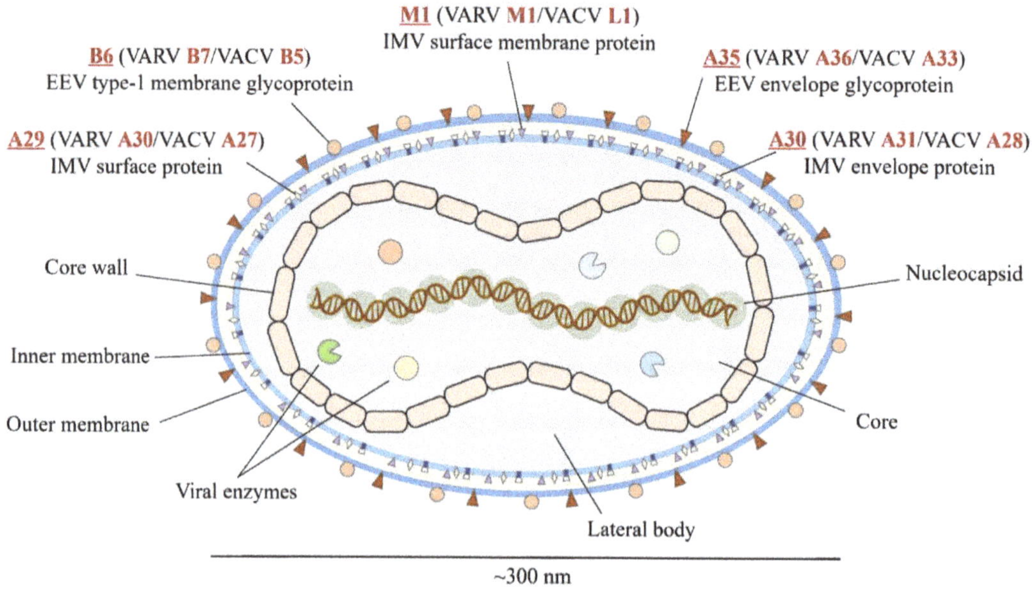

Figure 1. Structure of MPXV and selected viral antigens. Red font indicates the names of the antigens in MPXV, VARV, and VACV. Underline indicates the special emphasis on the name of antigen in MPXV. MPXV, monkeypox virus; VARV, variola virus; VACV, vaccinia virus.

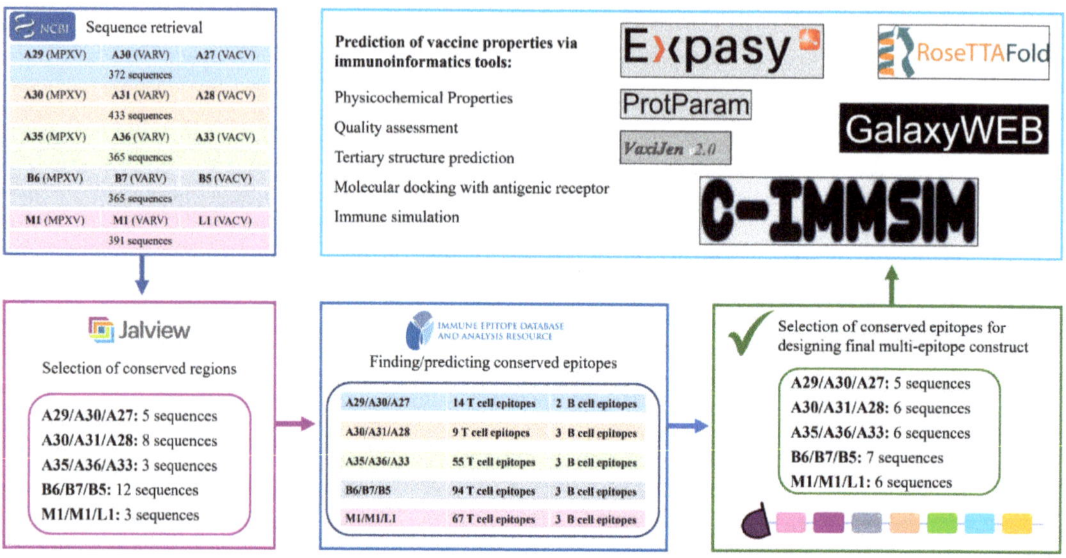

Figure 2. Strategy of designing a potentially universal MPXV mRNA vaccine. MPXV, monkeypox virus. MPXV, monkeypox virus; VARV, variola virus; VACV, vaccinia virus.

A

B cell epitopes (Predicted)	T cell epitopes and MHC ligands (Experimentally tested)
A29/A30/A27	
1. VQTGR 2. DGDDN	1. TLFPGDDDL 2. TLKQRLTNL 3. TLRAAMISL
A30/A31/A28	
1. SKSIGGTPA 2. QDVNDTISD 3. GPNNTR	1. FFIVVATAAVCLLFI 2. MNSLSIFFIVVATAAVCLLFI 3. ATAAVCLLFIQGYSIYENYGN
A35/A36/A33	
1. STHRKVASSTTQYDHK 2. PITKTTSDYQDSDVSQEV 3. EEQTS	1. LSMITMSAFLIVRLN 2. RVIGLCIRISMVISL 3. SAFLIVRLNQCMSAN
B6/B7/B5	
1. VSELYNKPLY 2. EEKNGNTSWNDTV 3. PDDETDLSKLSKD	1. ASYISCTANSWNVIP 2. GGVIHLSCKSGFILT 3. MKTISVVTLLCVLPA 4. GVIFLISVIVLVCSCN
M1/M1/L1	
1. EANAS 2. QTSVNT 3. APRQV	1. KIQNVIIDECY 2. AALFMYYAKR 3. NTLSERISSK

B

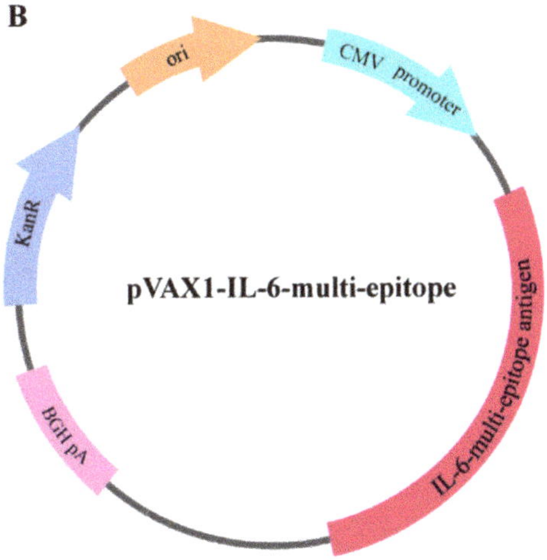

C

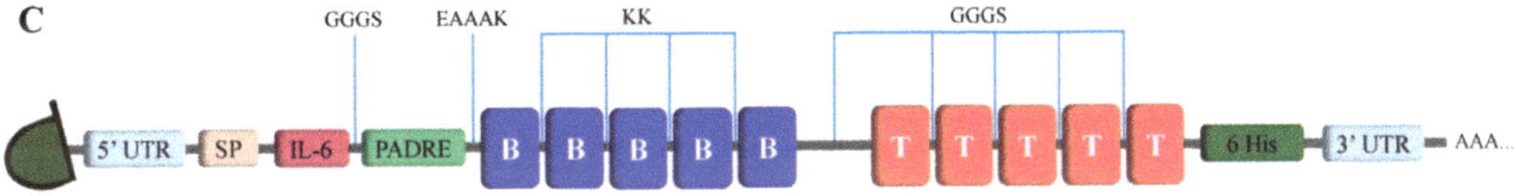

Figure 3. Schematic illustration of the designed mRNA and plasmid. (**A**) Table of selected conserved epitopes for the mRNA vaccine design. (**B**) Plasmid scheme. (**C**) Final design of the multi-epitope mRNA vaccine construct.

3.2. Assessment of Structure

Using RoseTTAFold, the tertiary structure of the protein of the vaccine was predicted and subsequently refined by GalaxyRefine (Figure 4). The tool generated five models out of which the one with the better characteristics was selected for the docking analysis. Of the amino acids, 93.0% were located in the most favored region of the Ramachandran, 6.3% were in the allowed region, 0.4% were in regions that were generously allowed, and 0.4% of the amino acid residues were in the outlier region. These parameters, along with the ERRAT with an overall quality factor of 92.9825, indicate the desired level of the quality of the vaccine product (Figure 5).

3.3. Physicochemical Properties, Allergenicity, and Antigenicity Analyses

According to the physicochemical properties predicted by ProtParam, the vaccine sequence contains 647 amino acid residues, and its MW is 68.97777 kDa. The theoretical pI was computed to be 9.31. The vaccine contains 55 negatively charged residues (Asp + Glu). The instability index was 39.82, which classifies the vaccine as stable. The estimated half-life in vitro was 100 h, while the estimated half-lives in vivo in yeast and *E. coli* were 20 h and 10 h, respectively. The aliphatic index was 82.32, which implies that the vaccine construct has a high level of thermostability. The vaccine's GRAVY score was −0.233, which suggests that the protein is hydrophilic.

The allergenicity prediction via the AlgPred server showed that the multi-epitope vaccine was non-allergen, while the server VaxiJen v2.0 with the default parameters and the threshold 0.4 [54] demonstrated its probable antigenicity with the antigenic score 0.4656.

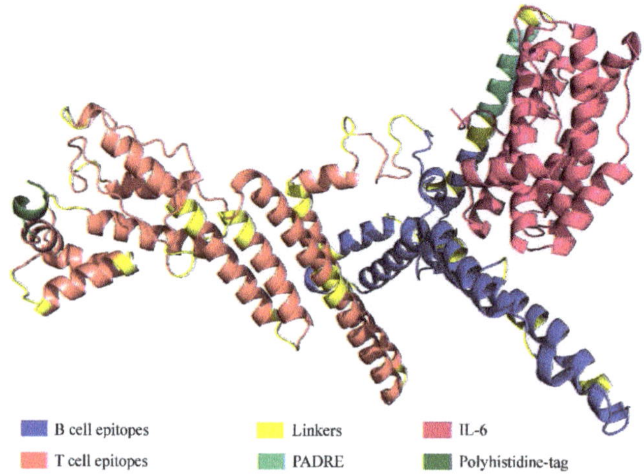

Figure 4. Tertiary structure of the protein comprising multiple epitopes.

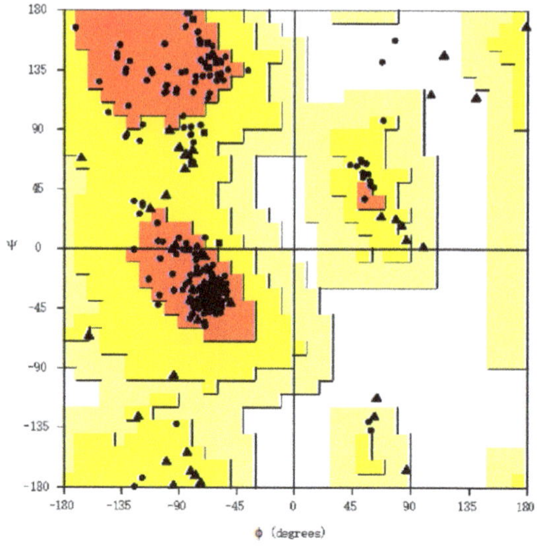

Figure 5. Ramachandran plot assessing vaccine structure quality. Each dot shows the amino acids. The positioning of these dots reflects the backbone dihedral angles ψ versus φ of the amino acids present in the vaccine product. Triangles indicate glycine, squares indicate prolines, while circles denote non-glycine and non-proline residues.

3.4. Conformational B Cell Epitopes

In total, four discontinuous B cell epitopes were predicted with the scores 0.69–0.859. The sizes varied between 44 and 109 residues. The presence of the conformational epitopes in the vaccine product indicates its potential capacity to induce a humoral immune response when the vaccine is administered in vivo. All the predicted discontinuous epitopes are given in Table 2 and illustrated in Figure 6.

Table 2. Discontinuous B cell epitopes predicted by ElliPro.

#	Epitopes	Residue Number	Score
1	A:K287, A:E289, A:Q290, A:T291, A:S292, A:K293, A:K294, A:V295, A:S296, A:E297, A:L298, A:Y299, A:N300, A:K301, A:P302, A:L303, A:Y304, A:K305, A:K306, A:E307, A:E308, A:K309, A:N310, A:G311, A:N312, A:T313, A:S314, A:W315, A:N316, A:D317, A:T318, A:V319, A:K320, A:K321, A:P322, A:D323, A:D324, A:E325, A:T326, A:D327, A:L328, A:S329, A:K330, A:L331	44	0.859
2	A:I511, A:V512, A:R513, A:L514, A:N515, A:Q516, A:C517, A:M518, A:S519, A:A520, A:N521, A:G522, A:G523, A:G524, A:S525, A:A526, A:S527, A:Y528, A:I529, A:S530, A:C531, A:T532, A:A533, A:N534, A:S535, A:N537, A:I539, A:G546, A:V547, A:I548, A:H549, A:L550, A:S551, A:C552, A:K553, A:S554, A:G555, A:F556, A:I557, A:L558, A:T559, A:G560, A:G561, A:G562, A:S563, A:M564, A:K565, A:T566, A:I567, A:S568, A:V569, A:V570, A:S589, A:V592, A:L593, A:V594, A:C595, A:S596, A:C597, A:N598, A:G599, A:G600, A:G601, A:S602, A:K603, A:I604, A:Q605, A:N606, A:V607, A:I608, A:I609, A:D610, A:E611, A:C612, A:Y613, A:G614, A:G615, A:G616, A:S617, A:A618, A:A619, A:L620, A:F621, A:M622, A:Y623, A:Y624, A:A625, A:K626, A:R627, A:G628, A:G629, A:G630, A:S631, A:N632, A:T633, A:L634, A:S635, A:E636, A:R637, A:I638, A:S639, A:S640, A:K641, A:H642, A:H643, A:H644, A:H645, A:H646, A:H647	109	0.761
3	A:K212, A:D216, A:N217, A:K218, A:K221, A:G224, A:G225, A:T226, A:P227, A:A228, A:K229, A:K230, A:Q231, A:D232, A:V233, A:N234, A:D235, A:T236, A:I237, A:S238, A:D239, A:K240, A:K241, A:G242, A:P243, A:N244, A:N245, A:T246, A:R247, A:K248, A:K249, A:S250, A:T251, A:H252, A:R253, A:K254, A:V255, A:F401, A:I402, A:V403, A:V404, A:A405, A:T406, A:A407, A:A408, A:V409, A:C410, A:L411, A:L412, A:F413, A:I414, A:G415, A:G416, A:G417, A:S418, A:M419, A:N420, A:S421, A:L422, A:S423, A:I424, A:F425, A:F426, A:I427, A:V428, A:V429, A:A430, A:T431, A:A432, A:A433	70	0.733
4	A:D9, A:V10, A:A11, A:A12, A:P13, A:H14, A:R15, A:Q16, A:P17, A:L18, A:T19, A:S20, A:S21, A:E22, A:R23, A:I24, A:D25, A:M66, A:A67, A:E68, A:K69, A:D70, A:G71, A:C72, A:F73, A:Q74, A:S75, A:G76, A:F77, A:N78, A:E79, A:E80, A:T81, A:C82, A:L83, A:V84, A:K85, A:I86, A:I87, A:T88, A:L91, A:V120, A:L121, A:I122, A:Q123, A:F124, A:L125, A:Q126, A:K128, A:A129, A:K130, A:N131, A:L132, A:D133, A:A134, A:I135, A:T136, A:T137, A:P138, A:D139, A:P140, A:T141, A:T142, A:A144, A:S145, A:T148, A:K149, A:S176, A:L177, A:R178, A:A179, A:L180, A:R181, A:Q182, A:M183, A:G184, A:G185, A:G186, A:S187, A:A188, A:K189, A:F190, A:V191, A:A192, A:A193, A:W194, A:T195, A:K197, A:A198, A:E201	90	0.69

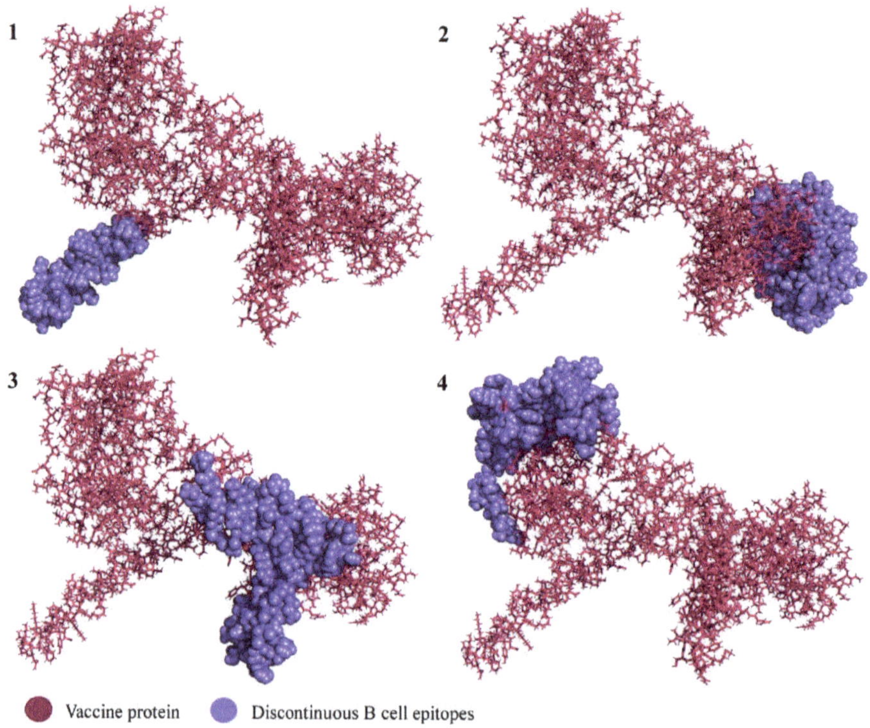

Figure 6. Predicted discontinuous epitopes of B cell epitopes in the vaccine product. The subfigure numbers (**1–4**) correspond to the conformational B cell epitope numbers listed in Table 2.

3.5. Molecular Docking and Immune Responses following Immune Simulation

The molecular docking analysis of epitopes with the MHC-I and MHC-II molecules revealed the stable binding (Figure 7). The molecular docking results are given in the Supplementary Materials (Tables S11 and S12). In total, three doses were administered four weeks apart for the immune simulation—day 0, day 28, and day 56 with an 8 h offset. As expected, the immune response was also compatible with the immune responses that are elicited in general after in vivo immunization. The immune responses after receiving additional and booster doses were more robust compared to the first shot. The immune responses following the second and third shots were stronger in comparison to the prime immunization. The antigen level was decreased while elevated levels of antibodies, including IgG1, IgG1 + IgG2, IgM, and IgG + IgM, were detected. Antigen abundance peaked at each vaccine injection (Figure 8A). The humoral response surged after each immunization, and the antibody levels remained high for weeks after the last vaccine injection. Following immunization, the activation of the B cell population, resulting in an increased number of B cells producing antibodies, was observed (Figure 8B). The counts of CD8 T-cytotoxic lymphocytes show the ignorable number of anergic cells and activation of T-cytotoxic cells (Figure 8C). The counts of CD4 T-helper lymphocytes demonstrate that the duplication phase starts immediately after each injection (Figure 8D). The cytokine production was also manifested upon immunization; particularly, high levels of IFN-γ, TGF-β, and IL-10 were induced by the infection (Figure 8E). The other parameters of the immune response elicited by the vaccine are presented in the Supplementary Materials (Figure S1).

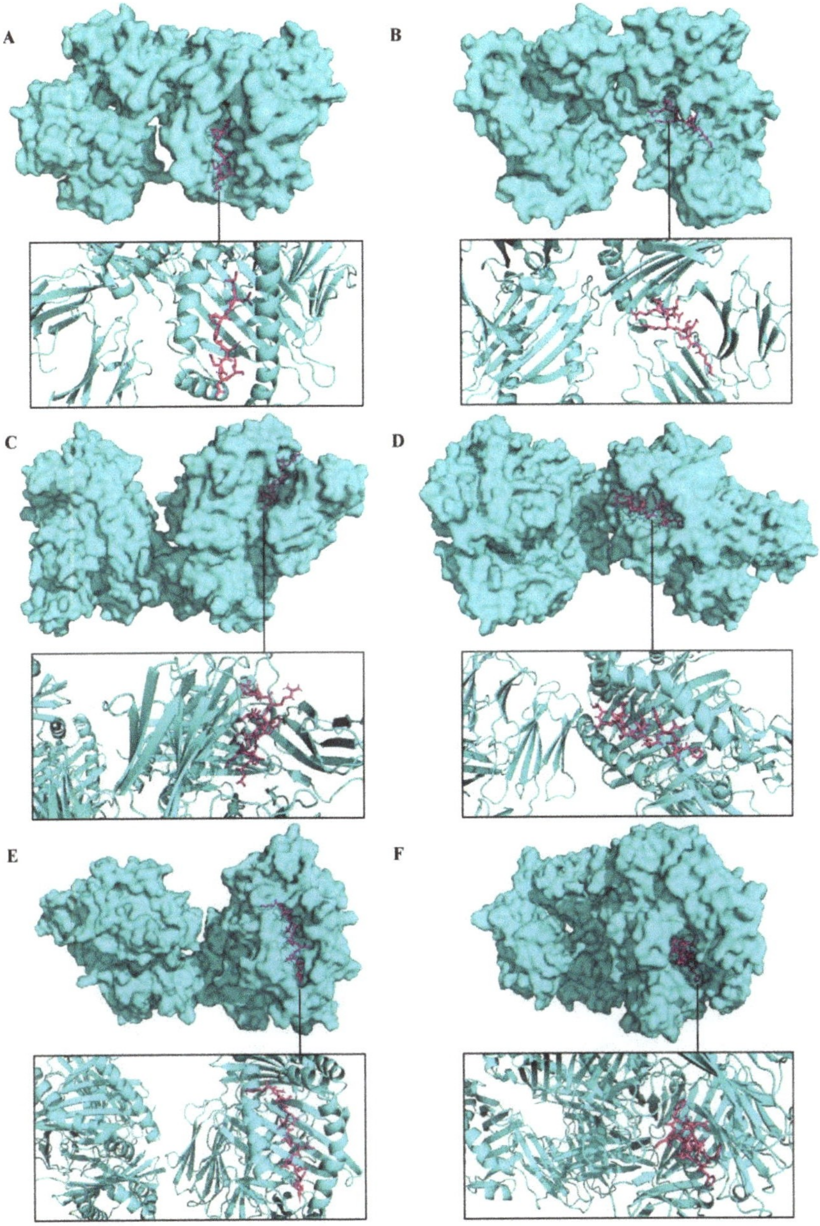

Figure 7. Representation of predicted molecular docking between epitopes of a multi-epitope vaccine construct and host receptors. The MHC molecules are displayed in cyan color and the epitope is shown in magenta. (**A**) Docked complex of epitope TLFPGDDDL (A29/A30/A27) and MHC-I. (**B**) Docked complex of epitope NTLSERISSK (M1/M1/L1) and MHC-I. (**C**) Docked complex of epitope LSMITMSAFLIVRLN (A35/A36/A33) and MHC-II. (**D**) Docked complex of epitope ASYISCTANSWNVIP (B6/B7/B5) and MHC-II. (**E**) Docked complex of epitope KIQNVIIDECY (M1/M1/L1) and MHC-II. (**F**) Docked complex of epitope FFIVVATAAVCLLFI (A30/A31/A28) and MHC-II.

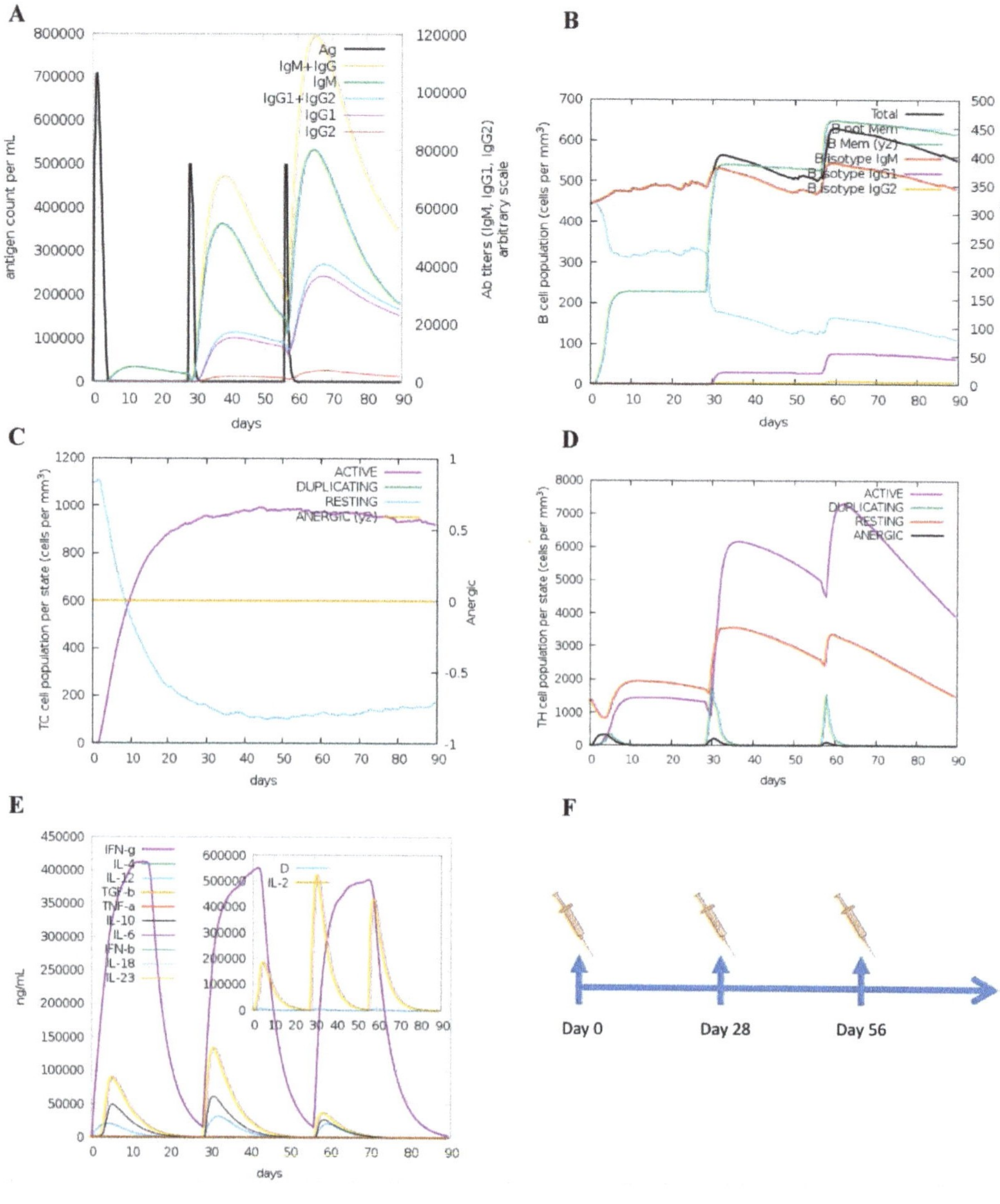

Figure 8. Immune responses elicited after in silico immunization. (**A**) Antibodies produced in response to immune simulation. Immunoglobulin subclasses are given in different colors. (**B**) Representation of evolution of B cell populations following the immunization with three doses. (**C**) Generation of CD8 T-cytotoxic lymphocytes induced by antigen exposure. The resting state denotes T-cytotoxic cells not actively engaged in fighting the infection, while the anergic state denotes the unresponsiveness of the CD8 T-cytotoxic cells due to repeated and prolonged exposure to the antigen. (**D**) Production of CD4 T-helper cells induced by exposure to the vaccine antigen. (**E**) The cytokine profile in response to the vaccine injections. The inserted graph shows the IL-2 level with the Simpson Index. D indicates a danger signal. (**F**) Schedule of in silico immunization.

4. Discussion

Mpox was known to be a rare zoonotic disease caused by the MPXV that usually emerged in rural rainforest regions of African countries [7]. The genus *Orthopoxvirus* contains four species that are pathogenic to humans: VARV (the causative agent of smallpox), MPXV, VACV, and cowpox. Smallpox and mpox are often life-threatening diseases, whereas vaccinia and cowpox are generally associated with local lesions. MPXV received this name as it was first isolated from laboratory macaque monkeys imported from Singapore to Copenhagen, Denmark, in 1958 [55]; however, it is considered that the host and the main source of MPXV are rodents such as squirrels, Gambian rats, etc. [56]. The first human case was documented in the Democratic Republic of the Congo (DRC) in 1970 while smallpox surveillance was taking place [7]. Particularly, a nine-month-old boy was admitted to the hospital with suspected smallpox. The specimens were sent to the Smallpox Reference Centre of the World Health Organization (WHO) where MPXV was isolated [57]. The symptoms of mpox include fever, malaise, respiratory symptoms, fatigue, lymphadenopathy, headache, and muscle aches, along with the main symptom—a rash similar to pimples and blisters that are itchy and painful and eventually crust over and are healed. The manifestation of mpox is similar to other rash diseases, hence, misdiagnoses often take place. Instead of mpox, mostly, chickenpox caused by *Varicella zoster* virus is mistakenly diagnosed [15,58,59]. The rush is usually concentrated on the face and extremities including palms and soles of the feet; however, according to the evidence during the current mpox outbreak, it is also localized in the perineal/perianal area as well as on the genitals [60,61]. Although mpox usually resolves by itself, the recent 2022 outbreak demonstrated that it is a life-threatening disease [62]. Indeed, after all, MPXV is categorized as a high-threat virus that belongs to biosafety level 3, according to EU regulations [63]. The possible reasons why MPXV is being spread more than usual include the increased exotic animal trade and international travel [58], the introduction from a single origin with super-spreader events [64], the long period of MPXV cryptic dissemination in humans as well as in animals in non-endemic countries [64], the affected human immunity levels due to the COVID-19 global pandemic while gaining the adaptability by the MPXV, and the waning of the immunity against smallpox in the world population which is caused by terminating the vaccination since 1980 when the disease was announced eradicated by WHO [7,55,56,65]. The latter one seems more reasonable as today only approximately 30% of the world population is vaccinated against smallpox [58] and the majority of infected people are younger than 50 and have never been vaccinated against smallpox. Vaccination to smallpox induced coincident immunity to MPXV, but smallpox eradication and halted vaccination as well as lack of vaccination triggered the MPXV to acquire clinical relevance [66]. Although the smallpox vaccine protects from MPXV [67], there are cases of MPXV infection in patients who are vaccinated against smallpox [18]. Mpox has the potential to grow as a global threat which necessitates the development of a specific vaccine. Moreover, if MPXV is spread easily due to the waned immunity against smallpox, smallpox itself might also re-emerge anytime soon, and that could be a huge plague. In addition, the "forgotten" viruses should get more attention as from time to time they manage to emerge, and in the case of the highly deadly virus, it would cause an unprecedentedly adverse outcome for the human population, e.g., Crimean-Congo hemorrhagic fever (CCHF) virus, Zika, Ebola, etc. Although they are considered to be limited geographically, the possibility of their spreading wider is high. MPXV was also reckoned as geographically limited and did not get much attention; however, in 2003, the outbreak took place outside of Africa for the first time which led to the increased attention to this virus. Moreover, there is a number of cases that are not reported due to the lack of surveillance systems in endemic areas, which most possibly resulted in the expansion of geographic areas [68]. According to the history of epidemics and pandemics, although the number of cases is not increasing recently, the new wave of MPXV can pose a higher risk to the health of the world population. Hence, it is important to design a more specific and universal vaccine against MPXV that will have fewer side effects. As MPXV, VARV, and VACV are from the same genus and their genomes are conserved, it is prudent to develop a vaccine based on the conserved elements. In this study, five antigens—A29, A30, A35, M1, and B6—were selected according to their importance for the viral life cycle. Importantly, these antigens have shown favorable outcomes when they were used in the line of studies and elicited protection from the MPXV challenge. Buchman et al., as well as Heraud et al., have challenged the non-human primates intradermally with the lethal dose of MPXV after the vaccination with a subunit vaccine consisting of VACV membrane proteins A33, B5, L1, and A27 [28,29] (see the names of the corresponding proteins of MPXV and VARV in Table 1), and all the animals survived. Hooper et al. used a DNA vaccine comprising the same antigens of VACV and demonstrated that the immunization with DNA vaccine encoding these four antigens protected rhesus macaques from a severe disease after the lethal MPXV challenge [27]. On the other hand, Hirao et al. used a DNA vaccine encoding eight antigens of VACV—A4, A56, F9, H3, A27, A33, B5, and L1—to immunize cynomolgus macaques and showed that the vaccination elicited protective immunity in the animals [24]. As to the fifth selected antigen in the current study, A28, it is one of at least eight transmembrane proteins in the VACV entry/fusion complex subunit and is evidenced to be a target of neutralizing and protective antibodies in rabbits [69].

The vaccine used for the eradication of smallpox was the live VACV-based vaccine that necessitated special precautions to protect from spreading the VACV from their vaccination spot [70]. Although the current VACV-based vaccine that is currently used against MPXV is a highly effective attenuated Modified Vaccinia Ankara-Bavarian Nordic (MVA-BN) virus [71,72], the development of a potentially universal next-generation vaccine against MPXV, VARV, and similar viruses from the same genus is reasonable. Nucleic acid vaccines, specifically mRNA-based vaccines, have recently garnered significant interest due to their remarkable advantages over other types of nucleic acid and conventional vaccines [73]. mRNA vaccines are highly potent, safe, cost-effective, and rapidly manufactured [19–21]. Indeed, the current COVID-19 pandemic has proven the superiority of mRNA vaccines over any other vaccines [74–76]. The advantages of mRNA-based immunization include efficacy, safety (no risk of genome integration as it is directly translated into the cytoplasm unlike DNA vaccines), cell-free manufacturing, and fast production. Hence, as mRNA technology represents a promising strategy, designing an mRNA vaccine against MPXV, VARV, and VACV is definitely reasonable [77,78].

Smallpox is the only infectious disease that is considered to be globally eradicated since 1980 [79]. However, the course of pandemic events in world history has revealed that a new pandemic can emerge unexpectedly despite the rapid advancement of science with numerous valuable studies. Thus, even though smallpox has been eliminated, there is a risk of re-emergence. Moreover, in the past, it killed millions of people until the effective vaccine was developed [80]. The current outbreak of mpox is an example of it. In addition, although VACV is not regarded as dangerous, evidently, it causes infection [81], and there is a possibility that it will become more virulent in the future. Accordingly, designing a potentially universal vaccine that can be protective to MPXV as a current emergency, VARV as a potential re-emergency, and VACV as its effective application as a vaccine seems prudent.

Noteworthily, a number of excellent immunoinformatics tools are available for finding the conserved sequences of the selected antigens [82–84], for finding experimentally tested epitopes or predicting the epitopes on the IEDB [31], and for the prediction of certain properties and immune response induced by the designed vaccine [30,40–47,49,51,85–87]. These in silico analyses save time and allow us to predict the potential outcome of the designed vaccine. Moreover, rather than commencing the time-consuming and expensive in vitro/in vivo experiments directly, using the immunoinformatics analysis that is favorable to formulate the universal MPXV mRNA vaccine and screen the multi-epitope construct containing conserved elements using in silico approaches seems very reasonable. Indeed, a line of research has been dedicated to developing vaccines for COVID-19 [41,88], influenza [89–91], and other viruses [92–94] using bioinformatics approaches. Until now, there is no study focusing on designing a multi-epitope universal mRNA vaccine against MPXV, VARV, and

VACV together. Hence, in the current study, epitopes of T cells that have already undergone experimental testing were chosen for their potential to enhance the favorable results of the proposed mRNA vaccine. On the other hand, B cell epitopes were predicted via IEDB prediction software [32] as there were no experimentally tested B cell epitopes found in the database. As the optimization of the final construct is of great importance and linkers that connect the epitopes to each other play a crucial role, flexible KK [41] and GGGS [35] linkers along with a rigid EAAAK linker [34,38] were used. The flexible linkers used here improved folding and stability in fusion proteins, while the EAAAK attached the peptides to each other to achieve an immunologically active multi-peptide [33].

In silico immunization showed that the vaccine designed for MPXV, VARV, and VACV in this study elicits optimal humoral and cellular immune responses following repeated immunization. In silico immunization with the proposed vaccine stimulated the B cell population, and increased the production of immunoglobulins, CD8 T-cytotoxic and CD4 T-helper cells, memory cells, as well as cytokines. In addition, the molecular docking showed favorable binding to the MHC molecules and other physicochemical parameters of the vaccine demonstrated stability and optimal characteristics of the vaccine.

The in vivo validation of the efficacy and safety of the potentially universal mRNA vaccine, which was designed and tested in silico in this study, is unquestionably necessary. For this, the multi-epitope antigen should be subcloned into an expression vector, transformed into DH5α competent cells, amplified, extracted and purified, linearized, in vitro transcribed, and encapsulated with lipid nanoparticles (LNPs). Then, the animal immunization experiment should be conducted for the assessment of immune responses. The schematic illustration of the steps needed for the development of a potentially universal mRNA vaccine candidate for MPXV, VARV, and VACV is shown in Figure 9.

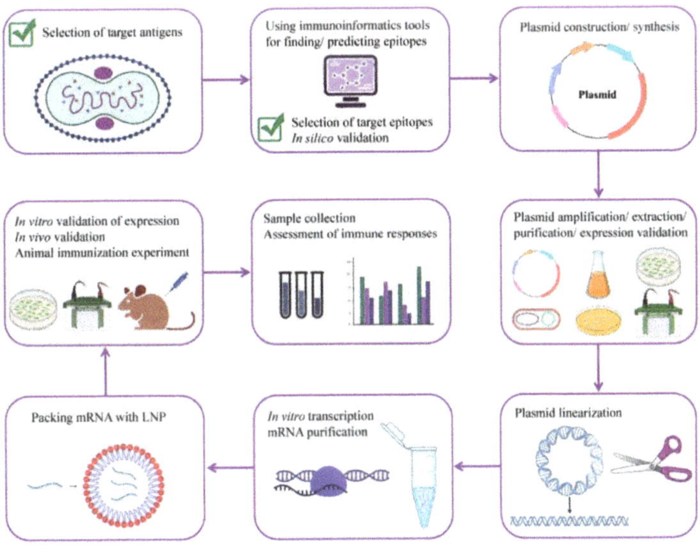

Figure 9. A roadmap for developing a potentially universal multi-epitope mRNA vaccine candidate against MPXV. MPXV, monkeypox virus.

5. Conclusions

Taken together, it is conceivable that mpox outbreaks will take place more frequently in the future [95]. Thus, more research is urgently needed for developing a specific and universal MPXV vaccine and therapeutics for the prevention and treatment of the disease, respectively. To the best of our knowledge, this is the first study focusing on designing of multi-epitope universal mRNA MPXV vaccine that will be potentially effective against

VARV and VACV as well. Indeed, immunoinformatics presents a promising strategy for contributing to the rapid development of mRNA vaccines, which could be achieved through subsequent pre-clinical and clinical investigations. This would enable the prevention, mitigation, and management of potential outbreaks before they escalate into deadly pandemics. Moreover, given that predicting the timing and location of the next pandemic is unfeasible, it is essential to have readily developed vaccine candidates that can be used swiftly to contain outbreaks before they escalate into worldwide pandemics. In summary, the current study demonstrated that the multi-epitope construct for mpox proposed here seems to be an auspicious candidate owing to the potential to induce an immune response against the two clades of MPXV and same family viruses—VACV and VARV—that might lay the groundwork to the alertness for epidemics and pandemics.

Supplementary Materials: The following supporting information can be downloaded at: https://www.mdpi.com/article/10.3390/v15051120/s1, Figure S1: Immune simulation results; Table S1: List of T cell epitopes (antigen A29L-A30L-A27L) containing conserved amino acid sequences (origin VACV); Table S2: List of B cell epitopes (antigen A29L-A30L-A27L); Table S3: List of T cell epitopes (antigen A30L/A31L/A28L) containing conserved amino acid sequences (origin VACV); Table S4: List of B cell epitopes (antigen A30L/A31L/A28L); Table S5: List of T cell epitopes (antigen A35R-A36R-A33R) containing conserved amino acid sequences (origin VACV); Table S6: List of B cell epitopes (antigen A35R-A36R-A33R); Table S7: List of T cell epitopes (antigen B6R-B7R-B5R) containing conserved amino acid sequences (origin VACV and VARV); Table S8: List of B cell epitopes (antigen B6R-B7R-B5R); Table S9: List of T cell epitopes (antigen M1R-M1R-L1R) containing conserved amino acid sequences (origin VACV); Table S10: List of B cell epitopes (antigen M1R-M1R-L1R); Table S11: Molecular docking results of the final vaccine constructs with MHC-I; Table S12: Molecular docking results of the final vaccine constructs with MHC-II.

Author Contributions: Conceptualization, N.R.; methodology, N.R. and J.M.; software, N.R. and J.M.; writing—original draft preparation, N.R.; writing—review and editing, D.P. and C.L.; visualization, D.P., S.F., X.Y. and J.L.; supervision, Y.H.; project administration, P.G.W. All authors have read and agreed to the published version of the manuscript.

Funding: This research was funded by SHENZHEN SCIENCE AND TECHNOLOGY INNOVATION PROGRAM, grant number KQTD20200909113758004.

Institutional Review Board Statement: Not applicable.

Informed Consent Statement: Not applicable.

Data Availability Statement: The datasets analyzed for this study can be found in the NCBI database [https://www.ncbi.nlm.nih.gov/; accessed on 19 August 2022] and the Immune Epitope Database (IEDB) [https://www.iedb.org/; accessed on 19 August 2022]. The data that support the findings of this study are available from the corresponding authors upon reasonable request. All data created during this study are included in this article. Further inquiries can be directed to the corresponding authors.

Conflicts of Interest: The authors declare no conflict of interest.

References

1. Centers for Disease Control and Prevention (CDC). 2022 Monkeypox Outbreak Global Map. Available online: https://www.cdc.gov/poxvirus/monkeypox/response/2022/world-map.html (accessed on 6 March 2023).
2. Vandenbogaert, M.; Kwasiborski, A.; Gonofio, E.; Descorps, S.; Selekon, B.; Andy, A.; Meyong, N.; Ouilibona, R.S.; Gessain, A.; Manuguerra, J.C.; et al. Nanopore Sequencing of a Monkeypox Virus Strain Isolated from a Pustular Lesion in the Central African Republic. *Sci. Rep.* **2022**, *12*, 10768. [CrossRef] [PubMed]
3. Shchelkunov, S.N.; Totmenin, A.V.; Safronov, P.F.; Mikheev, M.V.; Gutorov, V.V.; Ryazankina, O.I.; Petrov, N.A.; Babkin, I.V.; Uvarova, E.A.; Sandakhchiev, L.S.; et al. Analysis of the Monkeypox Virus Genome. *Virology* **2002**, *197*, 172–194. [CrossRef] [PubMed]
4. Zhu, M.; Ji, J.; Shi, D.; Lu, X.; Wang, B.; Wu, N.; Wu, J. Unusual Global Outbreak of Monkeypox: What Should We Do? *Front. Med.* **2022**, *16*, 507–517. [CrossRef] [PubMed]
5. Centers for Disease Control and Prevention (CDC). Past U.S. Cases and Outbreaks. Available online: https://www.cdc.gov/poxvirus/monkeypox/outbreak/us-outbreaks.html (accessed on 16 September 2022).

6. Gong, Q.; Wang, C.; Chuai, X.; Chiu, S. Monkeypox Virus: A Re-Emergent Threat to Humans. *Virol. Sin.* **2022**, *37*, 477–482. [CrossRef]
7. Sklenovská, N.; Ranst, M. Van Emergence of Monkeypox as the Most Important Orthopoxvirus Infection in Humans. *Front. Public Health* **2018**, *6*, 241. [CrossRef]
8. Rao, A.K.; Schulte, J.; Chen, T.-H.; Hughes, C.M.; Davidson, W.; Neff, J.M.; Markarian, M.; Delea, K.C.; Wada, S.; Liddell, A.; et al. Monkeypox in a Traveler Returning from Nigeria—Dallas, Texas, July 2021. *MMWR. Morb. Mortal. Wkly. Rep.* **2022**, *71*, 509–516. [CrossRef]
9. Costello, V.; Sowash, M.; Gaur, A.; Cardis, M.; Pasieka, H.; Wortmann, G.; Ramdeen, S. Imported Monkeypox from International Traveler, Maryland, USA, 2021. *Emerg. Infect. Dis.* **2022**, *28*, 1002–1005. [CrossRef]
10. Haider, N.; Guitian, J.; Simons, D.; Asogun, D.; Ansumana, R.; Honeyborne, I.; Velavan, T.P.; Ntoumi, F.; Valdoleiros, S.R.; Petersen, E.; et al. Increased Outbreaks of Monkeypox Highlight Gaps in Actual Disease Burden in Sub-Saharan Africa and in Animal Reservoirs. *Int. J. Infect. Dis.* **2022**, *122*, 107–111. [CrossRef]
11. Martínez, J.I.; Montalbán, E.G.; Bueno, S.J.; Martínez, F.M.; Juliá, A.N.; Díaz, S.; Marín, N.G.; Deorador, E.C.; Forte, A.N.; García, M.A.; et al. Monkeypox Outbreak Predominantly Affecting Men Who Have Sex with Men, Madrid, Spain, 26 April to 16 June 2022. *Eurosurveillance* **2022**, *27*, 2200471. [CrossRef]
12. Heskin, J.; Belfield, A.; Milne, C.; Brown, N.; Walters, Y.; Scott, C.; Bracchi, M.; Moore, L.S.; Mughal, N.; Rampling, T.; et al. Transmission of Monkeypox Virus through Sexual Contact—A Novel Route of Infection. *J. Infect.* **2022**, *85*, 334–363. [CrossRef]
13. Orviz, E.; Negredo, A.; Ayerdi, O.; Vázquez, A.; Muñoz-Gomez, A.; Monzón, S.; Clavo, P.; Zaballos, A.; Vera, M.; Sánchez, P.; et al. Monkeypox Outbreak in Madrid (Spain): Clinical and Virological Aspects. *J. Infect.* **2022**, *85*, 412–417. [CrossRef]
14. Beer, E.M.; Id, V.B.R. A Systematic Review of the Epidemiology of Human Monkeypox Outbreaks and Implications for Outbreak Strategy. *PLoS ONE* **2019**, *13*, e0007791. [CrossRef] [PubMed]
15. Jezek, Z.; Szczeniowski, M.; Paluku, K.M. Human Monkeypox: Clinical Features of 282 Patients. *J. Infect. Dis.* **1987**, *156*, 293–298. [CrossRef] [PubMed]
16. Anderson, M.G.; Frenkel, L.D.; Homann, S.; Guffey, J. A Case of Severe Monkeypox Virus Disease in an American Child: Emerging Infections and Changing Professional Values. *Pediatr. Infect. Dis. J.* **2003**, *22*, 1093–1096. [CrossRef]
17. Shchelkunov, S.N.; Totmenin, A.V.; Babkin, I.V.; Safronov, P.F.; Ryazankina, O.I.; Petrov, N.A.; Gutorov, V.V.; Uvarova, E.A.; Mikheev, M.V.; Sisler, J.R.; et al. Human Monkeypox and Smallpox Viruses: Genomic Comparison. *FEBS Lett.* **2001**, *509*, 66–70. [CrossRef] [PubMed]
18. Huhn, G.D.; Bauer, A.M.; Yorita, K.; Graham, M.B.; Sejvar, J.; Likos, A.; Damon, I.K.; Reynolds, M.G.; Kuehnert, M.J. Clinical Characteristics of Human Monkeypox, and Risk Factors for Severe Disease. *Clin. Infect. Dis.* **2005**, *41*, 1742–1751. [CrossRef] [PubMed]
19. Liu, C.; Rcheulishvili, N.; Shen, Z.; Papukashvili, D.; Xie, F.; Wang, Z.; Wang, X.; He, Y.; Wang, P.G. Development of an LNP-Encapsulated MRNA-RBD Vaccine against SARS-CoV-2 and Its Variants. *Pharmaceutics* **2022**, *14*, 1101. [CrossRef]
20. Pardi, N.; Hogan, M.J.; Porter, F.W.; Weissman, D. MRNA Vaccines-a New Era in Vaccinology. *Nat. Rev. Drug Discov.* **2018**, *17*, 261–279. [CrossRef]
21. Kowalzik, F.; Schreiner, D.; Jensen, C.; Teschner, D.; Gehring, S.; Zepp, F. MRNA-Based Vaccines. *Vaccines* **2021**, *9*, 390. [CrossRef]
22. Pütz, M.M.; Midgley, C.M.; Law, M.; Smith, G.L. Quantification of Antibody Responses against Multiple Antigens of the Two Infectious Forms of Vaccinia Virus Provides a Benchmark for Smallpox Vaccination. *Nat. Med.* **2006**, *12*, 1310–1315. [CrossRef]
23. Wolffe, E.J.; Vijaya, S.; Moss, B. A Myristylated Membrane Protein Encoded by the Vaccinia Virus L1R Open Reading Frame Is the Target of Potent Neutralizing Monoclonal Antibodies. *Virology* **1995**, *211*, 53–63. [CrossRef] [PubMed]
24. Hirao, L.A.; Draghia-akli, R.; Prigge, J.T.; Yang, M.; Satishchandran, A.; Wu, L.; Hammarlund, E.; Khan, A.S.; Babas, T.; Rhodes, L.; et al. Multivalent Smallpox DNA Vaccine Delivered by Intradermal Electroporation Drives Protective Immunity in Nonhuman Primates Against Lethal Monkeypox Challenge. *J. Infect. Dis.* **2011**, *203*, 95–102. [CrossRef]
25. Gilchuk, I.; Gilchuk, P.; Sapparapu, G.; Lampley, R.; Singh, V.; Kose, N.; Blum, D.L.; Hughes, L.J.; Satheshkumar, P.S.; Townsend, M.B.; et al. Cross-Neutralizing and Protective Human Antibody Specificities to Poxvirus Infections. *Cell* **2016**, *167*, 684–694.e9. [CrossRef] [PubMed]
26. Moss, B. Smallpox Vaccines: Targets of Protective Immunity. *Immunol. Rev.* **2011**, *239*, 8–26. [CrossRef] [PubMed]
27. Hooper, J.W.; Thompson, E.; Wilhelmsen, C.; Zimmerman, M.; Ichou, M.A.; Steffen, S.E.; Schmaljohn, C.S.; Schmaljohn, A.L.; Jahrling, P.B. Smallpox DNA Vaccine Protects Nonhuman Primates against Lethal Monkeypox. *J. Virol.* **2004**, *78*, 4433–4443. [CrossRef] [PubMed]
28. Buchman, G.W.; Cohen, M.E.; Xiao, Y.; Richardson-Harman, N.; Silvera, P.; DeTolla, L.J.; Davis, H.L.; Eisenberg, R.J.; Cohen, G.H.; Isaacs, S.N. A Protein-Based Smallpox Vaccine Protects Non-Human Primates from a Lethal Monkeypox Virus Challenge. *Vaccine* **2010**, *28*, 6627–6636. [CrossRef] [PubMed]
29. Heraud, J.; Edghill-smith, Y.; Ayala, V.; Kalisz, I.; Parrino, J.; Kalyanaraman, V.S.; Manischewitz, J.; King, L.R.; Hryniewicz, A.; Trindade, C.J.; et al. Subunit Recombinant Vaccine Protects against Monkeypox. *J. Immunol.* **2022**, *177*, 2552–2564. [CrossRef]
30. Waterhouse, A.M.; Procter, J.B.; Martin, D.M.A.; Clamp, M.; Barton, G.J. Jalview Version 2-a Multiple Sequence Alignment Editor and Analysis Workbench. *Bioinformatics* **2009**, *25*, 1189–1191. [CrossRef]
31. Vita, R.; Mahajan, S.; Overton, J.A.; Dhanda, S.K.; Martini, S.; Cantrell, J.R.; Wheeler, D.K.; Sette, A.; Peters, B.; Immune, T.; et al. The Immune Epitope Database (IEDB): 2018 Update. *Nucleic Acids Res.* **2019**, *47*, 339–343. [CrossRef]

32. Jespersen, M.C.; Peters, B.; Nielsen, M.; Marcatili, P. BepiPred-2.0: Improving Sequence-Based B-Cell Epitope Prediction Using Conformational Epitopes. *Nucleic Acids Res.* **2017**, *45*, W24–W29. [CrossRef]
33. Rahmani, A.; Baee, M.; Saleki, K.; Moradi, S.; Nouri, H.R. Applying High Throughput and Comprehensive Immunoinformatics Approaches to Design a Trivalent Subunit Vaccine for Induction of Immune Response against Emerging Human Coronaviruses SARS-CoV, MERS-CoV and SARS-CoV-2. *J. Biomol. Struct. Dyn.* **2021**, *40*, 6097–6113. [CrossRef] [PubMed]
34. Tarrahimofrad, H.; Rahimnahal, S.; Zamani, J.; Jahangirian, E.; Aminzadeh, S. Designing a Multi-epitope Vaccine to Provoke the Robust Immune Response against Influenza A H7N9. *Sci. Rep.* **2021**, *11*, 24485. [CrossRef] [PubMed]
35. Sanches, R.C.O.; Tiwari, S.; Ferreira, L.C.G.; Oliveira, F.M.; Lopes, M.D.; Passos, M.J.F.; Maia, E.H.B.; Taranto, A.G.; Kato, R.; Azevedo, V.A.C.; et al. Immunoinformatics Design of Multi-Epitope Peptide-Based Vaccine Against Schistosoma Mansoni Using Transmembrane Proteins as a Target. *Front. Immunol.* **2021**, *12*, 621706. [CrossRef] [PubMed]
36. Ahammad, I.; Lira, S.S. Designing a Novel MRNA Vaccine against SARS-CoV-2: An Immunoinformatics Approach. *Int. J. Biol. Macromol.* **2020**, *162*, 820–837. [CrossRef]
37. Yi, L.; Lee, Y.; Izzard, L.; Hurt, A.C. A Review of DNA Vaccines Against Influenza. *Front. Immunol.* **2018**, *9*, 1568. [CrossRef]
38. Nezafat, N.; Ghasemi, Y.; Javadi, G.; Khoshnoud, M.J.; Omidinia, E. A Novel Multi-Epitope Peptide Vaccine against Cancer: An in Silico Approach. *J. Theor. Biol.* **2014**, *349*, 121–134. [CrossRef]
39. Boyoglu-Barnum, S.; Ellis, D.; Gillespie, R.A.; Hutchinson, G.B.; Park, Y.; Moin, S.M.; Acton, O.J.; Ravichandran, R.; Murphy, M.; Pettie, D.; et al. Quadrivalent Influenza Nanoparticle Vaccines Induce Broad Protection. *Nature* **2021**, *592*, 623–628. [CrossRef]
40. Saha, S.; Raghava, G.P.S. AlgPred: Prediction of Allergenic Proteins and Mapping of IgE Epitopes. *Nucleic Acids Res.* **2006**, *34*, 202–209. [CrossRef]
41. Dong, R.; Chu, Z.; Yu, F.; Zha, Y. Contriving Multi-Epitope Subunit of Vaccine for COVID-19: Immunoinformatics Approaches. *Front. Immunol.* **2020**, *11*, 1784. [CrossRef]
42. Doytchinova, I.A.; Flower, D.R. Identifying Candidate Subunit Vaccines Using an Alignment-Independent Method Based on Principal Amino Acid Properties. *Vaccine* **2007**, *25*, 856–866. [CrossRef]
43. Doytchinova, I.A.; Flower, D.R. VaxiJen: A Server for Prediction of Protective Antigens, Tumour Antigens and Subunit Vaccines. *BMC Bioinform.* **2007**, *8*, 4. [CrossRef] [PubMed]
44. Gasteiger, E.; Hoogland, C.; Gattiker, A.; Duvaud, S.; Wilkins, M.R.; Appel, R.D.; Bairoch, A. Protein Identification and Analysis Tools on the ExPASy Server; The Proteomics Protocols Handbook. *Hum. Press* **2005**, *112*, 531–552. [CrossRef]
45. Baek, M.; Dimaio, F.; Anishchenko, I.; Dauparas, J.; Ovchinnikov, S.; Lee, G.R.; Wang, J.; Cong, Q.; Kinch, L.N.; Schaeffer, R.D.; et al. Accurate Prediction of Protein Structures and Interactions Using a 3-Track Neural Network. *Science* **2021**, *373*, 871–876. [CrossRef] [PubMed]
46. Heo, L.; Park, H.; Seok, C. GalaxyRefine: Protein Structure Refinement Driven by Side-Chain Repacking. *Nucleic Acids Res.* **2013**, *41*, 384–388. [CrossRef] [PubMed]
47. Nelson, D.L.; Cox, M.M.; Hoskins, A.A. *Lehninger Principles of Biochemistry*, 8th ed.; Macmillan Learning: Austin, TX, USA, 2021; ISBN 9781429234146.
48. Ponomarenko, J.; Bui, H.H.; Li, W.; Fusseder, N.; Bourne, P.E.; Sette, A.; Peters, B. ElliPro: A New Structure-Based Tool for the Prediction of Antibody Epitopes. *BMC Bioinform.* **2008**, *9*, 514. [CrossRef] [PubMed]
49. Rapin, N.; Lund, O.; Castiglione, F. Immune System Simulation Online. *Bioinformatics* **2011**, *27*, 2013–2014. [CrossRef] [PubMed]
50. Rapin, N.; Lund, O.; Bernaschi, M.; Castiglione, F. Computational Immunology Meets Bioinformatics: The Use of Prediction Tools for Molecular Binding in the Simulation of the Immune System. *PLoS ONE* **2010**, *5*, e9862. [CrossRef]
51. Kozakov, D.; Hall, D.R.; Xia, B.; Porter, K.A.; Padhorny, D.; Yueh, C.; Beglov, D.; Vajda, S.; Biology, Q. The ClusPro Web Server for Protein-Protein Docking. *Nat. Protoc.* **2017**, *12*, 255–278. [CrossRef]
52. Mirdita, M.; Schütze, K.; Moriwaki, Y.; Heo, L. ColabFold: Making Protein Folding Accessible to All. *Nat. Methods* **2022**, *19*, 679–682. [CrossRef]
53. Jumper, J.; Evans, R.; Pritzel, A.; Green, T.; Figurnov, M.; Ronneberger, O.; Tunyasuvunakool, K.; Bates, R.; Žídek, A.; Potapenko, A.; et al. Highly Accurate Protein Structure Prediction with AlphaFold. *Nature* **2021**, *596*, 583–589. [CrossRef]
54. Ghafouri, F.; Cohan, R.A.; Samimi, H.; Hosseini, A.; Naderi, M.; Noorbakhsh, F.; Haghpanah, V. Development of a Multiepitope Vaccine Against SARS-CoV-2: Immunoinformatics Study. *JMIR Bioinforma. Biotechnol.* **2022**, 3, e36100. [CrossRef] [PubMed]
55. Xiang, Y.; White, A. Monkeypox Virus Emerges from The Shadow of Its More Infamous Cousin: Family Biology Matters. *Emerg. Microbes Infect.* **2022**, *11*, 1768–1777. [CrossRef] [PubMed]
56. Durski, K.N.; Mccollum, A.M.; Nakazawa, Y.; Petersen, B.W.; Reynolds, M.G.; Briand, S.; Djingarey, M.H.; Olson, V.; Damon, I.K.; Khalakdina, A. Emergence of Monkeypox—West and Central Africa, 1970–2017. *MMWR. Morb. Mortal. Wkly. Rep.* **2018**, *67*, 306–310. [CrossRef] [PubMed]
57. Ladnyj, D.; Ziegler, P.; Kima, E. A Human Infection Caused by Monkeypox Virus in Basankusu Territory, Democratic Republic of the Congo. *Bull. World Health Organ.* **1972**, *46*, 593–597.
58. Kmiec, D.; Kirchhoff, F. Monkeypox: A New Threat? *Int. J. Mol. Sci.* **2022**, *23*, 7866. [CrossRef] [PubMed]
59. Osadebe, L.; Hughes, C.M.; Lushima, R.S.; Kabamba, J.; Nguete, B.; Malekani, J.; Pukuta, E.; Karhemere, S.; Tamfum, J.M.; Okitolonda, E.W.; et al. Enhancing Case Definitions for Surveillance of Human Monkeypox in the Democratic Republic of Congo. *PLoS Negl. Trop. Dis.* **2017**, *11*, e0005857. [CrossRef] [PubMed]

60. Pfäfflin, F.; Wendisch, D.; Scherer, R.; Jürgens, L.; Godzick, G.; Eva, N.; Tober, P.; Miriam, L.; Stegemann, S.; Max, V.; et al. Monkeypox In-Patients with Severe Anal Pain. *Infection* **2022**, *51*, 483–487. [CrossRef]
61. Jang, Y.R.; Lee, M.; Shin, H.; Kim, J.; Choi, M.; Kim, Y.M.; Lee, M.J.; Kim, J.; Na, H.K.; Kim, J.Y.; et al. The First Case of Monkeypox in the Republic of Korea. *Infect. Dis. Microbiol. Parasitol.* **2022**, *37*, e224. [CrossRef]
62. Meyer, H.; Perrichot, M.; Stemmler, M.; Emmerich, P.; Schmitz, H.; Varaine, F.; Shungu, R.; Tshioko, F.; Formenty, P. Outbreaks of Disease Suspected of Being Due to Human Monkeypox Virus Infection in the Democratic Republic of Congo in 2001. *J. Clin. Microbiol.* **2002**, *40*, 2919–2921. [CrossRef]
63. Tian, D.; Zheng, T. Comparison and Analysis of Biological Agent Category Lists Based on Biosafety and Biodefense. *PLoS ONE* **2014**, *9*, e101163. [CrossRef]
64. Isidro, J.; Borges, V.; Pinto, M.; Sobral, D.; Santos, J.D.; Nunes, A.; Mixão, V.; Ferreira, R.; Santos, D.; Duarte, S.; et al. Phylogenomic Characterization and Signs of Microevolution in the 2022 Multi-Country Outbreak of Monkeypox Virus. *Nat. Med.* **2022**, *28*, 1569–1572. [CrossRef] [PubMed]
65. Weaver, J.R.; Isaacs, S.N. Monkeypox Virus and Insights into Its Immunomodulatory Proteins. *Immunol. Rev.* **2009**, *225*, 96–113. [CrossRef] [PubMed]
66. Nguyen, P.; Ajisegiri, W.S.; Costantino, V.; Chughtai, A.A.; Macintyre, C.R. Reemergence of Human Monkeypox and Declining Population Immunity in the Context of Urbanization, Nigeria, 2017–2020. *Emerg. Infect. Dis.* **2021**, *27*, 1007–1014. [CrossRef]
67. Burki, T. Investigating Monkeypox. *Lancet* **2022**, *399*, 2254–2255. [CrossRef] [PubMed]
68. Yang, Z. Monkeypox: A Potential Global Threat? *J. Med. Virol.* **2022**, *94*, 4034–4036. [CrossRef] [PubMed]
69. Nelson, G.E.; Sisler, J.R.; Chandran, D.; Moss, B. Vaccinia Virus Entry/Fusion Complex Subunit A28 Is a Target of Neutralizing and Protective Antibodies. *Virology* **2008**, *380*, 394–401. [CrossRef]
70. Rizk, J.G.; Lippi, G.; Henry, B.M.; Forthal, D.N.; Rizk, Y. Prevention and Treatment of Monkeypox. *Drugs* **2022**, *82*, 957–963. [CrossRef] [PubMed]
71. Kandeel, M.; Morsy, M.A.; El-lateef, H.M.A.; Marzok, M.; El-Beltagi, H.S.; Khodair, K.M.A.; Albokhadaim, I.; Venugopala, K.N. Efficacy of the Modified Vaccinia Ankara Virus Vaccine and the Replication-Competent Vaccine ACAM2000 in Monkeypox Prevention Mahmoud. *Int. Immunopharmacol.* **2023**, *119*, 110206. [CrossRef]
72. Food and Drug Administration (FDA). Use of JYNNEOS (Smallpox and Monkeypox Vaccine, Live, Nonreplicating) for Pre-exposure Vaccination of Persons at Risk for Occupational Exposure to Orthopoxviruses: Recommendations of the Advisory Committee on Immunization Practices—United States. 2022. Available online: https://www.fda.gov/media/131078/download#:~:text=JYNNEOSisanattenuated%2Clive,preventionofsmallpoxandmonkeypox (accessed on 4 May 2023).
73. Turner, J.S.; Halloran, J.A.O.; Kalaidina, E.; Kim, W.; Schmitz, A.J.; Zhou, J.Q.; Lei, T.; Thapa, M.; Chen, R.E.; Case, J.B.; et al. SARS-CoV-2 MRNA Vaccines Induce Persistent Human Germinal Centre Responses. *Nature* **2021**, *596*, 109–113. [CrossRef]
74. Walsh, E.E.; Frenck, R.W.; Falsey, A.R.; Kitchin, N.; Absalon, J.; Gurtman, A.; Lockhart, S.; Neuzil, K.; Mulligan, M.J.; Bailey, R.; et al. Safety and Immunogenicity of Two RNA-Based COVID-19 Vaccine Candidates. *N. Engl. J. Med.* **2020**, *383*, 2439–2450. [CrossRef]
75. Haas, E.J.; Angulo, F.J.; McLaughlin, J.M.; Anis, E.; Singer, S.R.; Khan, F.; Brooks, N.; Smaja, M.; Mircus, G.; Pan, K.; et al. Impact and Effectiveness of MRNA BNT162b2 Vaccine against SARS-CoV-2 Infections and COVID-19 Cases, Hospitalisations, and Deaths Following a Nationwide Vaccination Campaign in Israel: An Observational Study Using National Surveillance Data. *Lancet* **2021**, *397*, 1819–1829. [CrossRef] [PubMed]
76. Corbett, K.S.; Flynn, B.; Foulds, K.E.; Francica, J.R.; Boyoglu-Barnum, S.; Werner, A.P.; Flach, B.; O'Connell, S.; Bock, K.W.; Minai, M.; et al. Evaluation of the MRNA-1273 Vaccine against SARS-CoV-2 in Nonhuman Primates. *N. Engl. J. Med.* **2020**, *383*, 1544–1555. [CrossRef]
77. Chaudhary, N.; Weissman, D.; Whitehead, K.A. MRNA Vaccines for Infectious Diseases: Principles, Delivery and Clinical Translation. *Nat. Rev. Drug Discov.* **2021**, *20*, 817–838. [CrossRef] [PubMed]
78. Damase, T.R.; Sukhovershin, R.; Boada, C.; Taraballi, F.; Pettigrew, R.I.; Cooke, J.P. The Limitless Future of RNA Therapeutics. *Front. Bioeng. Biotechnol.* **2021**, *9*, 628137. [CrossRef] [PubMed]
79. Osterholm, M.T.; Gellin, B. Confronting 21st-Century Monkeypox. *Science* **2022**, *377*, 349. [CrossRef]
80. Belongia, E.A.; Clinic, M. Smallpox Vaccine: The Good, the Bad, and the Ugly. *Clin. Med. Res.* **2003**, *1*, 87–92. [CrossRef] [PubMed]
81. Lu, B.; Cui, L.; Gu, M.; Shi, C.; Sun, C.; Zhao, K. Outbreak of Vaccinia Virus Infection. *Emerg. Infect. Dis.* **2019**, *25*, 2017–2020.
82. Sievers, F.; Wilm, A.; Dineen, D.; Gibson, T.J.; Karplus, K.; Li, W.; Lopez, R.; Mcwilliam, H.; Remmert, M.; Söding, J.; et al. Fast, Scalable Generation of High-Quality Protein Multiple Sequence Alignments Using Clustal Omega. *Mol. Syst. Biol.* **2011**, *7*, 539. [CrossRef] [PubMed]
83. Sievers, F.; Higgins, D.G. Clustal Omega for Making Accurate Alignments of Many Protein Sequences. *Protein Sci.* **2018**, *27*, 135–145. [CrossRef]
84. Sievers, F.; Barton, G.J.; Higgins, D.G. *Multiple Sequence Alignments. Bioinformatics*, 4th ed.; Baxevanis, A.D., Bader, G.D., Wishart, D.S., Eds.; Wiley: Hoboken, NJ, USA, 2020; Volume 227, ISBN 978-1-119-33558-0.
85. González-Pech, R.A.; Stephens, T.G.; Chan, C.X. Commonly Misunderstood Parameters of NCBI BLAST and Important Considerations for Users. *Bioinformatics* **2019**, *35*, 2697–2698. [CrossRef]
86. Zhang, J.; Liang, Y.; Zhang, Y. Atomic-Level Protein Structure Refinement Using Fragment Guided Molecular Dynamics Conformation Sampling. *Structure* **2011**, *19*, 1784–1795. [CrossRef] [PubMed]

87. Kozakov, D.; Brenke, R.; Comeau, S.R.; Vajda, S. PIPER: An FFT-Based Protein Docking Program with Pairwise Potentials. *Proteins* **2006**, *65*, 392–406. [CrossRef] [PubMed]
88. Oliveira, S.C.; Magalhães, M.T.Q. De; Homan, E.J. Immunoinformatic Analysis of SARS-CoV-2 Nucleocapsid Protein and Identi Fi Cation of COVID-19 Vaccine Targets. *Front. Immunol.* **2020**, *11*, 587615. [CrossRef]
89. Sharma, S.; Kumari, V.; Vasant, B.; Mukherjee, A.; Pandey, R.; Kondabagil, K. Immunoinformatics Approach for a Novel Multi-Epitope Subunit Vaccine Design against Various Subtypes of Influenza A Virus. *Immunobiology* **2021**, *226*, 152053. [CrossRef] [PubMed]
90. Behbahani, M.; Moradi, M.; Mohabatkar, H. In Silico Design of a Multi-epitope Peptide Construct as a Potential Vaccine Candidate for Influenza A Based on Neuraminidase Protein. *In Silico Pharmacol.* **2021**, *9*, 36. [CrossRef] [PubMed]
91. Rcheulishvili, N.; Mao, J.; Papukashvili, D.; Liu, C.; Wang, Z.; Zhao, J.; Xie, F.; Pan, X.; Ji, Y.; He, Y.; et al. Designing multi-epitope mRNA construct as a universal influenza vaccine candidate for future epidemic/pandemic preparedness. *Int J Biol Macromol.* **2023**, *226*, 885–899. [CrossRef] [PubMed]
92. Ros-Lucas, A.; Correa-Fiz, F.; Bosch-Camós, L.; Rodriguez, F.; Alonso-Padilla, J. Computational Analysis of African Swine Fever Virus Protein Space for the Design of an Epitope-Based Vaccine Ensemble. *Pathogens* **2020**, *9*, 1078. [CrossRef]
93. Mahmudul, H.; Shiful, I.; Sourav, C.; Abu, H.M.; Kazi, F.A.; Ziaul, F.J.; Nazmul, H.; Shakhawat, H.F.; Nazmul, H. Contriving a Chimeric Polyvalent Vaccine to Prevent Infections Caused by Herpes Simplex Virus (Type-1 and Type-2): An Exploratory Immunoinformatic Approach. *J. Biomol. Struct. Dyn.* **2020**, *38*, 2898–2915. [CrossRef]
94. Ali, A.; Khan, A.; Kaushik, A.C.; Wang, Y.; Ali, S.S.; Junaid, M.; Saleem, S.; Cho, W.C.S.; Mao, X.; Wei, D. Immunoinformatic and Systems Biology Approaches to Predict and Validate Peptide Vaccines against Epstein–Barr Virus (EBV). *Sci. Rep.* **2019**, *9*, 720. [CrossRef]
95. Papukashvili, D.; Rcheulishvili, N.; Liu, C.; Wang, X.; He, Y.; Wang, P.G. Strategy of developing nucleic acid-based universal monkeypox vaccine candidates. *Front Immunol.* **2022**, *13*, 1050309. [CrossRef]

MDPI
St. Alban-Anlage 66
4052 Basel
Switzerland
www.mdpi.com

Viruses Editorial Office
E-mail: viruses@mdpi.com
www.mdpi.com/journal/viruses

www.ingramcontent.com/pod-product-compliance
Lightning Source LLC
Chambersburg PA
CBHW061508210926
51850CB00008B/1853
9783036592336